AF581741

Modeling, Optimization and Design Method of Metal Manufacturing Processes

Modeling, Optimization and Design Method of Metal Manufacturing Processes

Editors

Guoqing Zhang
Zejia Zhao
Wai Sze YIP

MDPI • Basel • Beijing • Wuhan • Barcelona • Belgrade • Manchester • Tokyo • Cluj • Tianjin

Editors

Guoqing Zhang
College of Mechatronics and Control Engineering
Shenzhen University
Shen Zhen
China

Zejia Zhao
College of Mechatronics and Control Engineering
Shenzhen University
Shen Zhen
China

Wai Sze YIP
Industrial and Systems Engineering
The Hong Kong Polytechnic University
Hong Kong

Editorial Office
MDPI
St. Alban-Anlage 66
4052 Basel, Switzerland

This is a reprint of articles from the Special Issue published online in the open access journal *Processes* (ISSN 2227-9717) (available at: www.mdpi.com/journal/processes/special_issues/metal_manufacturing_processes).

For citation purposes, cite each article independently as indicated on the article page online and as indicated below:

LastName, A.A.; LastName, B.B.; LastName, C.C. Article Title. *Journal Name* **Year**, *Volume Number*, Page Range.

ISBN 978-3-0365-6034-2 (Hbk)
ISBN 978-3-0365-6033-5 (PDF)

Contents

About the Editors . . . . **vii**

Preface to "Modeling, Optimization and Design Method of Metal Manufacturing Processes" **ix**

Guoqing Zhang, Zejia Zhao and Wai Sze YIP
Special Issue on "Modeling, Optimization and Design Method of Metal Manufacturing Processes"
Reprinted from: *Processes* **2022**, *10*, 2461, doi:10.3390/pr10112461 . . . . **1**

Doru Stefan Andreiana, Luis Enrique Acevedo Galicia, Seppo Ollila, Carlos Leyva Guerrero, Álvaro Ojeda Roldán and Fernando Dorado Navas et al.
Steelmaking Process Optimised through a Decision Support System Aided by Self-Learning Machine Learning
Reprinted from: *Processes* **2022**, *10*, 434, doi:10.3390/pr10030434 . . . . **5**

Xingyu Chen, Jiayang Dai and Yasong Luo
Temperature Prediction Model for a Regenerative Aluminum Smelting Furnace by a Just-in-Time Learning JITL-Based Triple-Weighted Regularized Extreme Learning Machine
Reprinted from: *Processes* **2022**, *10*, 1972, doi:10.3390/pr10101972 . . . . **25**

Jiaqi Ran, Gangping Chen, Fuxing Zhong, Li Xu, Teng Xu and Feng Gong
The Influence of Size Effect to Deformation Mechanism of C5131 Bronze Structures of Negative Poisson's Ratio
Reprinted from: *Processes* **2022**, *10*, 652, doi:10.3390/pr10040652 . . . . **41**

Meng Wang, Yu Guo, Hongying Wang and Shengsheng Zhao
Characterization of Refining the Morphology of Al–Fe–Si in A380 Aluminum Alloy due to Ca Addition
Reprinted from: *Processes* **2022**, *10*, 672, doi:10.3390/pr10040672 . . . . **53**

Lang Cui, Shengmin Shao, Haitao Wang, Guoqing Zhang, Zejia Zhao and Chunyang Zhao
Recent Advances in the Equal Channel Angular Pressing of Metallic Materials
Reprinted from: *Processes* **2022**, *10*, 2181, doi:10.3390/pr10112181 . . . . **63**

Zakaria Ahmed M. Tagiuri, Thien-My Dao, Agnes Marie Samuel and Victor Songmene
Numerical Prediction of the Performance of Chamfered and Sharp Cutting Tools during Orthogonal Cutting of AISI 1045 Steel
Reprinted from: *Processes* **2022**, *10*, 2171, doi:10.3390/pr10112171 . . . . **93**

Ganggang Yin, Jianyun Shen, Ze Wu, Xian Wu and Feng Jiang
Experimental Investigation on the Machinability of PCBN Chamfered Tool in Dry Turning of Gray Cast Iron
Reprinted from: *Processes* **2022**, *10*, 1547, doi:10.3390/pr10081547 . . . . **119**

Fung Ming Kwok, Zhanwen Sun, Wai Sze Yip, Kwong Yu David Kwok and Suet To
Effects of Coating Parameters of Hot Filament Chemical Vapour Deposition on Tool Wear in Micro-Drilling of High-Frequency Printed Circuit Board
Reprinted from: *Processes* **2022**, *10*, 1466, doi:10.3390/pr10081466 . . . . **133**

Chuan-Zhi Jing, Ji-Lai Wang, Xue Li, Yi-Fei Li and Lu Han
Influence of Material Microstructure on Machining Characteristics of OFHC Copper C102 in Orthogonal Micro-Turning
Reprinted from: *Processes* **2022**, *10*, 741, doi:10.3390/pr10040741 . . . . **147**

Zexuan Huo, Guoqing Zhang, Junhong Han, Jianpeng Wang, Shuai Ma and Haitao Wang
A Review of the Preparation, Machining Performance, and Application of Fe-Based Amorphous Alloys
Reprinted from: *Processes* **2022**, *10*, 1203, doi:10.3390/pr10061203 . . . . **161**

About the Editors

Guoqing Zhang

Prof. Dr. Zhang Guoqing is currently a Research Professor in the College of Mechatronics and Control Engineering at Shenzhen University. He achieved his Bachelor's degree in Mechanical Engineering at Northeast Petroleum University, China, in 2005. In 2009, he achieved his Master's degree in Mechanical Engineering at Harbin Institute of Technology, China. Additionally, he achieved his Ph.D. degree at The Hong Kong Polytechnic University in 2014. He has been selected for the Shenzhen Peacock Program for Overseas/Professional High-Level Talents. He serves as a Deputy Director of Guangdong Key Laboratory of Electromagnetic Control and Intelligent Robots; Deputy Director of Shenzhen Key Laboratory of High Performance Nontraditional Manufacturing; Committee Member of the Chinese Mechanical Engineering Society Design Branch; and a Member of Advanced Optical Manufacturing Young Expert Committee of Chinese Optical Engineering Society. His research interests include ultra-precision machining technology and equipment, robots and intelligent equipment, etc. To date, he has hosted more than 10 research projects including 3 National Natural Science Foundation of China (NSFC) projects; he has published more than 80 peer-reviewed papers; he has been granted 14 US patents and 28 CN patents; and he has co-edited 2 books. He is now served as an Academic Editor for the international journal *Shock and Vibration*, and contributes as a reviewer for more than 10 other international journals. He has won more than five prizes, including the China Industry-University-Research Cooperation Promotion Award (individual) and a second prize in innovation achievement.

Zejia Zhao

Dr. Zhao Zejia is currently an Assistant Professor in the College of Mechatronics and Control Engineering at Shenzhen University. His research interests mainly include ultra-precision machining, the cutting of difficult-to-cut and brittle materials, and numerical modeling. As a principal investigator, he has taken charge of several projects, such as those organized by the National Natural Science Foundation of China (NSFC), the Department of Education of Guangdong Province, and Shenzhen Stable Support. Dr. Zhao is also a Member of the Advanced Optical Manufacturing Youth Expert Committee. He has been awarded as a talent of the Shenzhen Peacock Project.

Wai Sze YIP

Dr. Yip Wai Sze is currently a Research Assistant Professor at the Department of Industrial and Systems Engineering of The Hong Kong Polytechnic University. She achieved her Ph.D. degree in Ultra-Precision Machining at The Hong Kong Polytechnic University in 2018. She achieved a double degree of a BBA in Marketing and a BEng in Industrial and Systems Engineering at The Hong Kong Polytechnic University in 2014, and a BEng degree in Electronic and Communication Engineering at City University of Hong Kong in 2006. Before joining the ISE of PolyU, Dr Yip worked as a Research Fellow at the National University of Singapore. Her research focuses on the ultra-precision machining of difficult-to-cut materials, sustainable precision machining, and the sustainability development of precision manufacturing. Dr Yip has published research in top-tier SCI journals such as *Energy*, the *Journal of Cleaner Production*, the *Journal of Alloy and Compounds*, and *IEEE Access*. In 2018, one of her papers in *Scientific Reports* was recognized as being one of the top 100 papers in materials science. In 2019, she received the Excellent Thesis Award, 9th Hiwin Doctoral Dissertation, by the Chinese Mechanical Engineering Society.

Preface to "Modeling, Optimization and Design Method of Metal Manufacturing Processes"

Metal manufacturing process is an essential topic in the field of manufacturing, and is also key for the fabrication of precise components. Nowadays, complicated and high-performance metal products need smarter, more efficient, and more accurate manufacturing processes, which would require systematic improvements to the existing metal manufacturing techniques. Therefore, the optimization and improvement of metal manufacturing process is now key to preparing high-performance metal products and improving product performance.

This book presents studies of the manufacturing of difficult-to-deform metals, through the modeling, optimization, and design of the manufacturing processes, and provides an introduction to cutting-edge technology for improvements, including: (1) machine learning algorithms in manufacturing metal products; (2) the fabrication and mechanical property optimization of metals; and (3) numerical simulations and experiments in the machining of metals. The purpose of this book is to present a variety of feasible fabrication methods and advanced manufacturing techniques for precision metal products.

This book focuses on the modeling, optimization, and design methods of metal manufacturing processes, including the applications of artificial intelligence (AI) technology, the optimization of mechanical properties of metals, and numerical simulations in the machining of metals; as such, graduate students, engineering technicians, and researchers in related areas will benefit from the findings shown in this book, as would readers with interests in related application fields such as equipment manufacturing, advanced manufacturing, artificial intelligence, optics, etc.

We thank all the contributors and the Editors (Mr. Scott Pan and Ms. Sue Sun) for their enthusiastic support of the book, as well as the editorial staff of the journal *Processes* for their efforts.

Guoqing Zhang, Zejia Zhao, and Wai Sze YIP

Editors

Editorial

Special Issue on "Modeling, Optimization and Design Method of Metal Manufacturing Processes"

Guoqing Zhang [1], Zejia Zhao [1,*] and Wai Sze YIP [2]

[1] College of Mechatronics and Control Engineering, Shenzhen University, Shenzhen 518060, China
[2] Department of Industrial and Systems Engineering, The Hong Kong Polytechnic University, Hong Kong SAR 999077, China
* Correspondence: zhaozejia@szu.edu.cn

Citation: Zhang, G.; Zhao, Z.; YIP, W.S. Special Issue on "Modeling, Optimization and Design Method of Metal Manufacturing Processes". *Processes* **2022**, *10*, 2461. https://doi.org/10.3390/pr10112461

Received: 3 November 2022
Accepted: 18 November 2022
Published: 21 November 2022

Metal manufacturing processes are essential techniques to convert raw materials into desired metal products, which contributes significantly to the growth of industry and our society. In recent years, there has been a rapid increase in the demand for accurate modeling, proper optimization, and design methods prior to manufacturing processes, in order to achieve the sustainable, high-efficient, and low-cost production of metal products. The purposes of this Special Issue on "Modeling, Optimization and Design Method of Metal Manufacturing Processes" are to investigate and discuss the fundamental aspect of some metal manufacturing processes, as well as to highlight advancements in their development and applications. This Special Issue is available online at: https://www.mdpi.com/journal/processes/special_issues/metal_manufacturing_processes.

Machine learning algorithm in manufacturing metal products

In the production of large amounts of metal products, the industry is continuously seeking to improve the manufacturing efficiency and reduce energy consumption. As many manufacturing actions are selected based on operator experience, providing operators with sufficient technical support is recognized as one of the effective strategies for enhancing production efficiency. Recently, an application of machine learning algorithms in the manufacturing process obviously facilitated the working efficiency of the operators.

The paper by Andreiana et al. [1] introduces a reinforcement learning (RL) algorithm (concretely Q-Learning) as the core of a decision support system (DSS) in the steel manufacturing process of the Composition Adjustment by Sealed Argon Bubbling with Oxygen Blowing (CAS–OB), which provides significant benefits to operators in taking proper and correct decisions during manufacturing processes, especially for those who are less experienced. The proposed algorithm successfully learns the process using raw data from the historical database and recommends the same operation as those taken by the operator 69.23% of the time. Furthermore, incorporating the operator's experience into the DSS knowledge could facilitate the integration of operators with limited experience.

The paper by Chen et al. [2] presents a soft sensor modeling method called a just-in-time learning-based triple-weighted regularized extreme-learning machine (JITL-TWRELM) to measure the real-time liquid aluminum temperature in the manufacturing of aluminum on the regenerative aluminum smelting furnace. To address the process time-varying problem, a weighted JITL method (WJITL) is used to update the online local models, and a regularized extreme-learning machine model—with respect to sample similarities and variable correlations—is established as the local modeling method. The proposed models show satisfactory prediction accuracy in the manufacturing of aluminum.

Fabrication and mechanical property optimization of metals

The additive manufacturing of metal or alloys has attracted great attention in recent years. Selective laser melting (SLM) is a common additive manufacturing technology that has been widely used to fabricate some complex metal structures. The paper by Ran et al. [3] compares the mechanical properties of a metallic 3D auxetic structure produced by the SLM and micro-assembled (MA) methods, and investigates the influences of size effect on the deformation mechanism of C5131 bronze 3D auxetic structures with a negative Poisson's ratio, using both FEM simulations and experiment verification. This paper provides helpful instructions for establishing an accurate constitutive model for predicting the evolution of the mechanical behaviors of a 3D auxetic structure.

To expand the applications of metals, it is necessary to conduct optimization on their mechanical properties, which could be achieved by modifying their microstructures—particularly, grain size and phase composition. Reducing the grain size or modifying the phase composition are important approaches to achieve the superior mechanical strength of the metals. The paper by Wang et al. [4] introduces a method of element alloying to modify the microstructures of materials for obtaining high mechanical properties. The authors investigate the effects of Ca addition on the morphological modification of Al–Fe–Si alloys. Their results show that optimization of the amount of Ca addition (0.01–0.1 wt.%) is capable of refining α-AlFeSi and β-AlFeSi morphologies and transforming the β-AlFeSi phase into α-AlFeSi phase, thereby enhancing the mechanical performances of the alloys.

The review paper summarized by Cui et al. [5] provides state-of-the-art information regarding the effects of equal-channel angular-pressing (ECAP) processing parameters, such as passes, temperature and routes, on the microstructural evolution of metallic materials. Based on existing studies, the various parameters required to achieve submicron and nanoscale grain sizes as well as different phases are analyzed, providing practical guidance for optimizing the processing parameters during the ECAP of different metallic materials. A large strain rate and dynamic recrystallization are supposed to be the main mechanisms of grain refinement. The plastic deformation mechanism of the ECAP process is also discussed from the perspectives of dislocation slipping and twinning behavior. Additionally, some challenges and perspectives of the ECAP are presented at the end of this paper in an effort to enhance the ECAP manufacturing process.

Numerical simulation and experiments in the machining of metals

Mechanical machining is one of the primary techniques to manufacture metal components, especially to obtain the desired final parts with high precision and form accuracy, and it mainly includes milling, turning, drilling and grinding. The finite element method (FEM) has been widely used in the numerical simulation of the machining processes for parameter optimization and cost savings. The paper by Tagiuri et al. [6] uses the FEM-based DEFORM-2D software to simulate the machining performance in the orthogonal milling of AISI 1045 steel, and the authors study interactions between chamfer width, chamfer angle, sharp angle, cutting speed and feed rate. More specifically, the simulation results are statistically analyzed to determine significant parameters using the analysis of variance (ANOVA) test, indicating that both chamfer and sharp tools, feed rate, cutting speed, and their interactions are the most important parameters that influence machining temperature and stress.

To achieve highly efficient and low-cost machining, it is essential to reduce tool wear by optimizing the structure of cutting tools. The paper by Yin et al. [7] experimentally investigates the edge structure of polycrystalline cubic boron nitride (PCBN) tools and cutting parameters on machining performance in the dry turning of grey cast irons, and the results indicate that a minimum surface roughness could be obtained with a feed rate of 0.15 mm/r that exceeds the tool chamfer width in the turning process. The paper also analyzes the PCBN tool wear mechanism, revealing that the micro notches on the rake face and micro-chipping on the tool chamfer are the main tool wear modes. Another paper by Kwok et al. [8] introduces diamond protection films by hot-filament chemical vapor deposition (HFCVD) on the drilling tool for drilling high-frequency printed circuit boards (PCBs). The paper demonstrates that the tool wear with nanocrystalline protection films is almost 90% less than microcrystalline diamond-coated tools, resulting in a significant increase in tool life. In addition, their findings emphasize the significance of HFCVD parameters for coated drills that process high-frequency PCBs, thereby contributing to the highly efficient production of PCBs for industrial applications.

Aside from cutting parameters, the machining performance of metals can also be improved by modifying the microstructures of the materials, especially for microscale machining. The paper by Jing et al. [9] experimentally investigates the effects of grain size on the surface integrity, cutting forces and chip formation in the micro-turning of oxygen-free high-conductivity (OFHC) copper. By comparing the feed rates and material microstructures, it is found that a smooth surface and small width of the flake structure can be achieved when the feed rates are equivalent to the grain sizes.

The review paper by Huo et al. [10] summarizes the mechanical machining performance and applications of Fe-based amorphous alloys, which are known as difficult-to-machine materials due to their extreme hardness and severe chemical tool wear. The review also compares various assisted machining approaches, including tool-assisted machining, low-temperature lubrication-assisted machining and magnetic field-assisted machining, for improving the machining performance of Fe-based amorphous alloys. In the future, it is anticipated that difficult-to-machine amorphous alloys will be machined using a combination of the above assisted machining methods. This paper provides useful and practical instructions for the optimization and design of metal machining processes.

Funding: There are no funding supports.

Conflicts of Interest: The authors declare no conflict of interest.

References

1. Andreiana, D.S.; Acevedo Galicia, L.E.; Ollila, S.; Leyva Guerrero, C.; Ojeda Roldán, Á.; Dorado Navas, F.; del Real Torres, A. Steelmaking Process Optimised through a Decision Support System Aided by Self-Learning Machine Learning. *Processes* **2022**, *10*, 434. [CrossRef]
2. Chen, X.; Dai, J.; Luo, Y. Temperature Prediction Model for a Regenerative Aluminum Smelting Furnace by a Just-in-Time Learning-Based Triple-Weighted Regularized Extreme Learning Machine. *Processes* **2022**, *10*, 1972. [CrossRef]
3. Ran, J.; Chen, G.; Zhong, F.; Xu, L.; Xu, T.; Gong, F. The Influence of Size Effect to Deformation Mechanism of C5131 Bronze Structures of Negative Poisson's Ratio. *Processes* **2022**, *10*, 652. [CrossRef]
4. Wang, M.; Guo, Y.; Wang, H.; Zhao, S. Characterization of Refining the Morphology of Al–Fe–Si in A380 Aluminum Alloy due to Ca Addition. *Processes* **2022**, *10*, 672. [CrossRef]
5. Cui, L.; Shao, S.; Wang, H.; Zhang, G.; Zhao, Z.; Zhao, C. Recent Advances in the Equal Channel Angular Pressing of Metallic Materials. *Processes* **2022**, *10*, 2181. [CrossRef]
6. Tagiuri, Z.A.M.; Dao, T.-M.; Samuel, A.M.; Songmene, V. Numerical Prediction of the Performance of Chamfered and Sharp Cutting Tools during Orthogonal Cutting of AISI 1045 Steel. *Processes* **2022**, *10*, 2171. [CrossRef]
7. Yin, G.; Shen, J.; Wu, Z.; Wu, X.; Jiang, F. Experimental Investigation on the Machinability of PCBN Chamfered Tool in Dry Turning of Gray Cast Iron. *Processes* **2022**, *10*, 1547. [CrossRef]
8. Kwok, F.M.; Sun, Z.; Yip, W.S.; Kwok, K.Y.D.; To, S. Effects of Coating Parameters of Hot Filament Chemical Vapour Deposition on Tool Wear in Micro-Drilling of High-Frequency Printed Circuit Board. *Processes* **2022**, *10*, 1466. [CrossRef]
9. Jing, C.-Z.; Wang, J.-L.; Li, X.; Li, Y.-F.; Han, L. Influence of Material Microstructure on Machining Characteristics of OFHC Copper C102 in Orthogonal Micro-Turning. *Processes* **2022**, *10*, 741. [CrossRef]
10. Huo, Z.; Zhang, G.; Han, J.; Wang, J.; Ma, S.; Wang, H. A Review of the Preparation, Machining Performance, and Application of Fe-Based Amorphous Alloys. *Processes* **2022**, *10*, 1203. [CrossRef]

Article

Steelmaking Process Optimised through a Decision Support System Aided by Self-Learning Machine Learning

Doru Stefan Andreiana [1,*], Luis Enrique Acevedo Galicia [1], Seppo Ollila [2], Carlos Leyva Guerrero [1], Álvaro Ojeda Roldán [1], Fernando Dorado Navas [1] and Alejandro del Real Torres [1]

1 IDENER, IT Department, 41300 Sevilla, Spain; luisenrique.acevedo@idener.es (L.E.A.G.); carlos.leyva@idener.es (C.L.G.); alvaro.ojeda@idener.es (Á.O.R.); fernando.dorado@idener.es (F.D.N.); alejandro.delreal@idener.es (A.d.R.T.)

2 SSAB Europe Oy, Processes Development Steelmaking, 60100 Seinäjoki, Finland; seppo.ollila@ssab.com

* Correspondence: doru.stefan@idener.es

Abstract: This paper presents the application of a reinforcement learning (RL) algorithm, concretely Q-Learning, as the core of a decision support system (DSS) for a steelmaking subprocess, the Composition Adjustment by Sealed Argon-bubbling with Oxygen Blowing (CAS-OB) from the SSAB Raahe steel plant. Since many CAS-OB actions are selected based on operator experience, this research aims to develop a DSS to assist the operator in taking the proper decisions during the process, especially less experienced operators. The DSS is intended to supports the operators in real-time during the process to facilitate their work and optimise the process, improving material and energy efficiency, thus increasing the operation's sustainability. The objective is that the algorithm learns the process based only on raw data from the CAS-OB historical database, and on rewards set according to the objectives. Finally, the DSS was tested and validated by a developer engineer from the CAS-OB steelmaking plant. The results show that the algorithm successfully learns the process, recommending the same actions as those taken by the operator 69.23% of the time. The algorithm also suggests a better option in 30.76% of the remaining cases. Thanks to the DSS, the heat rejection due to wrong composition is reduced by 4%, and temperature accuracy is increased to 83.33%. These improvements resulted in an estimated reduction of 2% in CO_2 emissions, 0.5% in energy consumption and 1.5% in costs. Additionally, actions taken based on the operator's experience are incorporated into the DSS knowledge, facilitating the integration of operators with lower experience in the process.

Keywords: machine learning; reinforcement learning; Q-learning; steelmaking process CAS-OB; decision-support system; optimisation algorithm

Citation: Andreiana, D.S.; Acevedo Galicia, L.E.; Ollila, S.; Leyva Guerrero, C.; Ojeda Roldán, Á.; Dorado Navas, F.; del Real Torres, A. Steelmaking Process Optimised through a Decision Support System Aided by Self-Learning Machine Learning. *Processes* **2022**, *10*, 434. https://doi.org/10.3390/pr10030434

Academic Editors: Guoqing Zhang, Zejia Zhao and Wai Sze Yip

Received: 1 February 2022
Accepted: 18 February 2022
Published: 22 February 2022

1. Introduction

The Composition Adjustment by Sealed Argon-bubbling with Oxygen Blowing (CAS–OB) is a secondary steelmaking process developed in the 1980s by Nippon Steel Corporation [1]. The main goals during this process are homogenisation, temperature control and composition adjustment [2]. As a result, CAS-OB has become one of the relevant buffer stations in the secondary metallurgy of steelmaking thanks to its capability of good chemical composition control, steel homogeneity, and reheating [3]. Furthermore, the process enables the consistent correction of high alloy composition and the reheating of the steel using the exothermic reaction between oxygen and aluminium. The study presented has been done under the MORSE project (Model-based optimisation for efficient use of resources and energy) [4] funded by the European Commission within its Horizon 2020 framework programme. The case under investigation pertains to the SSAB Europe Oy Raahe steel plant with two CAS-OB stations [5], which provided the data for performing the training of the algorithms and then performed the validation of the final developed DSS.

1.1. Motivation

In large-scale production, such as steelmaking, even small changes in resource and energy consumption can make a difference. Hence, the steel industry is continuously looking to enhance sustainability, as stated in similar studies concerning aluminium reduction and energy efficiency [6,7].

Additionally, besides the efficiency of the plant, it is crucial to consider the high impact of those processes on the environment [8]. Energy consumption constitutes up to 40% of the cost of steel production [9]. Specific energy consumption at Raahe steelworks in 2016 was 17.8 GJ/t of steel and specific CO_2 emission 1650 kg/t of steel. One of the objectives of this study is to reduce those numbers and, in consequence, enhance the efficiency and sustainability of the process.

With that purpose in mind, the first step is to minimise the rejected products, and the second is to provide support to the operators. Many of the actions taken by the operators are based on their own experience, which makes the integration of the new staff into the workforce difficult. Therefore, another objective is to gather that knowledge based on operator experience into the DSS, for the ease of new workers' integration.

These kinds of processes usually lack digitalisation and integration of AI-based techniques. Thanks to the advances in those fields in the last years, the objectives can be achieved by replacing the classic control techniques with novel control techniques strengthened by machine learning algorithms, such as an adaptive controller based on Q-Learning [10]. This can speed up the transformation towards industry 4.0 and all its benefits [11], such as the enhancement of the steelmaking process's efficiency [12] and facilitating the work of the operators while accommodating them to digitalisation [13]. A review of the latest methods and tools for improving the steelmaking process was presented by T. Rotevatn et al. in 2015 [14]. Some of these methods have already been applied and have given excellent results [15], which also served as motivation, and provides evidence that this is the right path.

1.2. Methodology and Goals

This research aims to develop a decision support system (DSS) to assist the operator in taking the proper decisions since many CAS-OB actions decisions are made based on operator experience. The DSS includes self-learning ability thanks to the core based on Q-Learning, which has already been stated to give excellent results in similar studies of different metallurgic processes [16]. Furthermore, the algorithm will adapt and learn new features while working, improving its suggestions and achieving the goals defined, thanks to the high impact of the self-learning ability [17].

The guidance is expected to reduce additional corrective actions, reduce duration time and minimise failed operations. Furthermore, the Q-Learning algorithm used as the core of the DSS will learn the process from the historical data of two CAS-OB stations provided by SSAB Europe Oy. This way, it is possible to develop an RL algorithm based only on raw data. The steps followed during this research to achieve the objectives proposed are presented in the following list:

1. **Select the most suitable RL algorithm.** There are many RL algorithms [18], and each one suits a different kind of problem. Therefore, this step involves analysing the available algorithms and selecting the most suitable for the process's features.
2. **Transform CAS-OB process into a Markov Decision Process (MDP).** This task involves an exhaustive analysis of the process, including an analysis of historical data from it. Once done, the process's discretisation is undertaken, defining representative states and actions based on the analysis results.
3. **Train the algorithm.** In this study, the objective is to train an RL algorithm only with experimental data. Therefore, historical data is used from the process. The data will be transformed according to the discretisation set before, and episodes will be defined to "feed" the algorithm.

4. **User interface (UI).** The next step is to incorporate the DSS on the plant and facilitate access through a user interface.
5. **Test.** The DSS is tested in the CAS-OB steel plant station by development engineers. The test includes validating the correct behaviour of the DSS and gathering data for performance measurement.

2. Reinforcement Learning

2.1. Origin

Machine learning (ML) main areas are supervised learning and unsupervised learning [19,20]. Nevertheless, a third area has been gaining popularity in the last decade: Reinforcement Learning (RL). What distinguishes it from the other main areas is that it does not rely on correct behaviour examples, as supervised learning does. On the other hand, unlike supervised learning, its objective is not finding common or hidden patterns on data, neither classify.

Reinforcement learning's [21] origin rests on the idea of learning by interaction. RL algorithms aim to solve a problem by direct interaction with the environment, learning by trial and error, based on the consequences of the actions in terms of rewards. Practising and exploring new ways of doing something produces a wealth of information about each action's effects, and it helps define the best action path for solving the problem. Figure 1 presents a scheme with the RL methodology.

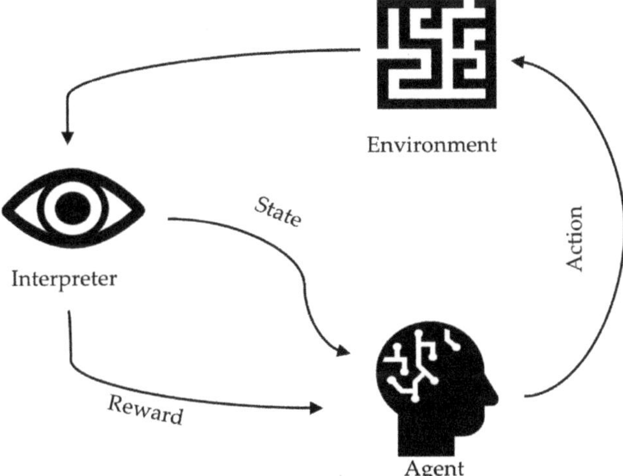

Figure 1. Reinforcement Learning methodology.

RL's popularity increased greatly thanks to its application on games defining agents capable of surpassing human abilities, such as Go [22] and Dota 2 [23]. Nevertheless, the RL algorithm can go beyond winning games [24]. Nowadays, thanks to the wide variety of algorithms developed, the applic ation area has greatly expanded due to its advantages over classic control techniques. The classic control techniques and RL algorithms are frequently compared [25].

2.1.1. Reinforcement Learning Elements

In Reinforcement Learning, the main elements are the agent and the environment, which interact with each other. However, the interaction involves some other subelements [21]:

- **Agent.** The learner and decision-maker.

- **Environment.** What the agent interacts with, comprising everything outside the agent. The interpreter defines the state of the environment and gives rewards as feedback to the agent.
- **State (S).** A sample of the environment's features involved in the case under study. It defines the situation of the environment and works as feedback.
- **Action (A).** A control signal of the environment. The defined actions must be those that change the states involved in the problem.
- **Episode.** A succession of steps, composed of the state of the environment and the action taken by the agent. In the case under study, the step ends when the action changes the environment status. At the end of the step, a reward is received based on the new state achieved. These episodes are used to train the RL agent. The format of an episode can be seen in Figure 2.
- **Policy (π).** The core of the RL agent, which defines how the agent behaves in each state, based on what has been learned as the best action. It can be a simple function or a lookup table, although it can involve extensive computation depending on the problem's complexity.
- **Reward (R).** The reward defines the goal of the problem. Each state gets a reward as feedback on how well the RL agent did when taking the previous actions. The RL agent will define a policy maximising the rewards obtained. Therefore, the rewards must be set according to the objectives of the process. Rewards assignment is a critical task because agent behaviour will rely on them.
- **Value function.** There are two types, value-functions and action-value functions. The difference is that the first one calculates the values of the states only, and the second one calculates the values for pairs of state and action. Moreover, in both cases, the formulas have modifications depending on the algorithm. The algorithm applied in this study, Q-Learning, uses an action-value function. The values calculated determine the best action, considering long-term consequences.

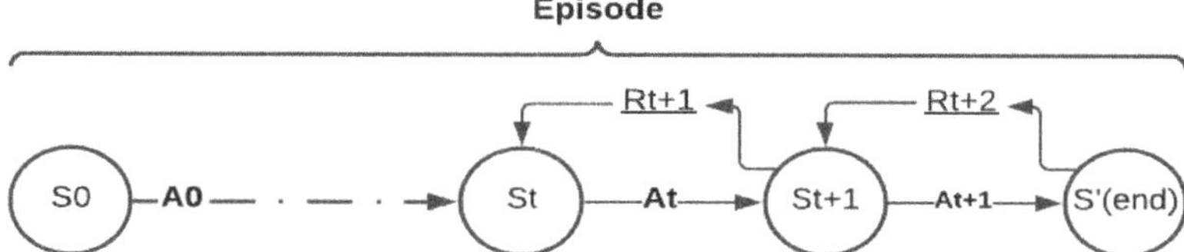

Figure 2. Episode structure.

The way the value function is updated is what distinguishes one RL method from another. It is done by following an update rule which depends on the RL method because each one estimates the new values differently. The generic update rule is formulated as in Equation (1):

$$NewEstimation \leftarrow OldEstimation + StepSize[Target - OldEstimation] \quad (1)$$

where *Target* is the estimated accumulative reward in each state, and the expression [*Target* − *OldEstimate*] is an error in the estimation. Moreover, the variable *StepSize*, known as alpha(α), controls how much the agent learns about its experience. This parameter is between 0 and 1, and it may be constant or variable.

2.1.2. Markov Decision Process

The Markov decision process is a classical definition of sequential decision-making, where actions influence immediate rewards and subsequent states, and therefore those future rewards. It is a substantial abstraction of the problem of goal-directed learning from interaction. It proposes that any problem can be reduced to three signals that provide communication between the agent and the environment. This concept is used as a mathematical form of the reinforcement learning problem [26]. However, it is used in many other fields, such as medicine [27] and robotics [28], among many others.

Aiming to frame the problem of learning from interaction, MPDs are the base and a requirement of most of the RL algorithms, amongst them the Q-Learning, used in this case study. In the MDP framework, the learner and decision-maker are called *agents*, and what they interact with is called the *environment*. The interaction is done through *actions*, and the environment responds with its new *state*, and the consequences, namely *rewards*, are set according to the goal. Further on, the agent will aim to maximise the *rewards* through its *actions* and the knowledge learned. Figure 3 illustrates the structure of an MDP.

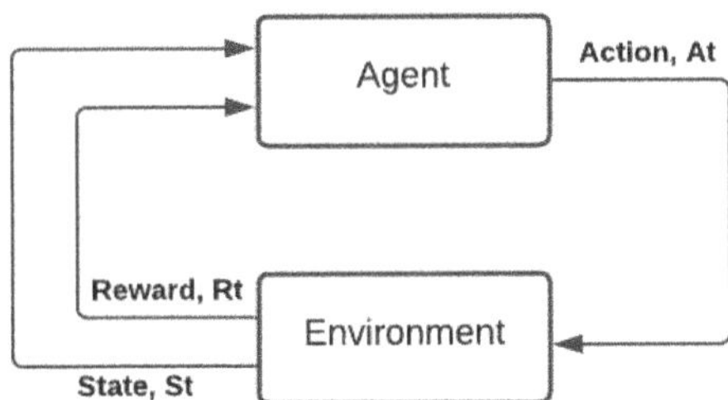

Figure 3. Agent–environment interaction in a Markov decision process.

The boundary between agent and environment represents the point beyond which the agent lacks absolute control and knowledge. However, the border does not have to be physical; it can be located at different places depending on the goal. This boundary is defined once the states, actions and rewards are defined. Then, the interaction is divided into *steps* (t), and the sequence generated depends on the actions selected and the dynamics of the environment.

2.2. Reinforcement Learning Algorithms

RL methods have the advantage that they can learn from direct experience of the process. If the system's operations are saved on a database, these can be used as training data. This also brings up probably the most significant advantage of the RL, continuous training. Moreover, this learning allows the agent to adapt to changes in the process without designing the method again.

The most relevant tabular methods have been studied to find the most suitable RL method for the problem. These methods are briefly described below. Their description also helps clarify the method used.

- Dynamic programming (DP) [29]: this is an assemblage of algorithms capable of computing optimal policies if a perfect environment model is given. This requirement, together with the tremendous computational cost, limits the utility of DP algorithms.
- Monte Carlo (MC) methods [30,31]: these methods are an improvement compared to dynamic programming because their computational expenses are lower, and a perfect model of the environment is not necessary. This last feature means that knowing the environment's dynamics is not required to attain the optimal policy; the agent learns from direct interaction. Instead, these methods only need a simulated or real experience of this interaction in the form of state-action-reward sequences. Another distinctive feature of these classes of techniques is that they estimate values functions by averaging sample returns. Thus, MC methods are very suitable for episodic tasks, since they base their estimations on the final outcome and not on intermediary outcomes.

2.3. Self-Learning Algorithm: Q–Learning

Temporal difference (TD) learning is the most significant thread in RL [32]. It is a combination of DP and MC. TD methods, just like MC, are model-free, and additionally, they can learn from raw experience [33]. On the other hand, they assimilate with DP on how the estimation is updated. The values are updated based on previously known estimations, performing the calculation on the fly. Consequently, in contrast with MC methods, TD methods do not have to wait until the end of the episode to determine the value states. TD methods only need to wait until the next time step, and the update is done in "transition". This is why TD methods are so popular. There is no need to know how the episode will end; the agent learns from its estimations. If the estimation were very wrong, it would learn much more. Further on, convergence is still guaranteed. Moreover, TD methods have been proved to be the fastest.

The Q–Learning [34] method is one of the most known and used methods of RL. It is a model-free method and uses an off-policy approach. A side effect of being model-free and a temporal difference method is the capacity of being used online. It can be used online because it is unnecessary to know the episode's ending to update the values. This feature is essential in the problem studied because, in the final version, the agent should be online and recommend actions in real-time during the steelmaking process, adapting to each situation. Besides, this method also can handle stochastic problems, as in the problem treated in this project.

Moreover, Q-Learning is a control method, so it is based on an action-value function, which suits perfectly the problem in hand. The function used approximates directly to the optimal action-value function. Furthermore, it converges faster because it always aims for taking the best action. The duration will primarily be determined by the size of the state and action space. The action-value function used in Q-Learning is shown in Equation (2). It is used to calculate the value of each pair of states and actions. Afterwards, based on these values, the most suitable action, the one with higher value is selected.

$$Q(S_t, A_t) = Q(S_t, A_t) + \alpha \left[\underbrace{R_{t+1} + \gamma \max_a Q(S_{t+1}, a)}_{Target} - Q(S_t, A_t) \right] \tag{2}$$

- The sub-index t indicates the step during the episodes.
- $\alpha \in (0, 1]$, the step size or learning rate determines to what extent newly acquired information overrides old information. The value could vary during the training, so we learn a large amount at the beginning, and at the end, when the optimal values are reached, there is no override.
- $\gamma \in [0, 1]$, the discount rate, indicates the importance of the next reward, or in this case, the value of the next possible state and action pair.

The pseudocode of Q-learning algorithm is presented in the Algorithm 1 [35]:

Algorithm 1. Q-learning (off-policy) for estimating $\pi \approx \pi_*$

Algorithm parameters: step size, $\alpha \in (0, 1]$, $\gamma \in (0, 1]$
Initialise Q (s, a) for all $s \in S^+$, a $\in A(s)$, arbitrarily except that Q (terminal,) = 0
Loop for each episode:
Initialise S
Loop for each step of the episode:
Choose A from S using policy derived from Q
Take action A, observe R, S′

$$Q(S_t, A_t) = Q(S_t, A_t) + \alpha[R_{t+1} + \gamma * max_a Q(S_{t+1}, a) - Q(S_t, A_t)]$$

$S \leftarrow S'$
Until S is terminal

2.4. Comparison with Classic Control Techniques

Classic control efficiency is mainly limited by the model reliability. The control method is designed to control the model. Moreover, the control design will not adapt to changes in the process. Any modification on the process means defining a new model and developing a new controller.

Meanwhile, RL algorithms, such as Q-Learning, fill the gap limitations of classical control. The algorithm's design only requires knowledge about the process, the states that define it, the actions (control signals), and the consequences of these actions. So, it is helpful to have a model; however, it is not a requirement. Figure 4 presents a summarised scheme of the comparison.

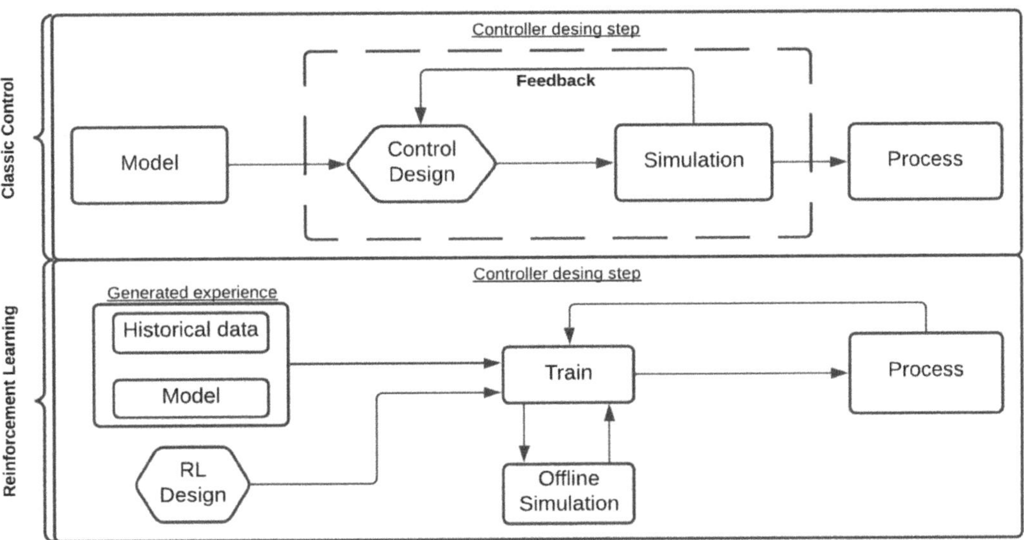

Figure 4. Reinforcement learning comparasion with classic control.

3. Development of the Solution

As already mentioned, the RL agent will use a Q-Learning algorithm. As its features have already been explained, this section will focus on the actual implementation. Figure 5 shows an overview of the whole developed and integrated solution.

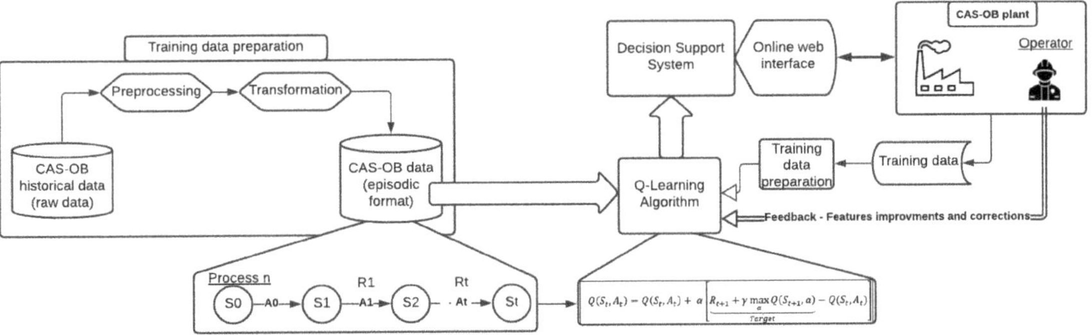

Figure 5. Scheme of solution developed and implemented.

In the first place, the historical data from the CAS-OB process must be processed, selecting only the valuable data since the database includes all the information regarding the CAS-OB process. Once the variables needed for the episode definition and the further training are designated, the next step is the transformation into an episodic format. As already explained, the episodic format is necessary due to the MDP nature of the self-learning algorithm. The algorithm receives the state of the environment as an input, and instead of following an action based on the policy, it selects the action defined in the episode. In other words, during the training, the agent will strictly follow the episodes while calculating the values of each state-action pair.

Once the algorithm finishes the training with the historical data and performs some validations, it can be introducted into the DSS implemented in the CAS-OB plant and used by a developer engineer. During this interaction, the algorithm still learns from the data gathered while it is used. Moreover, developer engineer feedback is crucial at the beginning for features improvements, corrections of the behaviour or states and action discretisation.

3.1. CAS-OB Steelmaking Process Description

CAS-OB works with liquid steel from a previous process, the *Basic Oxygen Furnace* (BOF) [36]. The main goals for this secondary metallurgy process are homogenisation, temperature control and adjustment of composition. After that, the liquid steel is transferred to the Continuous Casting Machine (CCM) (see Figure 6).

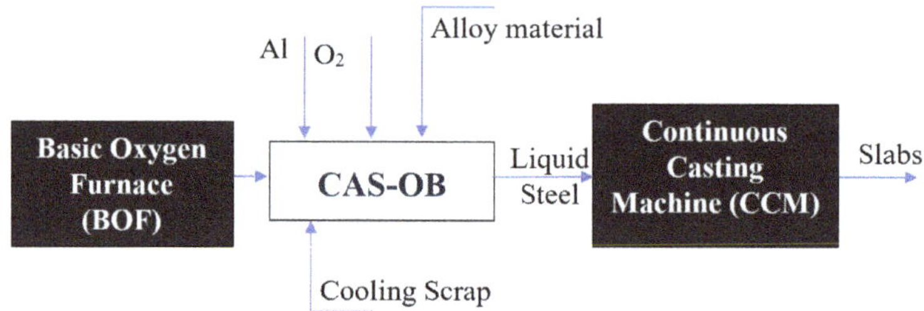

Figure 6. Diagram of the steelmaking process summarised.

Composition adjustment starts in the previous phase, the BOF tapping. Afterwards, during the CAS-OB process, the homogenisation and adjustment of the composition of molten steel must be maintained while the temperature is adjusted. Composition adjustment is achieved by adding a stoichiometric proportion of alloys which is calculated using a composition sample.

Temperature and dissolved oxygen measurements are made before alloying in order to verify whether these two variables need to be adjusted. Should there be a corrective action to be taken, it has to be carried out before alloying, in order to ensure no extra oxidation is needed, as it would alter the composition and therefore a subsequent adjustment would be needed. Once this alloying is done, composition is deemed to be satisfactory and no more alloying is necessary, unless a subsequent sample analysis may indicate the need. Liquid steel temperature can be increased by adding aluminium with oxygen blowing, using the exothermic reaction between them for reheating. Whenever refrigeration is needed, cooling scrap is added to decrease the temperature. Likewise, this technique can be used as well for minor temperature adjustments at the end of the process. Considering steel cleanliness, the time at which cooling scrap is added is not so critical, the reason for this being that it does not change the composition significantly, and therefore it can be added whenever it is needed. Once the goal is achieved, the liquid steel moves forward to the third process, the Continuous Casting Machine (CCM).

3.2. *Definition of Environment*

The process will be transformed into an MDP. This requires defining a finite representative number of states and actions. The definition of these states and actions is based on the historical data analysis and the process's objectives.

3.2.1. Historical Data Analysis and Treatment

The historical data includes all the information from a real CAS-OB plant, gathering 9720 steel treatments, also called "heats". Only the significant information for the formulation of the problem was extracted from this data, such as the measurements, targets and actions taken by the operators. Table 1 presents the parameters gathered from the historical data.

Table 1. Data gathered from CAS-OB historical data.

Treatment Data	Measurements	Actions	Targets
Heat number	Measuring time	Addition time	Composition
Start of treatment	Composition, % of each alloy	Material	Temperature
End of treatment	Temperature	Quantity, kg	
Start of reheating	Dissolved oxygen		
Reheating time, s			
End of reheating			
Count of reheating			
Total aluminium, including reheating, kg			
Cooling scrap, kg			
Steel amount			
The time when treatment should be ready			

Making use of the knowledge on how the plant works and the historical data provided by the steel factory the process was modelled as an MDP. The data was previously treated eliminating inconsistencies, outliers and corrupted information, and afterwards the states and actions were properly defined.

The heat number classifies the steel treatments in the database. Each heat contains information about measurements, actions, targets and what has been done during the process. Those are the most important information for the classification of the states and the actions. Timestamps will help set up the state and action transition chronologically and generate a complete episode when something has been done.

If some information is missing on some heat, e.g., missing essential measurement or misleading information, the first option will be to fill the gap if possible to preserve the episode and not lose data. As an example, there are instances in which steel temperature is recorded to be at 0 °C. This is obviously an impossible value for the process; as a result on these occasions temperature was removed from the log, but not the other variables. The missing values of temperature would be inferred manually afterwards.

After cleaning and analysing the data, the classification and definition of states can be done. The current database consists of 13,800 heats in total. The duration of the episodes varies depending on how many times reheating has been done. In addition, the data provide information about the features of the measurements, target, and actions, e.g., "adding Al".

The final dataset is divided based on how many times reheating has been done. This division is done because depending on the reheats done, the data will be treated differently. The distribution is presented in Figure 7. The idea of the RL is to generalise. Focusing excessively on exceptional cases would guide the agent too much. Hence it will not find new optimised ways of solving the problem. It must discover by itself those exceptional cases.

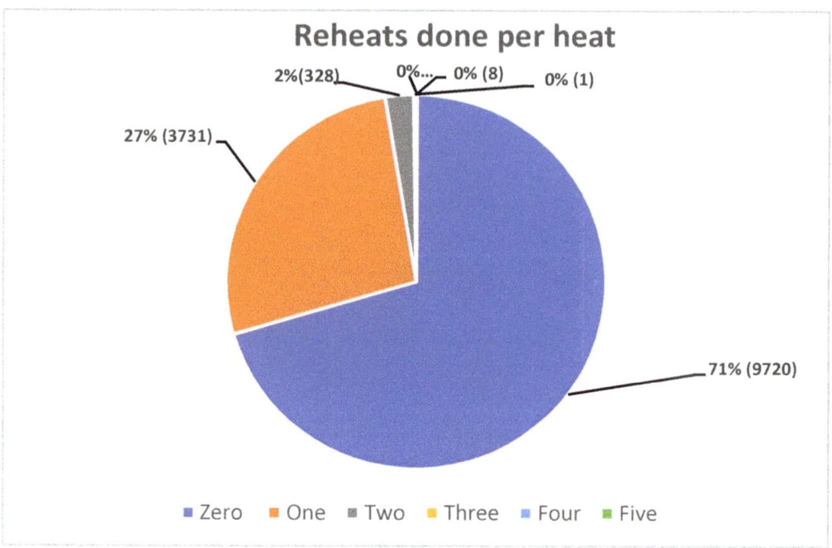

Figure 7. Pie chart representing the number of episodes per number of times reheating was done during the episode.

3.2.2. States

The states for the CAS-OB use case are defined by five parameters: the dissolved oxygen, the composition status, composition grade, steel mass, and the temperature difference. These parameters are continuous. Henceforth they must be discretised. If not, the state space will have infinite size. Below are listed and detailed the discretisation done to all the parameters.

- Mass of the steel:

The mass of steel is characterised into two categories, high and low. Its value for most of the steel heats is around 125 tons. Nevertheless, there are heats with reduced mass. The mass for these types of steel is 15–20% below the usual value. This must be taken into account because if the mass is considerably lower than usual, then the amounts of Al and O2 needed to heat the steel will be lower. A smaller mass requires reduced actions for the same results. So, the process with liquid steel mass below 110 tonnes will be considered low, and higher values will be regarded as high.

- Dissolved oxygen (DO):

DO is measured in ppm, and its target value is as close to zero as possible to ensure steel cleanliness. After the BOF process, the dissolved oxygen content is typically several hundred ppm. Dissolved oxygen is decreased by adding aluminium into steel first during BOF tapping and then during the CAS-OB process if needed, according to dissolved oxygen measurements. Aluminium reacts with oxygen forming aluminium oxide. Ideally, the exact amount of aluminium required for the reaction should be added.

Nevertheless, sometimes the amount added is not precise, or the reaction did not go on exactly as expected. From the historical data, the mean calculated value is equal to 5.18 ppm. The discretisation of the DO reduces the possible values to five ranges, all close to the mean. The discretisation of the DO is presented in Table 2.

- Temperature.

This state is defined as the difference between the target temperature during the process and the measured temperature. The discretisation is shown in Table 3 and is done based on the temperature difference distribution at the end of the process, considering any waiting time due to the scheduled ending time.

Table 2. Dissolved oxygen discretisation.

Dissolved Oxygen Intervals	Codification of the Intervals
DO < 1	0
1 ≤ DO < 2	1
2 ≤ DO < 3.5	2
3.5 ≤ DO < 5	3
5 ≤ DO < 7	4
7 ≤ DO < 10	5
10 ≤ DO	6

Table 3. Difference temperature states.

Temperature Difference Intervals	Codification of the Intervals
$\Delta T < -20$	−3
$-20 \leq \Delta T < -10$	−2
$-10 \leq \Delta T < -2$	−1
$-2 \leq \Delta T \leq 2$	0
$2 < \Delta T < 10$	1
$10 \leq \Delta T < 20$	2
$20 \leq \Delta T < 30$	3
$30 \leq \Delta T < 40$	4
$40 \leq \Delta T$	5

- Composition state.

The composition of the steel processed includes 16 elements: C, Si, Mn, P, S, Al, Nb, N, V, Ni, Ti, B, Ca, Cu, Cr, and Mo. The measurements taken during the process show the percentage of each element in the liquid steel, These are compared with the goals for those percentages, namely, minimum, target and/or maximum. Ideally, each percentage should reach its target if that is provided. However, if this value is not provided, then the composition should fit between the maximum and minimum percentage levels by setting the mean of both as the target percentage. The aimed result is to achieve the type of steel that is required.

Until the measurement is done, the composition may be considered incorrect. Once the measurements are taken, if any of the percentages are below the target, the composition is erroneous, and alloying is necessary., Otherwise, the composition is correct, and alloying is not necessary. So basically, there are **two possible states, correct and incorrect composition**. However, the incorrect composition state is disaggregated in four states: **Very Close, Close, Far and Very Far**. However, while the composition is incorrect, the action will always be "alloying", and how far the composition is from the target does not alter that. However, a higher resolution is required to estimate the temperature decrease accurately. The further the composition is from the target, the more alloys will be added, which could result in a higher drop in temperature. So, by making that division, the RL agent will learn that the alloying can reduce the temperature.

- Composition grade.

The grade of steel provides more information about the steel treated and its features. Some grades require different actions for the same consequences. Hence, the grades have been categorised into four groups, each with steels of different features.

3.2.3. Actions

CAS-OB process is complex and involves many actions. However, the RL agent will take into account only the actions, which are mainly based on operator experience and whose timing, and the amounts, directly affect the process's outcome. All the actions are independent from one another, and there can be one taken per state. The list below details the actions considered and how they have been discretised:

- Addition of aluminium:

  Aluminium is usually added to reduce the dissolved oxygen in the liquid steel or adjust the composition during the process. Nevertheless, it is also added besides O2 to increase the temperature, which is called reheating. However, only about 25% of all CAS-OB heats are reheated. The reheating serves to increase the temperature. Despite added aluminium, blown oxygen can also react with other alloying elements, consequently spoiling the composition. Therefore, reheating should always be performed before correcting the composition.

  The RL agent recommends when to add aluminium to reduce the dissolved oxygen and when to add it for reheating. Additionally, it suggests the amount in kilograms of aluminium that should be added. To recommend the amount of aluminium, this must be discretised to facilitate the estimation. To do so, it is necessary to know the usual amount of aluminium added during the process to make those adjustments. For that purpose, the historical data from the process was analysed. Figure 8 shows how many processes (episodes) there were for each amount of Al that has been added. It is seen that the usual quantity is a multiple of ten, and the bigger it is, the fewer times it is added.

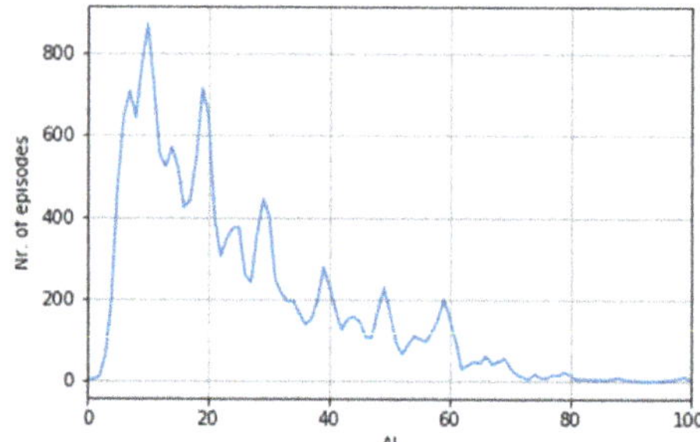

Figure 8. Number of episodes x Kg of Al that has been added.

- To perform the same analysis for the aluminium added for reheating, some pre-processing of the data was needed since the amount added for reheating is not specified in the historical data. Fortunately, the total amount of Al added during the process is specified. Hence, by subtracting the Al not added for reheating, the Al for reheating is obtained. The results of the analysis are similar to the previous analysis. The only difference is that the interval of amounts added is larger, and more significant amounts are usually added. Lastly, the RL agent does not specify the amount of O_2 added with the Al for the exothermic reaction, because the operator system directly calculates it.

A total of 26 discretised actions (15 for reheating, 11 for DO reduction) regarding the addition of aluminium will be considered, either for reheating or dissolved oxygen (Table 4).

Table 4. Aluminum addition discretisation.

Aluminium for Reheating	Aluminium for DO Correction
O_2 + Al 150 kg	Al 100 Kg
Between 135 and 145 kg of Al + O_2	Al between 85 and 95 Kg
Between 125 and 135 kg of Al + O_2	Al between 75 and 85 Kg
...	...
Between 15 and 25 kg of Al + O_2	Al between 6 and 15 Kg
Less than 15 kg of Al + O_2	Al less than 6 Kg

- Cooling Scrap:

  This action is used to reduce the temperature of the steel when needed. It is essential to mention that the RL agent may recommend cooling down the steel and ending the process faster. Nevertheless, the operator can refuse it if there is plenty of time for the steel to cool down by itself. The operator calculates the cooling scrap amount based on the temperature decrement desired.

- Alloying:

  This action corrects the composition and drastically reduces the temperature of the steel. The current control system of the plant provides the amount of alloying the process needs, and this action indicates to the operator the best time to add the alloys calculated by this control system.

3.2.4. Problem Size

Considering all the combinations between the sub-states of the parameters defined, the **size of the state space is 2520**. In other words, there are 2520 possible states. The size is quite large; however, many states will hardly ever occur if the physical nature of the problem is taken into consideration. On the other hand, summing all the possible actions, considering the discretisation of each one, there are **28 possible actions**. To conclude, this results in a Q table of 2520 $\times$ 28 size, hence **70.560 Q-values**. Many Q-values will probably not be calculated. Nevertheless, that is not a problem for the optimisation problem because, as mentioned with the states, many combinations of states and actions would make no sense in the real process, and just as the operator would do, the RL agent will directly discard those options.

3.2.5. Rewards

As explained before, rewards are the feedback for the agent indicating how well it is performing. This step is essential, as the behaviour of the agent will be based on how these rewards are assigned. The rewards must define the goal of the process correctly to achieve proper training and, in consequence, the correct functioning of the RL. If rewards are not appropriately set, the agent will not learn correctly how to optimise the process. Rewards must represent the goal of the process, hence the objective of the RL. As explained before, the RL will try to achieve the maximum value reward. With all this in mind, the rewards must be placed beneficially for the goal.

A reward will be set for all the possible end states, which in this case, means all of them as every state could be an end state. The reward will be selected depending on the end temperature, DO and composition status. Ideally, the temperature should be equal or very close to the target, DO approximately zero and the composition correct. Therefore, the reward for the ending status which such characteristics will be set with a maximum value, and rewards for approximate states with lower values. An incorrect composition will be determined directly as a bad ending, independently of the other states, and the RL agent will be penalised. The rewards will be placed as presented in Figure 9.

If the process ends with incorrect compositions, the "reward" is -100. Thus, almost all the rewards have their opposing pair for a negative end. By doing this, the categorisation and calculation of the values are more effective. Besides, for each incorrect action taken, there is a minor penalty (negative reward). The consequence of that constant minor negative reward will train the RL agent to finish faster and achieve the objectives in the minimum number of actions. Moreover, this penalty varies depending on the state. For example, the penalty is bigger if the temperature is much lower than the target, consequently, the RL agent will prioritise heating.

To summarise, the rewards will be set as follows:

- Maximum reward for the final states on the goal
- Intermediate reward for final states close to the goal, for example, good composition and temperature but a high dissolved oxygen level.

- High negative reward if the end composition is wrong.
- Intermediate negative reward for correct composition but wrong temperature or high DO.
- A small negative reward for each action taken to favour ending faster and reducing the number of actions taken.

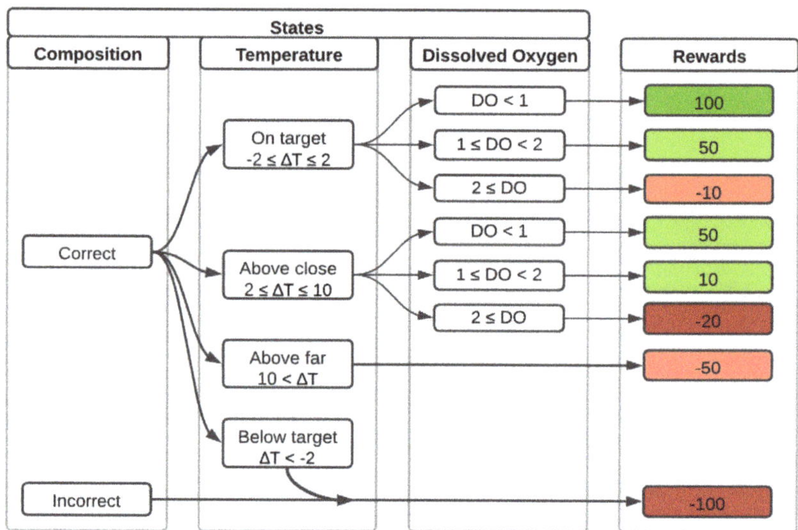

Figure 9. Rewards for the final states with the correct composition.

4. Implementation and Validation

4.1. Training Data

The training of the algorithm is done through the historical data of the process by SSAB. However, the data is not discretised and is not in the required format to be used as training data for the Q-learning algorithm. The first step is the discretisation of the data in state and actions. How is done has already been detailed above. The second step, detailed in this section, is the transformation of the data into episodic format and with the structure needed by the Q-learning formula (Figure 2).

Each episode will correspond to a heat, which is already classified. Regarding the discretisation, the data is transformed into the states and actions defined before. Once done each state must be associated with the action taken taken in that state, unless it is the ending state. The procedure followed is explained through an example of a real heat. After the discretisation, the format of the states of the heat is shown in Table 5 and of the actions in Table 6. However, some adjustments are required.

Firstly, the composition status is not entirely correct. In the process, once the alloying is performed, it is assumed that the composition is right and no further measurements of the composition are made, unless if the operator decides they are needed. Consequently, the composition status is not updated and, thus, the state is always "Close", hence incorrect composition. This first problem is solved by checking when the alloying action has been performed and uploading all the composition states that follow it to "Good".

Secondly, if we compare the addition times with the measuring times, it can be observed that several actions are taken between two state updates (those are marked on orange in the tables). Even if that can be done in practice, the configuration of the RL algorithm proposed does not allow taking several actions at the same time. Therefore, those states must be separated accordingly, associating an action to each one, so that they can be used as training data for the Q-learning algorithm. In order to correct this, the consequence of each action will be considered, in other words, how each action between those measurements would change the state. Based on that, the state will be divided, as many times as actions taken, and its substates will be updated according to the consequences of the actions, previously identified. In the example shown in the tables, the actions taken were "Al" and "Cooling Scrap". The purpose of these actions was to reduce the DO and the temperature. Therefore, the state is divided into two and is updated sequentially, as if the actions had been performed separately.

Table 5. Example of heat's state format after discretisation.

Measuring Time	T	T Difference	DO	Comp.	STATE
01-01-2019 01:32:53	$40 \leq \Delta T$	79	$5 \leq O2 < 10$	Close	S0
01-01-2019 01:42:29	$40 \leq \Delta T$	48	$5 \leq O2 < 10$	Close	S1
01-01-2019 01:51:15	$20 \leq \Delta T < 30$	23	$3.5 \leq O2 < 5$	Close	S2
01-01-2019 01:56:13	$10 \leq \Delta T < 20$	14	$3.5 \leq O2 < 5$	Close	S3
01-01-2019 01:59:53	$2 \leq \Delta T < 10$	4	$O2 < 3.5$	Close	S4

Table 6. Example of heat's actions after discretisation.

Addition Time	Material	Amount, kg	Action
01-01-2019 01:33:45	Al	39	Al between 35 and 45 kg
01-01-2019 01:33:45	Cooling Scrap	-	Cooling Scrap
01-01-2019 01:36:45	Cooling Scrap	-	Cooling Scrap
01-01-2019 01:44:54	Al	11	Al between 6 and 15 kg
01-01-2019 01:44:54	Cooling Scrap	-	Cooling Scrap
01-01-2019 01:47:42	Alloying	-	Alloying
01-01-2019 01:57:02	Al	10	Al between 6 and 15 kg
01-01-2019 01:57:02	Cooling Scrap	-	Cooling Scrap

Thirdly, there are also states without actions, intermediate measures done to check the status and if the actions taken have achieved their purpose. If this problem overlaps with the previous problem presented, the information gathered from the additional measurements is used as additional support in the procedure followed in the previous problem.

Finally, sometimes, although not frequently, there are actions taken after the last measurement. In these cases, the end state is estimated in function of the consequences of the action taken.

The example presented does not include a reheating sequence. However, same problems may occur and same procedure is followed. It is only necessary to add the reheating action between the already defined actions. The addition is done by comparing the sample times with the reheating start time.

In conclusion, all these problems leave gaps in the episode which must be filled. Once the gaps are filled, episodes with MDP structure are available and ready to be used as training data for the algorithm. The result of the procedures is presented in the following section, because they also serve as check on the work done so far.

Validation of Environment Transformation

The objective of this task was to validate if the CAS-OB process has been correctly transformed into an MDP and, hence, to validate the training data transformation. This validation is crucial since the agent's training is pointless if the data does not represent CAS-OB's process. To achieve the validation, the fact that every process gathered from the CAS-OB historical data was correctly transformed into an episodic format was checked.

Table 7 below includes the final result of heat transformation into MDP format. Therefore, the data can be used for training.

Table 7. Final result of episodes generation.

ΔT	DO	Comp.	Action	
40 ≤ ΔT	5 ≤ O2 < 10	Close	Al between 35 and 45 kg	S_0–A_{01}
		Close	Cooling Scrap	S_{01}–A_{02}
		Close	Al between 6 and 15 kg	S_1–A_{11}
	3.5 ≤ O2 < 5	Close	Cooling Scrap	S_{11}–A_{12}
20 ≤ ΔT < 30	3.5 ≤ O2 < 5	Close	Alloying	S_2–A_{13}
10 ≤ ΔT < 20	3.5 ≤ O2 < 5	Good	Al between 6 and 15 kg	S_3–A_{31}
	O2 < 3.5	Good	Cooling Scrap	S_{31}–A_{32}
2 ≤ ΔT < 10	O2 < 3.5	Good	F	S_4

4.2. Training and Validation of the Algorithm

Once the environment is validated, the next step is to validate the algorithm. RL algorithms are validated by achieving the convergence of the policy during the training stage. The policy of the Q-Learning algorithm converges if the Q-values remain constant between different trainings. The convergence proves that there is a solution for solving the problem. Hence the rewards are coherent with the environment, and the algorithm suits the problem.

The Q-Learning algorithm's training consists of following the historical data repeatedly until the solution is found, or in other words, it converges. The convergence is checked by calculating the maximum difference between the values from the previous training and the new ones. If the subtract is lower than the established accuracy parameter, set at 0.001, the convergence is achieved. Figure 10 illustrates the progress during the training, and the red line represents the accuracy parameter. Convergence is completed in 89 iterations.

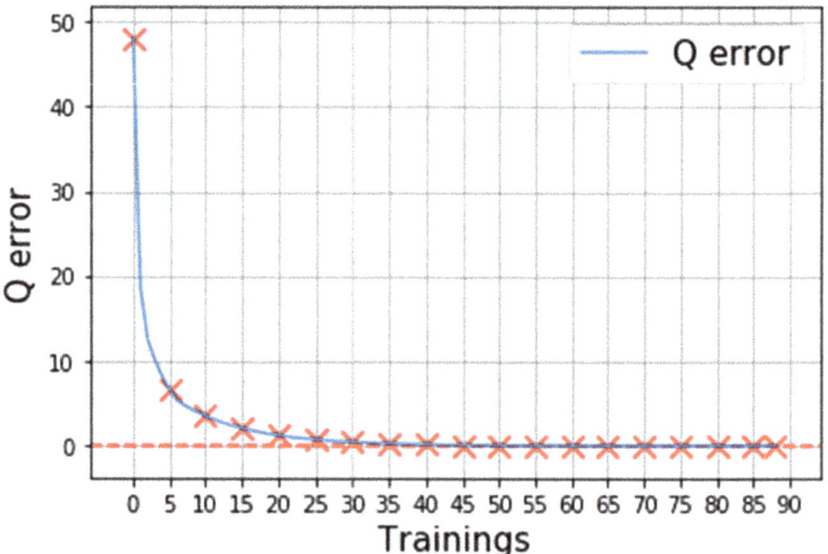

Figure 10. Training convergence.

Once the training ends, the program retains the Q values in a CSV file. Afterwards, the file is loaded and used to recommend actions. Moreover, it keeps learning and adjusts the values with the new experience gained. The training with the historical data is an initialisation of the Q values. Therefore, the DSS will recommend the right actions from the beginning while learning and enhancing its performance.

Nevertheless, before implementing the algorithm in the online DSS interacting with the real process, the policy must be validated. To accomplish this, the policy was validated by simulating episodes from the historical record following its recommendations. The validation continued by checking if the action recommended matched or was similar to the action taken in the historical dataset. The validation concludes when the DSS recommends coherent actions.

4.3. Online Interface

The online interface aims to facilitate the operator's work and to support him in real-time during the process operation. In Figure 11 the interface is shown. The RL UI will read from an OPC server the necessary data to define the state of the process. Therefore, it will not be necessary to set the inputs manually (although it is still possible to modify any input manually). Once the state is defined, the recommended action is shown, and every time the state changes, the action will change accordingly.

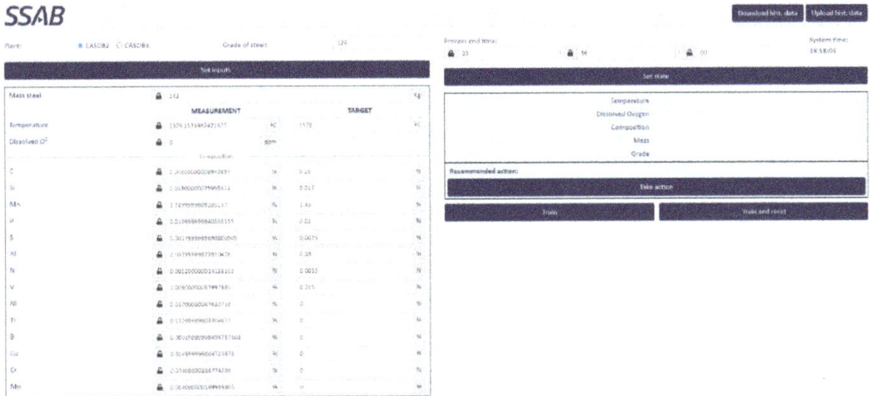

Figure 11. Decision support system online interface installed in CAS-OB process.

5. Testing and Results

The testing of the DSS and the gathering of results for performance measurement has been performed during two months at the CAS-OB plant. At the beginning of the testing, the focus was on the functionalities offered by the DSS and finding possible bugs. Thanks to the development engineers' feedback, it was possible to enhance the DSS by making the following adjustments to the environment definition and the learning algorithm:

- Adjust the rewards to be stricter on agent learning.
- Correction on composition status calculation.
- Addition of constant states during the process, such as the grade and mass of the steel.
- Higher resolution for dissolved oxygen states and aluminium addition actions.

In the case of the dissolved oxygen, some additional states closer to the target were added, which resulted in a better performance recommending aluminium to reduce the dissolved oxygen. Concerning the aluminium addition, the maximum amount that the system can suggest was increased. Before the change, the maximum amount that could be recommended was 50 Kg.

Thanks to these changes, the support offered by the DSS was more precise and helpful for the process. Additionally, it was observed that the RL agent prioritises firstly the composition and secondly the temperature, in worst scenarios aiming at least to maintain the temperature above target, which fits perfectly with the goals of the process. Further on, during all the testing periods, data has been gathered to measure the DSS's performance.

First, all the states processed by the DSS are stored together with the recommended action. Afterwards, these actions are compared with the actions taken by the operator, available in the historical data. By doing this, it was possible to analyse in detail the policy of the Q-learning algorithm. The analysis reflected that 69.23% of the time, the action concurred with the action taken by the operator.

Nevertheless, the 30.76% left does not mean the action recommended was incorrect, but the operators preferred to follow their experience. However, analysing the end of those heats, it was observed that many of them do not end ideally. In other words, the actions taken by the operator were not the best. Additionally, in order to ensure if the recommended action was better, a process engineer of CAS-OB process from SSAB, analysed the actions recommended and compared them with the actions taken by the operator. Estimating the consequences of the actions with the support of an external model, the process engineer from SSAB concluded that following the actions recommended, the heat would have ended perfectly in most of the cases. However, since CAS-OB is such a complex process and the DSS is a novel tool for the operator, it is understandable that experienced operators would follow their own experience instead of the recommendation of a tool as novel as the proposed DSS.

Finally, the efficiency of the process is measured with the following key performance indicators (KPIs) presented in Table 8:

Table 8. Key performance indicators for the RL agent application to the CAS-OB process.

KPI	Description	Units
Heat rejections	The percentage of heat rejections due to the wrong composition	%
Temperature hitting rate of CAS-OB heats	The percentage of heats in CAS-OB with the correct temperature in continuous casting	%

Analysing the episodes generated with the RL agent recommendations during the testing period, the heat rejections are reduced by **4%**. On the other side, **83.33%** of the heats end with the temperature on target or slightly above. The second KPI is expected to increase, at least to 89% over time with the continuous training of the agent. Additionally, both results, mainly the rejections reduction, reduced raw materials and energy consumption (~−0.5%) and, hence, the CO_2 emissions (~−2%) and costs (~−1.5%). The estimated percentages may seem low but given the huge scale of the steel production in the CAS-OB plant of SSAB, the impact is high. Those estimations were done by SSAB based on their internal data concerning their production, comparing results before and after the integration of the DSS.

6. Discussion

To conclude, it has been proved that a DSS using the Q-Learning algorithm can learn a complex process such as the steelmaking process CAS-OB. Moreover, it was possible to train an RL algorithm only with raw data from the historical database. Afterwards, it was validated and tested in the CAS-OB plant with direct interaction with the process. Additionally, thanks to the self-learning ability of the algorithm used as the core of the DSS, it will keep training, adapting and improving its performance while it is used.

The RL algorithm has learned how the process works, and the DSS can support the operator in real-time. Additionally, with constant interaction with the real process, the agent will learn every detail and improve its performance. In the future, it may even be possible to operate automatically while being supervised by the operator. Finally, with this research, the advantages of RL have been demonstrated. Moreover, it remains clear that RL algorithms, such as Q-Learning, are potent tools capable of learning and solving even complex steelmaking processes. This statement encourages further investigation in this direction and application in other steelmaking processes, thus opening a new way of enhancing the performance and sustainability of the steelmaking industry.

Author Contributions: D.S.A.: Conceptualisation, formal analysis, software, methodology and writing—original draft; L.E.A.G.: supervision, review and editing; S.O.: validation and review; C.L.G.: conceptualisation, supervision and review, project administration; A.d.R.T.: conceptualisation, funding acquisition; F.D.N.: conceptualisation, supervision and review; Á.O.R.: conceptualisation. All the authors have contributed to this manuscript. All authors have read and agreed to the published version of the manuscript.

Funding: This research has been conducted under the MORSE project, which has received funding from the European Union's Horizon 2020 research and innovation program under grant agreement No. 768652.

Institutional Review Board Statement: Not applicable.

Informed Consent Statement: Not applicable.

Data Availability Statement: Not applicable.

Conflicts of Interest: The authors declare no conflict of interest.

References

1. Kusunoki, S.; Nishihara, R.; Kato, K.; Sakagami, H.; Hirashima, S.F.N. Development of steelmaking processes for producing various high-quality steel grades at yawata works. *Nippon Steel Tech. Rep.* **2013**, *104*, 109–116.
2. Ghosh, A. *Secondary Steelmaking: Principles and Applications*; CRC Press: Boca Raton, FL, USA, 2000.
3. Ha, C.S.; Park, J.M. Start-up and some experience of CAS-OB at POSCO. *Arch. Metall. Mater.* **2008**, *53*, 2.
4. MORSE. Available online: https://www.spire2030.eu/morse (accessed on 13 July 2021).
5. Kanninen, K.; Lilja, J. Cost optimization system for full cycle metallurgical plants. *Chernye Met.* **2020**, *2020*, 3.
6. Kumar, S.; Singh, M.K.; Roy, S.; Kumar, V.; Kuma, M.; Anat, A. *Reducing Al Consumption in Steelmaking*; Digital-Feb'21; Steel Times International: Surrey, UK, 2021; pp. 41–45.
7. *DynStir. Dynamic Stirring for Improvement of Energy Efficiency in Secondary Steelmaking*; European Commission: Brussels, Belgium, 2019.
8. Janjua, R. *Energy Use in the Steel Industry*; World Steel Association: Brussels, Belgium, 2013; pp. 1–3.
9. Hasanbeigi, A.; Price, L.K.; McKane, A.T. *The State-of-the-Art Clean Technologies (SOACT) for Steelmaking Handbook*; Asia-Pacific Partnership on Clean Development and Climate: Sydney, Australia, 2010.
10. Shi, Q.; Lam, H.; Xiao, B.; Tsai, S. Adaptive PID controller based on Q-learning algorithm. *CAAI Trans. Intell. Technol.* **2018**, *3*, 235–244. [CrossRef]
11. Oztemel, E.; Gursev, S. Literature review of Industry 4.0 and related technologies. *J. Intell. Manuf.* **2018**, *31*, 127–182. [CrossRef]
12. Marcos, M.; Pitarch, J.; de Prada, C. Integrated Process Re-Design with Operation in the Digital Era: Illustration through an Industrial Case Study. *Processes* **2021**, *9*, 1203. [CrossRef]
13. Szabó-Szentgróti, G.; Végvári, B.; Varga, J. Impact of Industry 4.0 and digitization on labor market for 2030-verification of Keynes' prediction. *Sustainability* **2021**, *13*, 7703. [CrossRef]
14. Rotevatn, T.; Wasbø, S.O.; Järvinen, M.; Fabritius, T.; Ollila, S.; Hammervold, A.; De Blasio, C. A Model of the CAS-OB Process for Online Applications. *IFAC-PapersOnLine* **2015**, *48*, 6–11. [CrossRef]
15. Backman, J.; Kyllönen, V.; Helaakoski, H. Methods and Tools of Improving Steel Manufacturing Processes: Current State and Future Methods. *IFAC-PapersOnLine* **2019**, *52*, 1174–1179. [CrossRef]
16. Lee, S.; Cho, Y.; Lee, Y.H. Injection Mold Production Sustainable Scheduling Using Deep Reinforcement Learning. *Sustainability* **2020**, *12*, 8718. [CrossRef]
17. Shin, S.-J.; Kim, Y.-M.; Meilanitasari, P. A Holonic-Based Self-Learning Mechanism for Energy-Predictive Planning in Machining Processes. *Processes* **2019**, *7*, 739. [CrossRef]
18. Mehta, D. Panjab University (UIET) State-of-the-Art Reinforcement Learning Algorithms. *Int. J. Eng. Res.* **2020**. [CrossRef]
19. Kubat, M. *An Introduction to Machine Learning*; Springer: Cham, Switzerland, 2017.
20. Yang, S.-S.; Yu, X.-L.; Ding, M.-Q.; He, L.; Cao, G.-L.; Zhao, L.; Tao, Y.; Pang, J.-W.; Bai, S.-W.; Ding, J.; et al. Simulating a combined lysis-cryptic and biological nitrogen removal system treating domestic wastewater at low C/N ratios using artificial neural network. *Water Res.* **2020**, *189*, 116576. [CrossRef] [PubMed]
21. Sutton, R.S.; Barto, A.G. *Reinforcement Learning: An Introduction*; MIT Press: Cambridge, MA, USA, 2018.
22. Silver, D.; Huang, A.; Maddison, C.J.; Guez, A.; Sifre, L.; van den Driessche, G.; Schrittwieser, J.; Antonoglou, I.; Panneershelvam, V.; Lanctot, M.; et al. Mastering the game of Go with deep neural networks and tree search. *Nature* **2016**, *529*, 484–489. [CrossRef]

23. Berner, C.; Brockman, G.; Chan, B.; Cheung, V.; Dębiak, P.; Dennison, C.; Farhi, D.; Fischer, Q.; Hashme, S.; Hesse, C.; et al. Dota 2 with large scale deep reinforcement learning. *arXiv* **2019**, arXiv:1912.06680.
24. Pang, J.-W.; Yang, S.-S.; He, L.; Chen, Y.-D.; Cao, G.-L.; Zhao, L.; Wang, X.-Y.; Ren, N.-Q. An influent responsive control strategy with machine learning: Q-learning based optimization method for a biological phosphorus removal system. *Chemosphere* **2019**, *234*, 893–901. [CrossRef]
25. Gottschalk, S.; Burger, M. Differences and similarities between reinforcement learning and the classical optimal control framework. *PAMM* **2019**, *19*, e201900390. [CrossRef]
26. Wang, H.; Dong, S.; Shao, L. Measuring Structural Similarities in Finite MDPs. *IJCAI Int. Jt. Conf. Artif. Intell.* **2019**, *2019*, 3684–3690. [CrossRef]
27. Maass, K.; Kim, M. A Markov decision process approach to optimizing cancer therapy using multiple modalities. *Math. Med. Biol.* **2019**, *37*, 22–39. [CrossRef]
28. Roy, N.; Thrun, S. Coastal navigation with mobile robots. *Adv. Neural Inf. Process. Syst.* **1999**. [CrossRef]
29. Wang, F.-Y.; Zhang, J.; Wei, Q.; Zheng, X.; Li, L. PDP: Parallel dynamic programming. *IEEE/CAA J. Autom. Sin.* **2017**, *4*, 1–5. [CrossRef]
30. Kővári, B.; Hegedüs, F.; Bécsi, T. Design of a Reinforcement Learning-Based Lane Keeping Planning Agent for Automated Vehicles. *Appl. Sci.* **2020**, *10*, 7171. [CrossRef]
31. Song, S.; Chen, H.; Sun, H.; Liu, M. Data Efficient Reinforcement Learning for Integrated Lateral Planning and Control in Automated Parking System. *Sensors* **2020**, *20*, 7297. [CrossRef] [PubMed]
32. Jang, B.; Kim, M.; Harerimana, G.; Kim, J.W. Q-Learning Algorithms: A Comprehensive Classification and Applications. *IEEE Access* **2019**, *7*, 133653–133667. [CrossRef]
33. Silver, D.; Sutton, R.S.; Müller, M. Temporal-difference search in computer Go. *Mach. Learn.* **2012**, *87*, 183–219. [CrossRef]
34. Watkins, C.J.C.H.; Dayan, P. Technical Note: Q-Learning. *Mach. Learn.* **1992**, *8*, 279–292. [CrossRef]
35. Zhang, L.; Tang, L.; Zhang, S.; Wang, Z.; Shen, X.; Zhang, Z. A Self-Adaptive Reinforcement-Exploration Q-Learning Algorithm. *Symmetry* **2021**, *13*, 1057. [CrossRef]
36. Singh, R. Production of steel. *Appl. Weld. Eng.* **2020**, 35–52. [CrossRef]

 processes 

Article

Temperature Prediction Model for a Regenerative Aluminum Smelting Furnace by a Just-in-Time Learning-Based Triple-Weighted Regularized Extreme Learning Machine

Xingyu Chen, Jiayang Dai * and Yasong Luo

Guangxi Key Laboratory of Intelligent Control and Maintenance of Power Equipment, School of Electrical Engineering, Guangxi University, Nanning 530004, China
* Correspondence: daijiayang@gxu.edu.cn; Tel.: +86-185-7439-5495

Abstract: In a regenerative aluminum smelting furnace, real-time liquid aluminum temperature measurements are essential for process control. However, it is often very expensive to achieve accurate temperature measurements. To address this issue, a just-in-time learning-based triple-weighted regularized extreme learning machine (JITL-TWRELM) soft sensor modeling method is proposed for liquid aluminum temperature prediction. In this method, a weighted JITL method (WJITL) is adopted for updating the online local models to deal with the process time-varying problem. Moreover, a regularized extreme learning machine model considering both the sample similarities and the variable correlations was established as the local modeling method. The effectiveness of the proposed method is demonstrated in an industrial aluminum smelting process. The results show that the proposed method can meet the requirements of prediction accuracy of the regenerative aluminum smelting furnace.

Keywords: temperature prediction; weighted regularized extreme learning machine; just-in-time learning; sample similarities; variable correlations

Citation: Chen, X.; Dai, J.; Luo, Y. Temperature Prediction Model for a Regenerative Aluminum Smelting Furnace by a Just-in-Time Learning-Based Triple-Weighted Regularized Extreme Learning Machine. *Processes* **2022**, *10*, 1972. https://doi.org/10.3390/pr10101972

Academic Editors: Guoqing Zhang, Zejia Zhao and Wai Sze YIP

Received: 1 September 2022
Accepted: 23 September 2022
Published: 30 September 2022

1. Introduction

Aluminum can be made into alloys with various metals; it is widely used in automotive, aviation, and military industries due to its good ductility, plasticity, recyclability, and oxidation resistance. A regenerative aluminum smelting furnace is important for the aluminum smelting process, in which the real-time measurement and control of liquid aluminum temperatures influence the quality of the aluminum. However, on industrial sites, there are many influencing factors, such as the aging of temperature-measuring thermocouples and fluctuations in the operating voltage, which bring difficulties to the real-time measurements of the aluminum liquid temperature. Hence, it is essential to develop a modeling method to predict the liquid aluminum temperature for quality improvement of the aluminum. The aluminum smelting process is a typical complex industrial furnace production process. In recent decades, many studies on industrial furnaces have been performed (regarding 'mechanism modeling') [1–3]. Although the physical meaning of 'mechanism modeling' is clear, there are some problems, such as complicated calculations for industrial furnace systems. At the same time, mechanism models may not be reliable enough since they usually make simplified assumptions. The furnace temperature, airflow rate, etc., fluctuate greatly in different working states due to the intermittent working characteristics of the regenerative aluminum smelting furnace. The real-time update of the model for the regenerative aluminum smelting furnace is also a problem that needs to be considered.

To overcome the shortcomings of mechanism modeling, a soft-sensor that makes full use of the industrial data is proposed [4]. There are many researchers working on the data-driven modeling of industrial furnaces and similar processes, such as partial least squares (PLS) [5], the kernel principal component regression (KPCR) [6], and kernel partial least squares (KPLS) [7], which have been successfully applied with good results. However, these methods are generally considered to be global modeling (and trained offline). Moreover, after these models are put into application, they will face problems, such as difficulties in model updating. Consequently, to deal with the adaptive update problem of the model, the moving window technique [8,9], recursive models [10,11], and the just-in-time learning (JITL) strategy [12,13] are usually used as online adaptive update strategies. The JITL strategy trains an online local model to predict the query samples by selecting similar samples from historical samples, so it is more suitable for processes such as industrial furnaces with state mutations. For example, Chen et al. [14] proposed a least squares support vector machine temperature prediction model based on JITL to deal with large temperature change lags in roller kilns. Dai et al. [15] combined the moving window technique and the JITL strategy as an update strategy to select similar samples in both time and space dimensions, and they verified the effectiveness of the proposed method on an industrial kiln. In [16], a locally weighted partial least squares regression (LWPLS) model was proposed by JITL-based local modeling. In LWPLS, the samples most similar to the query sample are assigned different weights and selected for local modeling. The current model will be discarded when the next query sample is available. Then, a new local PLS model will be established for the model's online update. However, LWPLS only considers the sample similarities, not the variable correlations. The data of the aluminum smelting process often present high-dimensional characteristics and each input variable has a different degree of influence on the liquid aluminum temperature. Hence, except for the sample similarities, it is necessary to consider the variable correlations [17–19]. Furthermore, the accuracy of the JITL strategy depends on the quality of the selected samples. However, the traditional similarity measurement criteria, such as Euclidean distance and Mahalanobis distance, only consider the input information without considering the output information, and often cannot obtain accurate similar samples. Thus, investigating new similarity measurement criteria is important for the JITL strategy.

In recent years, artificial intelligence algorithms, such as long short-term memory networks (LSTM) [20–22] and extreme learning machine(s) (ELM) [23–26] have also been used in soft sensor modeling. The basic assumption for LSTM is that process data are sampled at even and unified frequencies; it is very difficult to meet these conditions for 'process data measurements' in industrial processes, especially for quality variables. Hence, LSTM is unsuitable for some processes with irregular sampling frequencies. ELM is a single hidden layer neural network with a low algorithm complexity, which does not need backpropagation to solve iteratively, and has been used in the temperature prediction of regenerative aluminum smelting processes. Huang et al. [27] proposed an extreme learning machine furnace temperature prediction model based on the kernel principal component analysis and showed that ELM has a better effect than the traditional BP neural network. Liu et al. [28] proposed an ELM model optimized by the restricted Boltzmann machine (RBM) to solve the random initialization of the input weights and biases in the ELM. Moreover, ELM has a fast learning speed and is suitable as an online prediction model. For example, Li et al. [29] built a local online ELM model in combination with a JITL strategy, allowing the online prediction of polyethylene terephthalate (PET) viscosity without relying on time-consuming laboratory analysis procedures. However, this ELM-based online prediction model neither considers sample similarities nor variable correlations, which is unreasonable in local modeling. Moreover, the original ELM runs the risk of model overfitting. Hence, a regularized extreme learning machine (RELM) [30] was proposed to solve the model's overfitting problem.

Although some research studies have been carried out on ELM, there are few discussions about sample similarities and variable correlations in RELM, especially in temperature prediction. Based on the above discussions, a soft sensor modeling method of the JITL-based triple-weighted regularized extreme learning machine (JITL-TWRELM) was proposed to solve the above problems. Compared with the traditional data-driven modeling method described above, the method proposed in this paper not only allows real-time updating of the model but also obtains more accurate local modeling samples due to the use of the WJITL strategy, which uses correlation information between the input and output variables in the sample selection stage. Meanwhile, in the local modeling stage, the proposed method overcomes the shortcomings of the traditional local modeling method, which only considers the sample similarities and analyzes the variable correlations, highlighting the influences of different variables on the output. The remainder of this article is structured as follows. Firstly, the regenerative aluminum smelting furnace is briefly introduced. Secondly, the regularized extreme learning machine (RELM), sample weighted regularized extreme learning machine (SWRELM), and variable weighted regularized extreme learning machine (VWRELM) are introduced, respectively. Then, the JITL-based triple-weighted regularized extreme learning machine (JITL-TWRELM) is described. Next, the flexibility and effectiveness of the proposed method are validated in the industrial aluminum smelting processing. Finally, we present the conclusions.

2. Related Methods

Since ELM runs the risk of model overfitting, the regularization method is used to solve the overfitting problem. Considering sample similarities and variable correlations, the sample weighted regularized extreme learning machine (SWRELM) and the variable weighted extreme learning machine (VWRELM) are introduced, respectively. Three related methods are discussed next. To better understand the derivation of the relevant equations, the definition of symbols in this paper is shown in Table 1.

Table 1. Definition of symbols in this paper.

Symbols	Definition
x_n, t_n	the nth historical input and output variable vectors
β_i	the output weight of the ith hidden layer unit
$\beta, \beta^S, \beta^V, \beta^t$	the output weight vectors in RELM, SWRELM, VWRELM, JITL-TWRELM
T	the output vector of RELM
ω_i, b_i	the input weight and bias connecting input layer and ith hidden layer unit
$t_j, \hat{t}_q^t$	the output corresponding to x_j, the output of the query sample in JITL-TWRELM
N	the number of training samples
C	the regularization coefficient
ξ	the training error vector
H, H^V, H^t	the hidden layer output matrices in RELM, VWRELM, JITL-TWRELM
$\Omega_{sn}, \Omega_s, \Omega_s^t$	the sample weight of the nth sample, the sample weighted matrix, the sample weighted matrix in JITL-TWRELM
λ	the Lagrange multiplier vector
ρ	the Pearson correlation coefficient
$E(x), E(t)$	the expectation of the single input variable and output variable
v_i	the contribution of each variable
V	the variable contribution matrix
x_n^v, x_n^w	the variable weighted input sample, the variable weighted local modeling sample
$d_{on}, d_{on}^w, d_{on}^{tw}$	the original Euclidean distance and weighted Euclidean distance, the weighted Euclidean distance in JITL-TWRELM
x_q, x_q^w	the query sample, the variable weighted query sample in JITL-TWRELM
$\Omega_v, \Omega_v^l, \Omega_v^g$	the correlation coefficient matrix, the local correlation coefficient matrix, the global correlation coefficient matrix
φ	the adjusted parameter
X, X^w	the local modeling sample matrix, the variable weighted local modeling sample matrix in JITL-TWRELM

2.1. RELM

As shown in Figure 1, the structure of ELM consists of three parts, which are the input layer, hidden layer, and output layer [31]. The core idea of ELM is to randomly select the input weights and hidden layer biases of the network. The output weights between the hidden layer and output layer are obtained by minimizing the loss function and solving the Moore–Penrose generalized inverse operation. Owing to the particularity of the single hidden-layer structure, ELM has a faster learning speed, minimal human interference, and it is easier to implement than traditional networks. However, the original ELM model only considers the empirical risk minimization (ERM) principle, which tends to result in an overfitting model. To overcome this deficiency, a regularized extreme learning machine (RELM) was proposed based on empirical risk minimization and structural risk minimization (SRM) principles and has proven to be a better generalization performance than ELM.

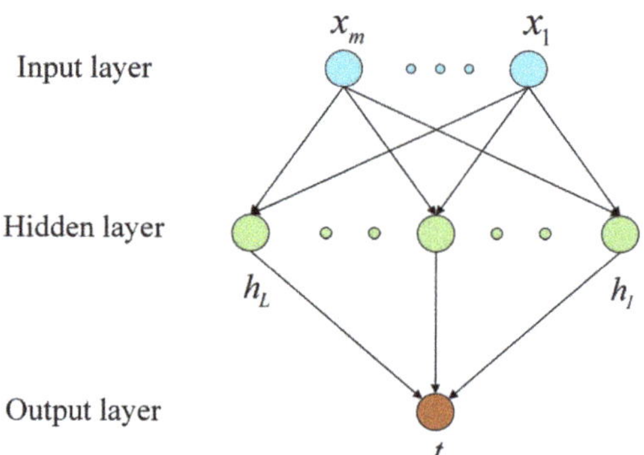

Figure 1. The structure of ELM.

It is assumed that the nth historical input variable vector and the output variable are denoted as $x_n = [x_{n1}, x_{n2}, \ldots, x_{nm}]$ and t_n, respectively, where m is the number of input variables. (x_n, t_n) is the nth historical sample composed of x_n and t_n. The output function of the RELM with L hidden layer neurons can be represented as

$$\sum_{i=1}^{L} \beta_i g(\omega_i x_j^T + b_i) = t_j, j = 1, \ldots, N \tag{1}$$

where β_i is the output weight of the ith hidden layer unit, $\omega_i = [\omega_{j1}, \ldots, \omega_{jm}]$, and b_i are the input weight, bias connecting input layer, and ith hidden layer unit, respectively. $x_j = [x_{j1}, x_{j2}, \ldots, x_{jm}]$ is the input variable vector, t_j denotes the output corresponding to x_j, N is the number of training samples. $g(.)$ is the activation function. Usually, $g(.)$ is set as the sigmoid function. We re-write Equation (1) in matrix form

$$H\beta = T \tag{2}$$

where

$$H = [h(x_1^T)^T, \ldots, h(x_N^T)^T]^T = \begin{pmatrix} g(\omega_1 x_1^T + b_1) & \cdots & g(\omega_L x_1^T + b_L) \\ \vdots & \ddots & \vdots \\ g(\omega_1 x_N^T + b_1) & \cdots & g(\omega_L x_N^T + b_L) \end{pmatrix} \tag{3}$$

$$\beta = [\beta_1, \ldots, \beta_L]^T \tag{4}$$

$$T = [t_1, \ldots, t_N]^T \tag{5}$$

Due to ω_i and b_i being randomly given, to obtain the output weight vector β, the optimization equation can be represented as

$$\begin{aligned} &\min \frac{1}{2}\|\beta\|^2 + \frac{C}{2}\|\xi\|^2, \\ &s.t. h(x_j^T)\beta = t_j + \xi_j, j = 1, \ldots, N \end{aligned} \tag{6}$$

where C represents the regularization coefficient, which can adjust the empirical risk and structural risk. $\xi = [\xi_1, \ldots, \xi_N]^T$ is the training error vector. By constructing the Lagrange function, the solution of Equation (6) is

$$\beta = \begin{cases} (H^T H + \frac{I_L}{C})^{-1} H^T T, L < N \\ H^T (H^T H + \frac{I_N}{C})^{-1} T, L > N \end{cases} \tag{7}$$

where $I_L \in R^{L \times L}$, $I_N \in R^{N \times N}$.

2.2. SWRELM

Not all samples have the same contribution to the output; moreover, the original RELM considers all samples equally important and does not consider the differences between different samples. Thus, to obtain a more realistic result, the sample weighted matrix $\Omega_s = diag(\Omega_{s1}, \ldots, \Omega_{sN})$ is added to Equation (6), which is expressed as

$$\begin{aligned} &\min \frac{1}{2}\left\|\beta^S\right\|^2 + \frac{C}{2}\|\Omega_s \xi\|^2, \\ &s.t. h(x_j^T)\beta^S = t_j + \xi_j, j = 1, \ldots, N \end{aligned} \tag{8}$$

The Lagrange function can be represented as follows:

$$\begin{aligned} &L(\beta^S, \Omega_s, \lambda) \\ &= \frac{1}{2}\left\|\beta^S\right\|^2 + \frac{C}{2}\|\Omega_s \xi\|^2 - \sum_{j=1}^{N} \lambda (\sum_{i=1}^{L} h(x_j^T)\beta^S - t_j - \xi_j) \\ &= \frac{1}{2}\left\|\beta^S\right\|^2 + \frac{C}{2}\|\Omega_s \xi\|^2 - \lambda (H\beta^S - T - \xi) \end{aligned} \tag{9}$$

where $\lambda = [\lambda_1, \ldots, \lambda_N]$ denotes the Lagrange multiplier vector. According to the KKT condition, taking the derivative of Equation (9) and setting the derivative to zero, we have

$$\frac{\partial L}{\partial \beta^S} = 0 \rightarrow (\beta^S)^T = \lambda H \tag{10}$$

$$\frac{\partial L}{\partial \xi} = 0 \rightarrow C\xi^T \Omega_s^2 + \lambda = 0 \tag{11}$$

$$\frac{\partial L}{\partial \lambda} = 0 \rightarrow H\beta^S = T + \xi \tag{12}$$

With Equations (11) and (12), the Lagrange multiplier vector λ can be expressed as

$$\lambda = -C(H\beta^S - T)^T \Omega_s^2 \tag{13}$$

Similarly, with Equations (10) and (13), the expression of the output weight vector of the sample weighted regularized extreme learning machine (SWRELM) is

$$\beta^S = (H^T \Omega_s^2 H + \frac{I_L}{C})^{-1} H^T \Omega_s^2 T, L < N \tag{14}$$

Equation (14) is suitable when the number of modeling samples is greater than the number of hidden neurons. Moreover, in this case, β^S has a faster calculation speed [32].

2.3. VWRELM

The original RELM treats all input variables with equal importance, while not all input variables have the same effect on the output variable, some input variables are more strongly correlated with the output variable than others. Thus, to reflect the differences of input variables and obtain better quality-related features, a variable contribution method based on the Pearson correlation coefficient was adopted. On this basis, the variable weighted extreme learning machine (VWRELM) was proposed. The Pearson correlation coefficient is defined as

$$\rho = \frac{E(xt) - E(x)E(t)}{\sqrt{E(x^2) - E^2(x)}\sqrt{E(t^2) - E^2(t)}} \tag{15}$$

where $E(x)$ and $E(t)$ are the expectations of the single input variable and output variable, respectively. ρ represents the degree of correlation between the two variables; two highly correlated variables will also have a larger ρ. As a result, the variable contribution can be defined by ρ. For a training sample $(x_n, t_n), n = 1, \ldots, k$, where each input sample x_n has m dimensions, the contribution of each variable can be defined as

$$v_i = \frac{|\rho_i|}{\sum\limits_{j=1}^{m} |\rho_i|}, i = 1, \ldots, m \tag{16}$$

where ρ_i represents the Pearson correlation coefficient between the ith input variable and the output variable. The variable contribution matrix can be written as

$$V = diag(v_1, \ldots, v_m) \tag{17}$$

Hence, taking the variable contribution as the variable weights, and applying the variable weights to the input sample x_n, the weighted input sample can be expressed as

$$x_n^v = x_n V = x_n diag(v_1, \ldots, v_m) = (x_{n1}v_1, \ldots, x_{nm}v_m) \tag{18}$$

where x_n^v represents the input sample weighted by variable weights. It can be seen from Equation (18) that each dimension of the input sample is given a different weight, reflecting the differences between variables. By variable weighting, Equation (3) can be rewritten as

$$\begin{aligned} H^V &= [h((x_1^v)^T)^T, \ldots, h((x_N^v)^T)^T]^T \\ &= \begin{pmatrix} g(\omega_1((x_1^v)^T)^T + b_1) & \ldots & g(\omega_L((x_1^v)^T)^T + b_L) \\ \vdots & \ddots & \vdots \\ g(\omega_1((x_N^v)^T)^T + b_1) & \cdots & g(\omega_L((x_N^v)^T)^T + b_L) \end{pmatrix} \end{aligned} \tag{19}$$

when $L < N$, the output weight vector is

$$\beta^V = ((H^V)^T H^V + \frac{I_L}{C})^{-1}(H^V)^T T, L < N \tag{20}$$

3. The Proposed JITL-TWRELM Model

In the previous analysis, the RELM, SWRELM, and VWRELM models have been established. However, in a multi-data, multivariate prediction model, the different samples and variables to the predicted outputs are different, especially in the aluminum smelting process. Table 2 shows the shortcomings of the three methods. Both sample similarities and variable correlations should be taken into account in RELM. Hence, to obtain a better model, combined with the weighted JITL strategy (WJITL), a JITL-based triple-weighted regularized extreme learning machine is proposed.

Table 2. Shortcomings of the three methods.

Method	Shortcomings
RELM	Neither sample similarities nor variable correlations are considered, and the model cannot be updated in real-time.
SWRELM	Only the sample similarities are considered, no variable correlations are considered, and the model cannot be updated in real-time.
VWRELM	Only the variable correlations are considered, no sample similarities are considered, and the model cannot be updated in real-time.

3.1. Weighted Similarity Measurement Criterion

The original Euclidean distance is usually used as a similarity measurement criterion, expressed as

$$d_{on} = \sqrt{(x_q - x_n)(x_q - x_n)^T} \tag{21}$$

where $x_q \in R^{1\times m}$ is the current query sample, $x_n \in R^{1\times m}$ is the nth historical sample, and d_{on} indicates the Euclidean distance between the current query sample and the nth historical sample. The more similar the historical sample is to the query sample, the smaller the distance d_{on}. However, Equation (21) only uses the input information of the historical sample and query sample, while the information of the output is not taken into consideration. Moreover, the calculation of the Euclidean distance can be regarded as the accumulation of each dimension of the sample. It is easy to see that the importance of each dimension may be different, with some dimensions contributing more to distance than others. Hence, inspired by Equation (15), the connections between the input variables and output variables are established through the correlation analysis. We define a weighted Euclidean distance as a weighted similarity measure criterion, expressed as

$$d_{on}^{w} = \sqrt{(x_q - x_n)\Omega_v(x_q - x_n)^T} \tag{22}$$

where $\Omega_v = diag(\rho_1, \ldots, \rho_m)$. Then, the sample weight is expressed as

$$\Omega_{sn} = \exp(\frac{d_{on}^{w}}{\varphi^2}) \tag{23}$$

where φ is the adjust parameter, which can adjust the change rate of weight value with the sample distance. For a better expression, the JITL strategy that applied this weighted similarity measurement criterion is called WJITL.

3.2. JITL-TWRELM

A JITL-based triple-weighted regularized extreme learning machine (JITL-TWRELM) soft sensor method, combined with the WJITL strategy, was established to simultaneously incorporate sample weights and variable weights. The detailed derivation steps are as follows.

$N(N < H)$ samples $(x_n, t_n), n = 1, \ldots, H$ from historical samples were selected for each query sample to local modeling. First, Pearson correlation coefficients between input variables and output variables of all historical samples were calculated to obtain the correlation coefficient matrix

$$\Omega_v^g = diag(\rho_1^g, \ldots, \rho_m^g) \tag{24}$$

To distinguish it from the subsequent derivation, we call Ω_v^g the global correlation coefficient matrix, where $\rho_i^g, i = 1, \ldots, m$ is the global correlation coefficient. As a result, the weighted Euclidean distance between the query samples and the historical samples can be obtained by Equation (25).

$$d_{on}^{tw} = \sqrt{(x_q - x_n)\Omega_v^g (x_q - x_n)^T}, n = 1, \ldots, H \tag{25}$$

We sort $d_{on}^{tw}, n = 1, \ldots, H$ from small to large, and the first N samples are selected as modeling samples. The sample weighted matrix is obtained as

$$\Omega_s^t = diag(\exp(\frac{d_{o1}^{tw}}{\varphi^2}), \ldots, \exp(\frac{d_{oN}^{tw}}{\varphi^2})) = diag(\Omega_{s1}^t, \ldots, \Omega_{sN}^t) \tag{26}$$

Then, the Pearson correlation coefficient of N local modeling samples is calculated, and the local correlation coefficient matrix is obtained as

$$\Omega_v^l = diag(\rho_1^l, \ldots, \rho_m^l) \tag{27}$$

where Ω_v^l is used as the local variable weighted matrix for local modeling samples and the query sample

$$X^w = X\Omega_v^l = \{x_n^w\}, n = 1, \ldots, N \tag{28}$$

$$x_q^w = x_q \Omega_v^l \tag{29}$$

where $X \in R^{N \times m}$ consists of local modeling samples, X^w and x_q^w are the variable weighted local modeling sample and variable weighted query sample, respectively. Thus, the new local modeling dataset $(x_n^w, t_n), n = 1, \ldots, N$ is used to build the local model. The optimization equation for the output weight vector is established as Equation (30)

$$\begin{aligned} &\min \frac{1}{2}\|\beta^t\|^2 + \frac{C}{2}\|\Omega_s^t \xi\|^2, \\ &s.t. h((x_j^w)^T)\beta^t = t_j + \xi_j, j = 1, \ldots, N \end{aligned} \tag{30}$$

The output matrix of the hidden layer is

$$H^t = \begin{pmatrix} g(\omega_1 (x_1^w)^T + b_1) & \cdots & g(\omega_L (x_1^w)^T + b_L) \\ \vdots & \ddots & \vdots \\ g(\omega_1 (x_N^w)^T + b_1) & \cdots & g(\omega_L (x_N^w)^T + b_L) \end{pmatrix} \tag{31}$$

According to Equations (9)–(14), the output weight vector of JITL-TWRELM is

$$\beta^t = ((H^t)^T (\Omega_s^t)^2 H^t + \frac{I_L}{C})^{-1} (H^t)^T (\Omega_s^t)^2 T, L < N \tag{32}$$

Finally, the prediction output of the query sample is

$$\overset{\wedge}{t_q^t} = \sum_{i=1}^{L} \beta^t g(\omega_i (x_q^w)^T + b_i) \tag{33}$$

4. Industrial Case

4.1. Process Description of the Regenerative Aluminum Smelting Furnace

An industrial regenerative aluminum smelting furnace and its internal structure are shown in Figure 2a and Figure 2b, respectively. The regenerative aluminum smelting furnace consists of a furnace chamber, regenerative burner (including burner and ceramic sphere accumulator), reversing valve, flue gas pipe, etc. The regenerative burners are arranged in pairs, and the two opposite burners are a group (A and B). Normal temperature air from the blower enters burner B through the reversing valve and is heated as it flows through the hot ceramic sphere accumulator. Then, the normal temperature air is heated to a temperature close to the furnace chamber (generally 80% to 90% of the furnace chamber temperature). The heated high-temperature air enters the furnace chamber and then rolls up the flue gas around the furnace to form a thin oxygen-poor high-temperature airflow with an oxygen content lower than 21%. Then, the mixture of the oxygen-poor high-temperature air and the injected flue gas is ignited to smelt the aluminum material. At the same time, the high-temperature flue gas passes through burner A, the heat is stored in the cold ceramic sphere accumulator, and then the flue gas is discharged at a temperature lower than 150 °C through the flue gas pipe. When the stored heat reaches saturation, the reversing valve is reversed, and the regenerative burner A and B change their combustion and heat storage working states, and so on, resulting in energy savings (and reducing emissions).

(a)

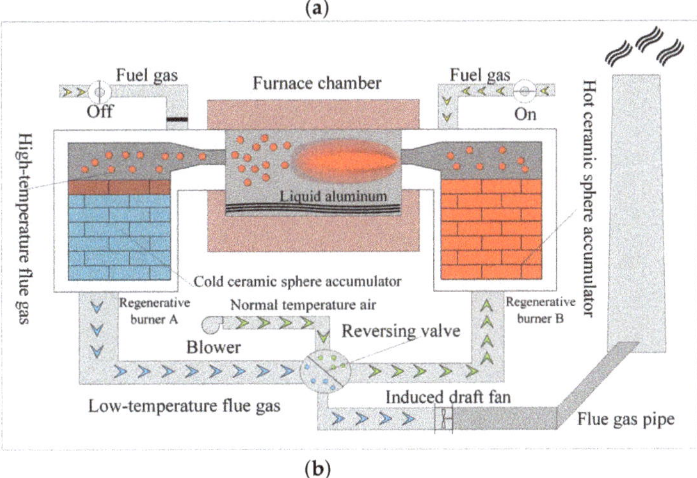

(b)

Figure 2. (**a**) An industrial regenerative aluminum smelting furnace; (**b**) the internal structure of the regenerative aluminum smelting furnace.

4.2. Model Establishment

To construct the model for the prediction of the liquid aluminum temperature, 12 secondary variables were chosen as the input variables, which are shown in Table 3. These input variables were measured by the sensor. The measurement ranges and errors of the sensors are shown in Table 4. The sampling interval of each sampling point was five minutes. There were 4400 data samples collected for modeling, of which, 4000 samples were used as historical data for training, and 400 samples for model testing. To better test the effectiveness of the proposed method, two groups of data (D1 and D2) from different periods were used as the testing dataset, with 200 samples in each group.

Table 3. Input variables for the model in the aluminum smelting process.

Input	Variable
1	Material temperature
2	Furnace pressure
3	12 # combustion airflow
4	12 # combustion air pressure difference
5	34 # combustion airflow
6	34 # combustion air temperature
7	34 # combustion air pressure difference
8	34 # gas air-fuel ratio
9	B1 # exhaust gas temperature
10	B2 # exhaust gas temperature
11	B3 # exhaust gas temperature
12	B4 # combustion air temperature

Table 4. Sensor measurement range and error.

Sensor Type	Measurement Range	Measurement Error
Pressure meter	0–15,000 Pa	1%
Flow meter	0–15 m^3/h	1.5%
Thermocouple	0–1300 °C	1%

The flowchart of JITL-TWRELM model is shown in Figure 3. To validate the performance of JITL-TWRELM, the six methods listed below were employed for comparison.

- Method 1: JITL-RELM (it applies the original JITL strategy and original RELM).
- Method 2: JITL-SWRELM (it applies the original JITL strategy and sample weights on RELM).
- Method 3: WJITL-RELM (it applies the WJITL strategy and original RELM).
- Method 4: JITL-VWRELM (it applies the original JITL strategy and local variable weights on RELM).
- Method 5: JITL-DWRELM (it applies the WJITL strategy and sample weights on RELM).
- Method 6: JITL-TWRELM (it applies the WJITL strategy, sample weights, and local variable weights on RELM).

The detailed step-by-step procedure of the proposed method is as follows.

Step 1: Prepare the input and output variables of the historical samples and perform the standardization.

Step 2: Determine the number N of training samples selected from the total historical samples, the parameter φ for the sample weight calculation, the hidden neuron number L, and the regularization coefficient C of the regularized extreme learning machine.

Step 3: Analyze the global correlation between the input variables and output variables of all historical samples. The global correlation coefficient matrix Ω_v^g is calculated for the sample similarity measurement.

Step 4: Calculate the weighted Euclidean distances between the current query samples and the training samples; N samples closest to the current query sample are selected as local modeling samples.

Step 5: Analyze the local correlation between input variables and output variables of the local modeling samples. The local correlation coefficient matrix Ω_v^l is determined.

Step 6: The JITL-TWRELM model is established, and the output of the current query sample is predicted.

Step 7: Before the next query sample arrives, the previous model is discarded and a new model is constructed based on the next query sample, enabling real-time updating of the model.

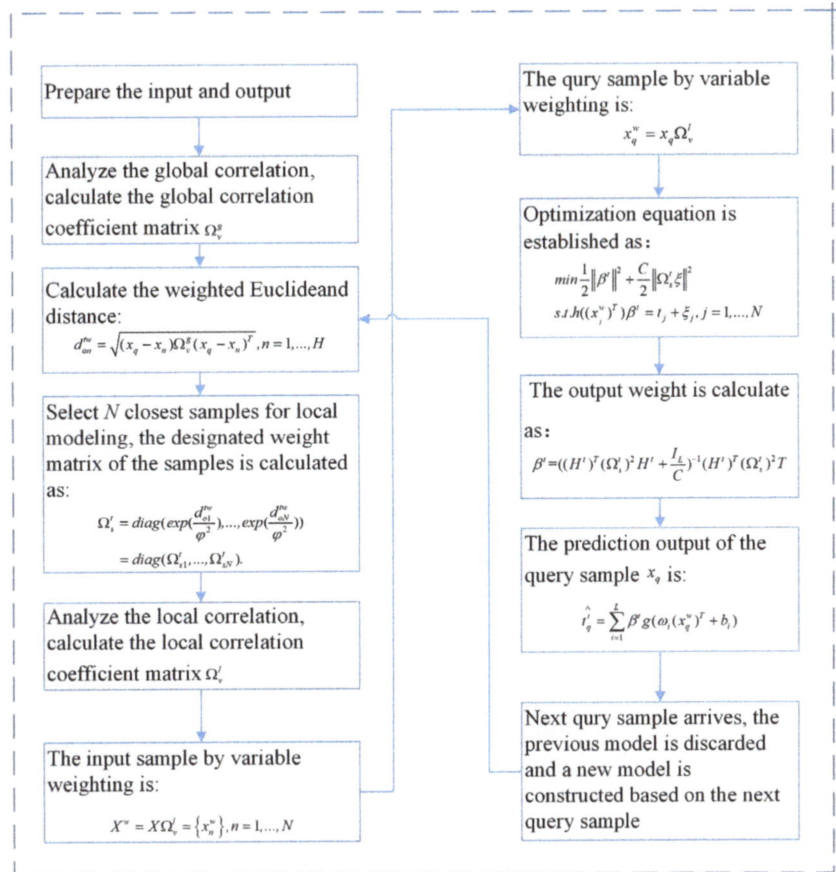

Figure 3. The flowchart of JITL-TWRELM model.

To evaluate the performance of the proposed method, four indices, including mean absolute error (MAE), root mean squared error ($RMSE$), mean absolute percentage error ($MAPE$), and coefficient of determination (R^2) are used in the performance evaluation, which are as follows:

$$MAE = \frac{1}{N_T} \sum_{i=1}^{N_T} |y_i - \hat{y}_i| \tag{34}$$

$$RMSE = \sqrt{\frac{1}{N_T} \sum_{i=1}^{N_T} (y_i - \hat{y}_i)^2} \tag{35}$$

$$MAPE = \frac{1}{N_T} \sum_{i=1}^{N_T} |\frac{y_i - \hat{y}_i}{y_i}| \tag{36}$$

$$R^2 = 1 - \frac{\sum_{i=1}^{N_T} (y_i - \hat{y}_i)^2}{\sum_{i=1}^{N_T} (y_i - \bar{y}_i)^2} \quad (37)$$

where N_T denotes the number of samples used for testing, y_i and $\hat{y}_i$ denote the values of the actual output variable and predicted output, respectively, $\bar{y}_i$ denotes the mean value of the actual output variable. It is essential to have small MAE, $RMSE$, and $MAPE$, and large R^2 for a prediction model.

Before establishing the JITL-TWRELM model, four parameters need to be determined. By trial and error experiments on dataset D1, N was set as a proper value of 200, which has a good prediction accuracy without increasing the computational burden. Similarly, the parameters φ and L are set to 0.3 and 20, respectively. Table 5 shows the prediction accuracy of the model under the different regularization coefficients. It can be seen that when $C = 150$, the model has a better effect.

Table 5. Comparison of the modeling accuracy with C.

C	MAE	$RMSE$	$MAPE$	R^2
140	15.0884	18.6443	0.020354	0.98666
150	14.7273	17.9456	0.019897	0.98764
160	15.3797	19.5019	0.020879	0.98541
170	15.5217	20.4519	0.021023	0.98395
180	15.8944	20.1194	0.021629	0.98447
190	16.0543	19.8318	0.021839	0.98491
200	17.1959	22.2015	0.023265	0.98109

4.3. Results and Discussion

To reduce the effect of randomness on the results, we took the average of ten tests as the final result. The prediction error indices of the six methods on two groups of the testing samples are shown in Table 6. We use testing dataset D1 as an example; in general, the proposed method (method 6) performed better than the other five methods on all four indices. Despite using the JITL strategy, the original method (method 1) had the worst performance on all indices. Methods 2, 3, 4, and 5 also achieved higher prediction accuracies than method 1, as neither the sample weights nor variable weights were used in method 1. Method 2 emphasizes the importance of the samples and introduces sample weights to reflect the effects of different samples on the output. Contrasted with the original JITL strategy, method 3 uses a weighted similarity measurement criterion; samples that are more similar to the query sample were selected to set up the local model, resulting in a more accurate prediction. Different from the previous methods, method 4 considers the local variable weights before establishing the model; the variable weights can be used to improve the influence of output-related variables and reduce that of irrelevant variables in feature extraction [33]. Although methods 2 to 4 have good prediction accuracy improvements, these methods only consider certain types of weighting strategies, such as individual sample weights or variable weights. Hence, method 5 introduces the sample weights and the WJITL strategy, and the R^2 is improved from 0.89453 to 0.97690 compared with method 1. Meanwhile, based on methods 4 and 5—method 6 has the smallest MAE, $RMSE$, and $MAPE$, and the highest R^2 among all methods. The R^2 of method 6 is improved from 0.97690 to 0.98764 compared with method 5. Correspondingly, the proposed method 6 has good prediction accuracy on D2, the R^2 reached 0.97427, which is 0.072 higher than method 1.

Table 6. The indices of the six methods of two groups of testing datasets.

Dataset	Method	*MAE*	*RMSE*	*MAPE*	R^2
D1	JITL-RELM	43.0278	52.4279	0.062519	0.89453
	JITL-SWRELM	38.1265	47.9605	0.052589	0.91174
	WJITL-RELM	32.7444	42.7606	0.044443	0.92984
	JITL-VWRELM	38.0149	46.1055	0.054197	0.91843
	JITL-DWRELM	20.7980	24.5347	0.029768	0.97690
	JITL-TWRELM	14.7273	17.9456	0.019897	0.98764
D2	JITL-RELM	26.7981	36.1509	0.035878	0.90223
	JITL-SWRELM	26.3624	33.526	0.03657	0.91174
	WJITL-RELM	27.4121	34.5595	0.044443	0.91065
	JITL-VWRELM	24.3605	31.4511	0.032461	0.92600
	JITL-DWRELM	16.2472	22.2734	0.021632	0.96289
	JITL-TWRELM	14.8733	18.5463	0.019646	0.97427

To more intuitively demonstrate the performances of these six methods, the detailed prediction results for each method on D1 and D2 are shown in Figures 4 and 5, in which (a–f) shows the prediction results of the six methods, respectively. It is easy to see that the prediction of JITL-TWRELM matches well with the curve of the actual measurement of the furnace temperature, while the prediction curve of JITL-RELM cannot track with the real output curve in some samples. In addition, although the other four methods have certain improvements, they still do not achieve the desired effects. In summary, the flexibility and effectiveness of the proposed methods are validated.

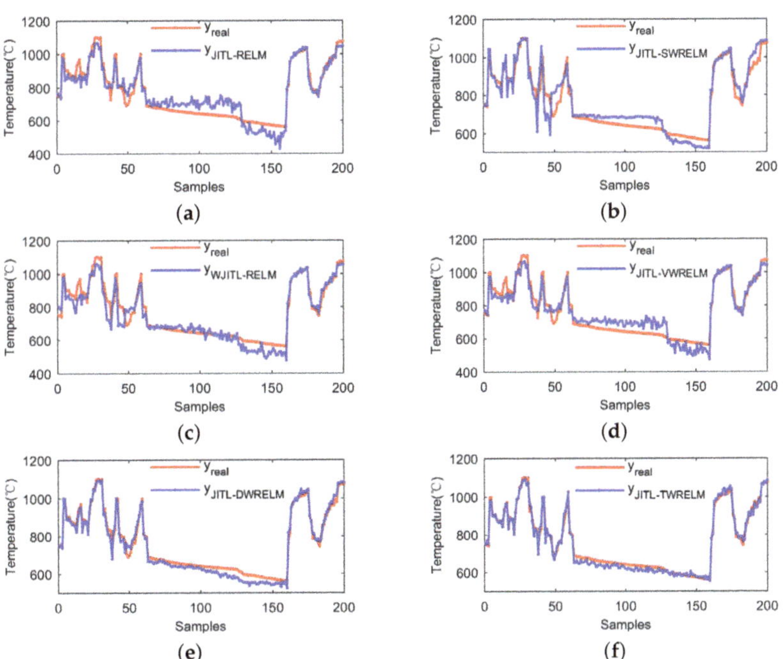

Figure 4. The detailed prediction results of six methods on D1; (**a**) method 1; (**b**) method 2; (**c**) method 3; (**d**) method 4; (**e**) method 5; (**f**) method 6.

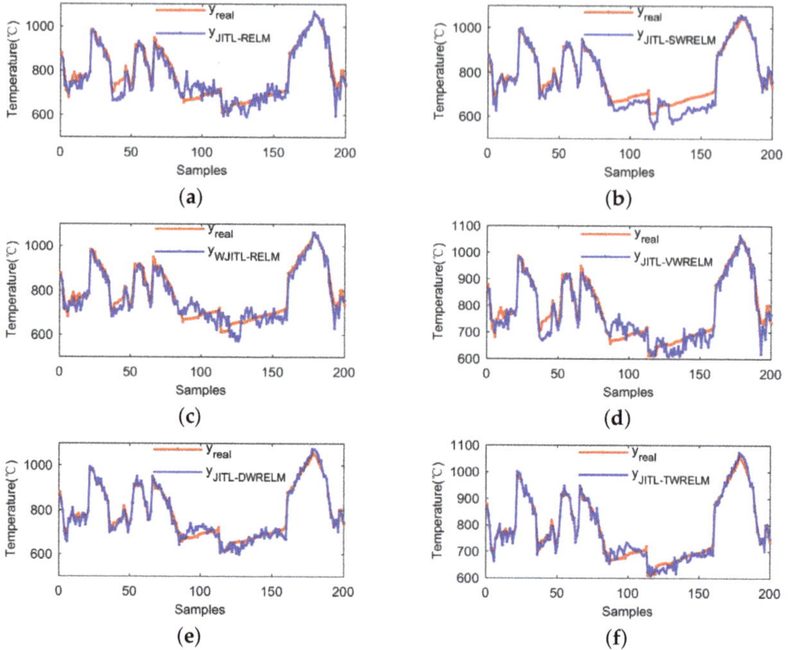

Figure 5. The detailed prediction results of six methods on D2; (**a**) method 1; (**b**) method 2; (**c**) method 3; (**d**) method 4; (**e**) method 5; (**f**) method 6.

The liquid aluminum temperature of the regenerative smelting furnace is generally controlled by feedback. The thermocouple is set in the furnace chamber, and if the temperature is detected to be lower than the set value, the regenerative burner starts to work. In a real industrial site, the temperature measurement performance of the thermocouple used to measure the temperature of aluminum liquids is often affected by voltage fluctuations and the aging of the protective jacket. Old thermocouples need to be replaced frequently, resulting in increased costs. The proposed method in this paper only requires the establishment of a historical database in the industrial site, and whenever a new query sample arrives, the modeling sample is selected from the historical database for modeling, and the prediction results of the aluminum liquid temperature can be obtained. As can be seen from Table 6, the MAEs of the proposed method 6 are 14.7273 and 14.8733 for the two test sets, respectively. Comparing the temperature range and measurement error of the thermocouple in Table 4, the accuracy of the proposed soft measurement model is close to the actual sensor, with a close to 2% error at the maximum temperature measurement range, but the efficiency and costs are more advantageous than the sensor. Therefore, the method proposed in this paper is significant for reducing production costs and improving product quality.

5. Conclusions

This paper mainly deals with the estimation of the liquid aluminum temperature in the regenerative aluminum smelting furnace. A JITL-TWRELM soft sensor modeling method is proposed. In this method, both the sample similarities and the variable correlations are considered in RELM to deal with the differences between samples and variables. Each modeling sample is assigned different weights according to the similarity calculation, and each dimension of the sample is also assigned a corresponding weight according to the correlation analysis, which improves the accuracy of the modeling compared with the original RELM. Furthermore, a weighted similarity measurement criterion is proposed for

JITL to select similar samples for local modeling. Compared with the original JITL strategy, more similar modeling samples are selected for each query sample, enhancing the accuracy and reliability of the local modeling dataset. The flexibility and effectiveness of JITL-TWRELM were validated through the industrial aluminum smelting process. The industrial applications show that the proposed method can effectively deal with the nonlinear and time-varying problems in the regenerative aluminum smelting process and achieve a higher accuracy of temperature prediction compared with the other five methods.

For each query sample, the model needs to be updated once, although some adjacent query samples do not need to update the model so frequently. Selective updating of the model will improve the modeling efficiency. Therefore, developing a selective update strategy will be the focus of future work.

Author Contributions: Data curation, Y.L.; methodology, J.D.; supervision, J.D.; validation, X.C.; writing—original draft, X.C.; writing—review and editing, Y.L. All authors have read and agreed to the published version of the manuscript.

Funding: This research received no external funding.

Data Availability Statement: Not applicable.

Conflicts of Interest: The authors declare that there are no conflict of interest regarding the publication of this paper.

References

1. Froehlich, C.; Strommer, S.; Steinboeck, A.; Niederer, M.; Kugi, A. Modeling of the media supply of gas burners of an industrial furnace. *IEEE Trans. Ind. Appl.* **2016**, *52*, 2664–2672. [CrossRef]
2. Strommer, S.; Niederer, M.; Steinboeck, A.; Kugi, A. A mathematical model of a direct-fired continuous strip annealing furnace. *Int. J. Heat Mass Trans.* **2014**, *69*, 375–389. [CrossRef]
3. Qiu, L.; Feng, Y.H.; Chen, Z.G.; Li, Y.L.; Zhang, X.X. Numerical simulation and optimization of the melting process for the regenerative aluminum melting furnace. *Appl. Therm. Eng.* **2018**, *145*, 315–327. [CrossRef]
4. Li, D.Y.; Song, Z.H. A novel incremental gaussian mixture regression and its application for time-varying multimodal process quality prediction. In Proceedings of the 2020 IEEE 9th Data Driven Control and Learning Systems Conference, Liuzhou, China, 19–21 June 2020.
5. Bao, L.; Yuan, X.F.; Ge, Z.Q. Co-training partial least squares model for semi-supervised soft sensor development. *Chemometr. Intell. Lab.* **2015**, *147*, 75–85. [CrossRef]
6. Wang, G.; Luo, H.; Peng, K.X. Quality-related fault detection using linear and nonlinear principal component regression. *J. Franklin. Inst.* **2016**, *353*, 2159–2177. [CrossRef]
7. Liu, H.B.; Yang, C.; Huang, M.Z.; ChangKyoo, Y. Soft sensor modeling of industrial process data using kernel latent variables-based relevance vector machine. *Appl. Soft. Comput.* **2020**, *90*, 106149. [CrossRef]
8. Shi, X.; Huang, G.L.; Hao, X.C.; Yang, Y.; Li, Z. A Synchronous Prediction Model Based on Multi-Channel CNN with Moving Window for Coal and Electricity Consumption in Cement Calcination Process. *Sensors* **2021**, *21*, 4284. [CrossRef]
9. Zhe, L.; Yi-Shan, L.; Chen, J.H.; Qian, Y.W. Developing variable moving window PLS models: Using case of NOx emission prediction of coal-fired power plants. *Fuel* **2021**, *296*, 120441–120456.
10. Mai, T.; Yu, X.L.; Gao, S.; Frejinger, E. Routing policy choice prediction in a stochastic network: Recursive model and solution algorithm. *Transport. Res. B-Methodol.* **2021**, *151*, 42–58. [CrossRef]
11. Doshi, P.; Gmytrasiewicz, P.; Durfee, E. Recursively modeling other agents for decision making: A research perspective. *Artif. Intell.* **2020**, *279*, 103202–103220. [CrossRef]
12. Qiu, K.P.; Wang, J.L.; Zhou, X.J.; Guo, Y.Q.; Wang, R.T. Soft sensor framework based on semisupervised just-in-time relevance vector regression for multiphase batch processes with unlabeled data. *Ind. Eng. Chem. Res.* **2020**, *59*, 19633–19642. [CrossRef]
13. Wu, Y.J.; Liu, D.J.; Yuan, X.F.; Wang, Y.L. A just-in-time fine-tuning framework for deep learning of SAE in adaptive data-driven modeling of time-varying industrial processes. *IEEE. Sens. J.* **2020**, *21*, 3497–3505. [CrossRef]
14. Chen, N.; Luo, L.H.; Gui, W.H.; Guo, Y.Q. Integrated modeling for roller kiln temperature prediction. In Proceedings of the 2017 Chinese Automation Congress, Shandong, China, 20–22 October 2017.
15. Dai, J.Y.; Chen, N.; Yuan, X.F.; Gui W.H.; Luo, L.H. Temperature prediction for roller kiln based on hybrid first-principle model and data-driven MW-DLWKPCR model. *ISA Trans.* **2020**, *98*, 403–417. [CrossRef]
16. Yao, L.; Ge, Z.Q. Locally weighted prediction methods for latent factor analysis with supervised and semisupervised process data. *IEEE Trans. Autom. Sci. Eng.* **2016**, *14*, 126–138. [CrossRef]
17. Chen, J.M.; Yang, C.H.; Zhou, C.; Li, Y.G.; Zhu, H.Q.; Gui, W.H. Multivariate regression model for industrial process measurement based on double locally weighted partial least squares. *IEEE. Trans. Instrum. Meas.* **2019**, *69*, 3962–3971. [CrossRef]

18. Yuan, X.F.; Huang, B.; Ge, Z.Q.; Song, Z.H. Double locally weighted principal component regression for soft sensor with sample selection under supervised latent structure. *Chemometr. Intell. Lab.* **2018**, *153*, 116–125. [CrossRef]
19. Lin, W.L.; Hang, H.F.; Zhuang, Y.P.; Zhang, S.L. Variable selection in partial least squares with the weighted variable contribution to the first singular value of the covariance matrix. *Chemometr. Intell. Lab.* **2018**, *183*, 113–121. [CrossRef]
20. Yuan, X.F.; Li, L.; Shardt, Y.; Wang, Y.L.; Yang, C.H. Deep learning with spatiotemporal attention-based LSTM for industrial soft sensor model development. *IEEE Trans. Ind. Electron.* **2020**, *65*, 4404–4414. [CrossRef]
21. Yuan, X.F.; Li, L.; Wang, Y.L. Nonlinear dynamic soft sensor modeling with supervised long short-term memory network. *IEEE Trans. Ind. Inform.* **2019**, *16*, 3168–3176. [CrossRef]
22. He, Z.S.; Chen, Y.H.; Xu, J. A Combined Model Based on the Social Cognitive Optimization Algorithm for Wind Speed Forecasting. *Processes* **2022**, *10*, 689. [CrossRef]
23. Wang, X.L.; Zhang, H.; Wang, Y.L.; Yang, S.M. ELM-Based AFL–SLFN Modeling and Multiscale Model-Modification Strategy for Online Prediction. *Processes* **2019**, *7*, 893. [CrossRef]
24. Li, G.Q.; Chen, B.; Qi, X.B.; Zhang, L. Circular convolution parallel extreme learning machine for modeling boiler efficiency for a 300 MW CFBB. *Soft. Comput.* **2019**, *23*, 6567–6577. [CrossRef]
25. Hu, P.H.; Tang, C.X.; Zhao, L.C.; Liu, S.L.; Dang, X.M. Research on measurement method of spherical joint rotation angle based on ELM artificial neural network and eddy current sensor. *IEEE Sens. J.* **2021**, *21*, 12269–12275. [CrossRef]
26. Su, X.L.; Zhang, S.; Yin, Y.X.; Xiao, W.D. Prediction model of hot metal temperature for blast furnace based on improved multi-layer extreme learning machine. *Int. J. Mach. Learn. Cybern.* **2019**, *10*, 2739–2752. [CrossRef]
27. Huang, Q.B.; Lei, S.N.; Jiang, C.L.; Xu, C.H. Furnace Temperature Prediction of Aluminum Smelting Furnace Based on KPCA-ELM. In Proceedings of the 2018 Chinese Automation Congress, Xi'an, China, 30 November–2 December 2018.
28. Liu, Q.; Wei, J.; Lei, S.; Huang, Q.B.; Zhang, M.Q.; Zhou, X.B. Temperature prediction modeling and control parameter optimization based on data driven. In Proceedings of the 2020 IEEE Fifth International Conference on Data Science in Cyberspace, Hong Kong, China, 24–27 June 2020.
29. Li, Z.X.; Hao, K.R.; Chen, L.; Ding, Y.S.; Huang, B. Pet viscosity prediction using jit-based extreme learning machine. In Proceedings of the 2018 IFAC Conference, Changchun, China, 20–22 September 2018.
30. Wang, Y.; Li, Y.G.; Wu, B.Y. Improved regularized extreme learning machine short-term wind speed prediction based on gray correlation analysis. *Wind. Eng.* **2021**, *45*, 667–679.
31. Pan, Z.Z.; Meng, Z.; Chen, Z. J; Gao, W.Q.; Shi, Y. A two-stage method based on extreme learning machine for predicting the remaining useful life of rolling-element bearings. *Mech. Syst. Signal Process.* **2020**, *144*, 106899–106916. [CrossRef]
32. Zong, W.W.; Huang, G.B.; Chen, Y.Q. Weighted extreme learning machine for imbalance learning. *Neurocomputing* **2013**, *101*, 229–242. [CrossRef]
33. Chen, N.; Dai, J.Y.; Yuan, X.F.; Gui, W.H.; Ren, W.T.; Koivo, H. Temperature prediction model for roller kiln by ALD-based double locally weighted kernel principal component regression. *IEEE Trans. Instrum. Meas.* **2018**, *67*, 2001–2010. [CrossRef]

Article

The Influence of Size Effect to Deformation Mechanism of C5131 Bronze Structures of Negative Poisson's Ratio

Jiaqi Ran, Gangping Chen, Fuxing Zhong, Li Xu, Teng Xu * and Feng Gong *

Shenzhen Key Laboratory of High Performance Nontraditional Manufacturing, College of Mechatronics and Control Engineering, Shenzhen University, Shenzhen 518060, China; ranjiaqi26@szu.edu.cn (J.R.); 1910293003@email.szu.edu.cn (G.C.); zhongfuxing2020@email.szu.edu.cn (F.Z.); Jerson.l.xu@foxconn.com (L.X.)
* Correspondence: tengxu@szu.edu.cn (T.X.); gongfeng@szu.edu.cn (F.G.)

Abstract: 3D auxetic structures, which present negative Poisson's ratio in the uniaxial compression deformation, is an ideal artificial material for meta-implants because of its lightweight, good material property and suitable porosity for bone recovery compared with conventional meta-biomaterials. Selective laser melting (SLM) is commonly used to produce metallic 3D auxetic structures but limited by the melting temperature and reflect rate of the material, and micro assembled (MA) structures is an alternative manufacturing process. However, the influence of size effect in 3D auxetic structures and the difference of the constitutive model of 3D auxetic structure produced by SLM and MA have not been discussed. In tandem of this, the mechanical property comparison of 3D auxetic structures produced by SLM and MA is conducted and a structural surface layer model for 3D auxetic structures is proposed. The result indicated that both SLM and MA structure can achieve auxetic effect. It is found that the Poisson's ratio of the SLM and MA structures decrease when increasing the size factor of the structure, and the negative Poisson's ratio effect is more obvious when the Young's modulus is relatively small. FE simulation result of Poisson's ratio is closer to experimental result of MA structures due to complexity of 3D auxetic structures. This paper thus provides a relatively helpful constitutive model for the prediction of the mechanical behavior of 3D auxetic structure.

Keywords: 3D auxetic structures; selective laser melting; micro assembled; structural surface layer model

Citation: Ran, J.; Chen, G.; Zhong, F.; Xu, L.; Xu, T.; Gong, F. The Influence of Size Effect to Deformation Mechanism of C5131 Bronze Structures of Negative Poisson's Ratio. *Processes* **2022**, *10*, 652. https://doi.org/10.3390/pr10040652

Academic Editor: Antonino Recca

Received: 6 February 2022
Accepted: 24 March 2022
Published: 28 March 2022

1. Introduction

Auxetic material and structure has been widely studied in modern industry [1]. In contrast with the conventional material and structure, the auxetic structure is contracted when compression force is applied in lateral direction because of the negative Poisson's ratio effect. Lakes created a foam structure with negative Poisson's ratio by conducting heat treatment to polyurethane foam, and successfully obtained a structure with the Poisson's ratio of −0.7 [2]. Evans et al. discovered that negative Poisson's ratio effect existed when porous Teflon structure was under compression stress [3], and a simplified microstructure model was established to explain this phenomenon. Later, Evans et al. named the material with microstructure of negative Poisson's ratio as the "auxetic materials" [4].

In the field of the computational study of auxetic material, Wojciechowski was the first to introduce a constant thermodynamic tension Monte Carlo method to study a two-dimensional system of hard cyclic hexamers. He demonstrated the existence of a crystal-crystal phase transition in the system and proved that Poisson's ratio of the densest phase can be negative [5]. Later, he provided a 2D lattice model of hexagonal molecules and proved that the Poisson's ratio of an isotropic system can be equal to any negative value in the range between minus one and zero [6]. The simulation results were confirmed in Ref. [7] by using another simulation method. Kolpakov presented a model indicated that the Poisson coefficient of a macroscopic structure of periodic configuration can be less than zero under certain conditions [8]. Hoover et al. provided the earliest approach on dynamic analyses of macroscopic periodic cellular structures and the maximum negative Poisson's ratio was obtained [9]. Pozniak et al. studied the size effects in auxetics for non-periodic structure [10]. For auxetic star structure, Bilski et al. proposed research on the elastic properties of 2D crystalline structures, and the phenomenon that Poisson's ratio decreased with increasing pressure was revealed [11]. Deformations and damages in wood-based sandwich beams with three dimensional structures similar to the one studied in the present paper were recently investigated by Smardzewski, and the ability of energy absorption in multilayer panel was studied [12].

As for the auxeticity of other mechanism except the mechanical relations of beams in rigid bodies, Baughman et al. indicated that pairwise central forces are the main reason of the negative Poisson's ratio for the (110) stretch [13]. Piglowski et al. studied the size polydispersity on auxetic properties of Yukawa crystals and the conclusion that the decrease of the Poisson's ratio is caused by particle size polydispersity was carried out [14]. Narojczyk and Wojciechowski found that the Poisson's ratio increased with increasing disorder at a certain number density of soft sphere systems [15]. Later, they proposed a study on the influence of inclusion particles' size to the Poisson's ratio of nanocomposites [16]. As one of the most common negative Poisson's ratio structures, the concave structure achieves the negative Poisson's ratio of the material by unfolding both sides of the unit cell. The concave structure was first transformed from the two-dimensional honeycomb structure. Almgren obtained the concave hexagonal structure by connecting rigid bars with hinges, and its isotropic elastic property in macro scale has been discovered [17]. Larsen et al. used numerical topology optimization to design a novel negative Poisson's ratio structure composed of double-arrow unit cells which was fabricated by chemical vapor deposition [18]. Grima established a two-dimensional periodic star structure, and the conclusion that the negative Poisson's ratio effect doesn't exist in every kind of star structure has been made [19]. Shokri et al. proposed a 3D auxetic structure based on 2D structure. By comparing the FE simulation and experiment result, the influence of the pillar thickness of the unit cell to the mechanical property has been revealed [20]. Hengsbach et al. used direct laser writing to obtain 3D auxetic structure with complex shape [21]. Wang et al. presented a 3D negative Poisson's ratio structure by applying interlocking assemble technic [22]. The influence of the length and angle between the vertical and oblique strut of each unit cell was discussed, and the negative Poisson's ratio effect was quantified. Li et al. prepared an auxetic structure of Ti-6Al-4V using electron beam melting (EBM) and the influence of the Poisson's ratio to structural mechanical property was discovered [23]. This previous research has provided the theoretical foundation for the application of the auxetic structure.

For complex 3D structure, most of the negative Poisson's ratio structures were prepared by 3D printing, which was able to manufacture arbitrarily complex parts, shorten product development cycle, and improve material utilization. The disadvantages of this technic were the relatively high cost, low production efficiency and dependence on materials. In previous research, little attention has been paid to the influence of the pillar thickness of the unit cell to the Poisson's ratio [22]. For this reasons, microforming and micro assembled 3D auxetic structure is carried out in this research, and the mechanical properties are compared with the structures made by SLM. The quasi-static compression test is conducted on both SLM and MA auxetic structure, and the stress-strain relationship obtained from the experiment is compared with the corresponding FE simulation result. The influence of pillar thickness of the unit cell is revealed and a structural-surface-layer (SSL) model is thus established.

2. Experiment

According to the previous research [22,24], the auxetic structure shown in Figure 1 is combined with several unit cells of umbrella shape, and the auxetic effect is caused by the enfoldment of the structure. In this research, the mechanical property of the bronze SLM structure and MA structure are studied. As the unit cell and the total structure is symmetrical, the mechanical behavior of the structure along the vertical direction during the compression process is important. The coefficient of the 3D auxetic structure has been shown: t is the thickness of the pillar, θ is the angle of the vertical and oblique strut, l is the length of oblique strut and H is the length of vertical pillar.

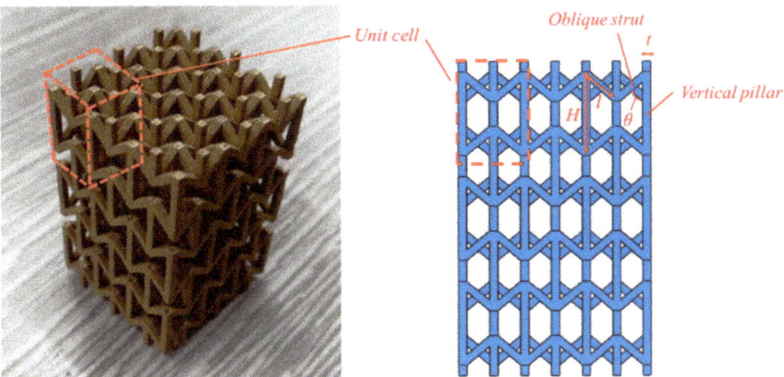

Figure 1. 3D auxetic structure.

The SLM 3D auxetic structure is prepared via a Riton R120 metal 3D printer. It is a 3 × 3 × 3 layer 3D structure contains 27 unit cells. Side holders are applied to prevent the material falling from the structure during the 3D printing process, and then removed after 3D printing is finished. The MA 3D auxetic structure is produced by microforming and micro assemble process. A bronze foil is delivered in the progressive molding tool and the blanking process is conducted at each stage. Four kinds of 2D parts are produced via microforming and then assembled into a 3D auxetic structure, as shown in Figure 2.

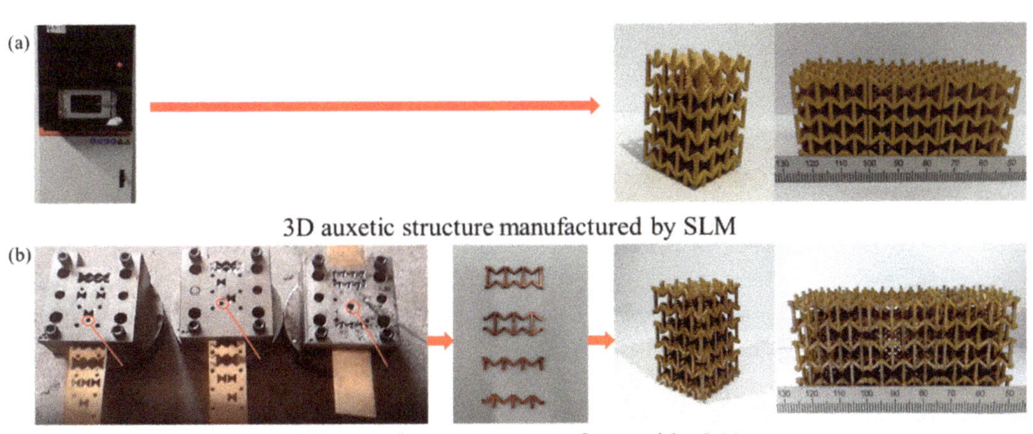

Figure 2. 3D auxetic strcutre manufactured by (**a**) SLM and (**b**) MA.

In order to compare the mechanical property of the SLM with the MA structure, bronze C5131 is applied in both of these two processes. The compression test is conducted on a Zwick/Roell Z050 testing machine, which is connected with a VideoXtens extensometer. Laser tags are bound on the top, bottom and both sides of the structure, and the maximum resolution of the extensometer is 0.01 μm. The punch speed of the testing machine is set to 0.1 mm/min to avoid the influence of strain rate, as shown in Figure 3. The horizontal and vertical strain is obtained via the video extensometer and the Poisson's ratio is thus calculated.

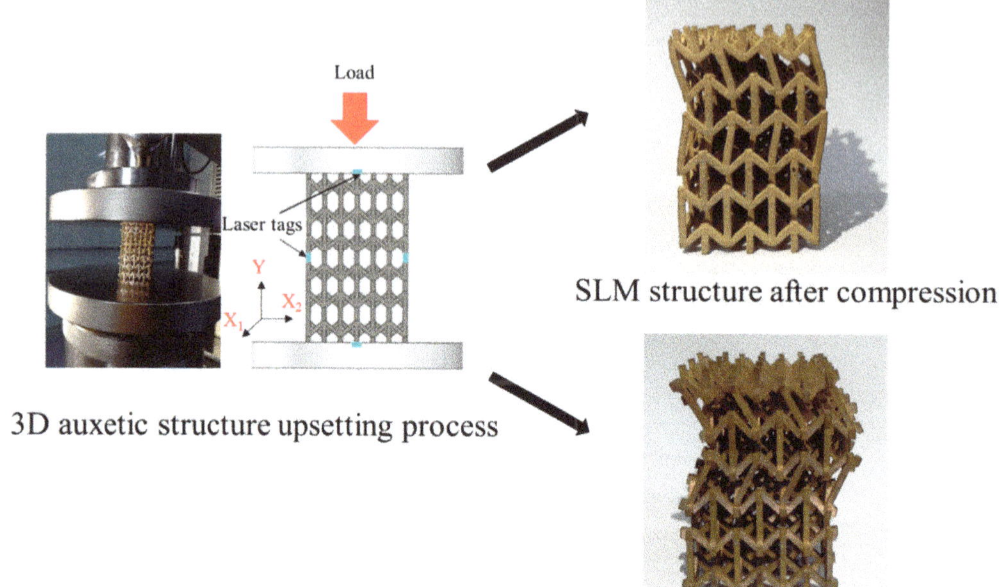

Figure 3. Quasi-static compression test for SLM and MA 3D auxetic structure.

3. Result Analysis

By conducting FE simulation, it is found that the stress concentration is on the oblique strut of the structure, and the bending process of the oblique strut is the main reason of the phenomenon of negative Poisson's ratio [25], as shown in Figure 4. During the compression test, the vertical pillars of both sides of the auxetic structure move towards the central of the structure and thus achieve the lateral shrinkage of total structure.

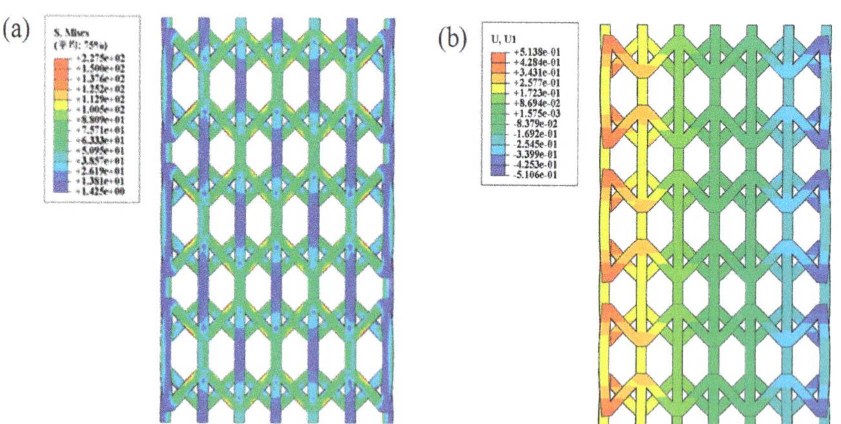

Figure 4. FE simulation of compression test of the 3D auxetic structure: (**a**) Stress distribution (**b**) Stroke distribution.

3.1. Comparison of SLM and MA Structure

In this research, the Poisson's ratio of different thickness of the vertical pillar is measured. For SLM structure, the pillar thickness of the inner layer of the structure is 1 mm, and the change of the dimension along horizontal direction is studied. As the 3D auxetic structure is symmetric, X_1 and X_2 direction are not distinguished, which means X_2 will not be considered. The laser tags at the bottom and top of the structure will measure the displacement along the loading direction Y, and the laser tags at both sides will measure the displacement along the X direction. The slope of the stress-strain curve is the Young's modulus and the slope of the ε_x and ε_y is the Poisson's ratio of the structure. The data of ε_x and ε_y is obtained via the two laser extensometers of horizontal and vertical direction and the ε_x-ε_y curve is demonstrated in Figure 5a. It is found that the absolute value of the Poisson's ratio has decreased with the pillar thickness. When the pillar thickness reach 1.5 mm, the slope of ε_x-ε_y curve is near 0, which means the auxetic effect of the structure no longer exists. As the thickness increases, more overlap areas exist at the joint point of the vertical pillar and oblique strut, and thus reduce the effective length of the oblique strut. From Figure 5b, it is found that the Young's modulus of the 1.5 mm structure is the largest.

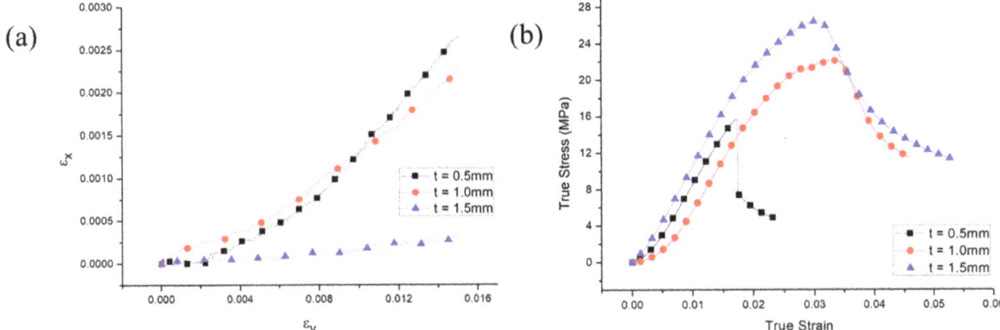

Figure 5. The ε_x-ε_y curve and stress-strain curve of SLM auxetic structure of different pillar thickness. (**a**) ε_x-ε_y curve (**b**) Stress-strain curve.

For MA structure, the Poisson's ratio curve is shown in Figure 6a. It is found that the auxetic effect exists in the structure with all the three different pillar thickness, but the influence of the thickness is slightly different with the SLM structure. The Poisson's ratio of the MA structure with 1.5 mm pillar thickness is much larger than the SLM structure, and the reason of this phenomenon may be due to the destabilization of the structure and has thus increased ε_x. For the specimens with the pillar thickness of 0.5 mm, the pillar contact area is relatively small. During the compression test, some of the bottom pillars may have "stick" on the die of the testing machine due to the friction force. Therefore, it is possible for these bottom pillars to buckling down and thus cause the Young's modulus of 0.5 mm higher than 1 mm. In Figure 6b the curve of Young's modulus of MA structure has been shown. The result is consistent with the one for SLM structure, which means the largest Young's modulus is for the 1.5 mm thickness structure.

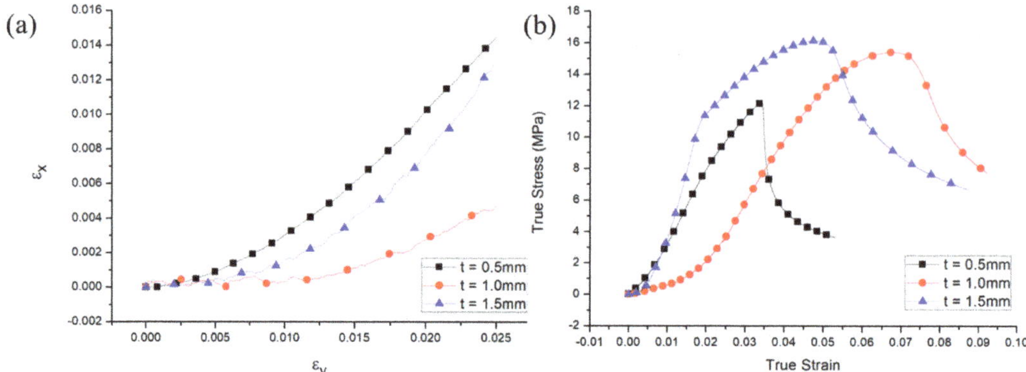

Figure 6. The Poisson's ratio εx-εy curve and stress-strain curve of MA auxetic structure of different pillar thickness. (**a**) ε_x-ε_y curve (**b**) Stress-strain curve.

At the beginning stage of the compression test, the flow stress of the MA structure increases slowly, because each component of MA structure has not been fully contacted in this stage, which is the reason that ε_y of MA structure is larger than SLM structure. When the components are fully contact, the flow stress of the structure increases rapidly with the increasing strain.

By comparing the mechanical properties of SLM and MA structure, it is found that the auxetic effect exists in both structures during the compression test. The auxetic effect is more obvious in MA structure as the absolute value of the Poisson's ratio is larger than SLM structure, as shown in Figure 7a. Although the accuracy and clearance of each component has reduced the Young's modulus of the MA structure, the comparison of the mechanical property between SLM and MA structure still proves that microforming and micro-assemble process is an effective way for manufacturing complex 3D auxetic structure. Meanwhile, it is found that another reason of the difference of the Young's modulus of MA and SLM structure may be caused by the coarse grains due to sudden temperature change during the manufacturing process of SLM structure. The Young's modulus of both structures decreases and then increase when the thickness of the pillar increases, as shown in Figure 7b.

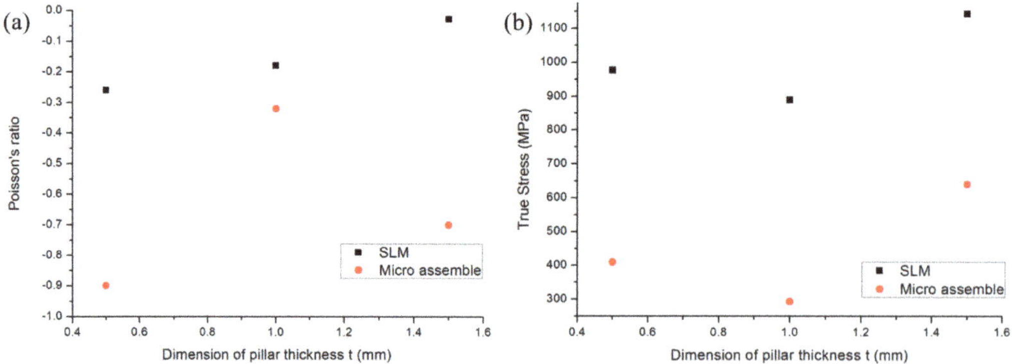

Figure 7. Mechanical properties of SLM and MA 3D auxetic structure: (**a**) Poisson's ratio (**b**) Young's Modulus.

According to the result of the compression test, it is found that the upper layer of the auxetic structure will first destabilized from the elastic stage to plastic stage. This may cause by the friction between the structure and the MTS testing machine. The structure is taking axial compression stress and shear stress of the top and bottom surface because of the friction stress, which makes the structure destabilized. Micro cracks thus exist after structure destabilization, as shown in the SEM photo of Figure 8.

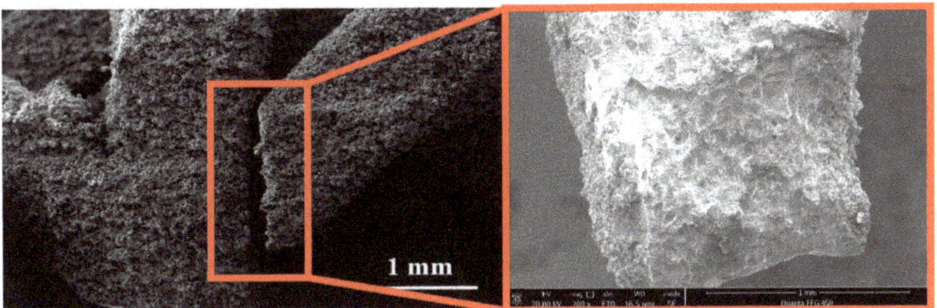

Figure 8. The micro cracks in the oblique strut of the 3D auxetic structure.

3.2. Structural Size Effect and Structural Surface Layer Model

In macro and micro plastic deformation process, surface layer model has been widely used to reveal the stress contribution of different layer of the specimen. As the surface layer grains has free surface, its mechanical property is better than the inner layer grain. The size factor η, which is the proportion of the surface grains number to total grain number, is introduced into the surface layer model [26].

$$\eta = \frac{N_{Surface}}{N_{Total}} \tag{1}$$

The basic expression of the surface layer model is:

$$\sigma = \eta\sigma_{Surface} + (1 - \eta)\sigma_{inner} \tag{2}$$

In macro scaled plastic deformation process, the total grain number is very large and the size factor is thus small and the size effect is not obvious. When the size of the specimen decreases, or the grain size of the specimen increases, the size factor becomes larger and the stress contribution of the surface layer grain cannot be ignored. By using Equation (2), the stress strain relationship of a specimen affect by size effect can thus be obtained.

In this research, the so-called structural size effect is discovered and the structural surface layer model is provided to explain this phenomenon. As the 3D auxetic structure is consisted of several unit cells with the same size, the unit cells can be classified into corner unit cell, surface layer unit cells and inner layer unit cells, which is similar to previous research [26]. The corner unit cells and the surface layer unit cells have free surfaces while the inner layer unit cells haven't, as shown in Figure 9. The unit cells in the central area, which are marked with red, are defined as the inner layer unit cells. The corner unit cells and surface layer unit cells have at least one free surface, which are marked with green and blue.

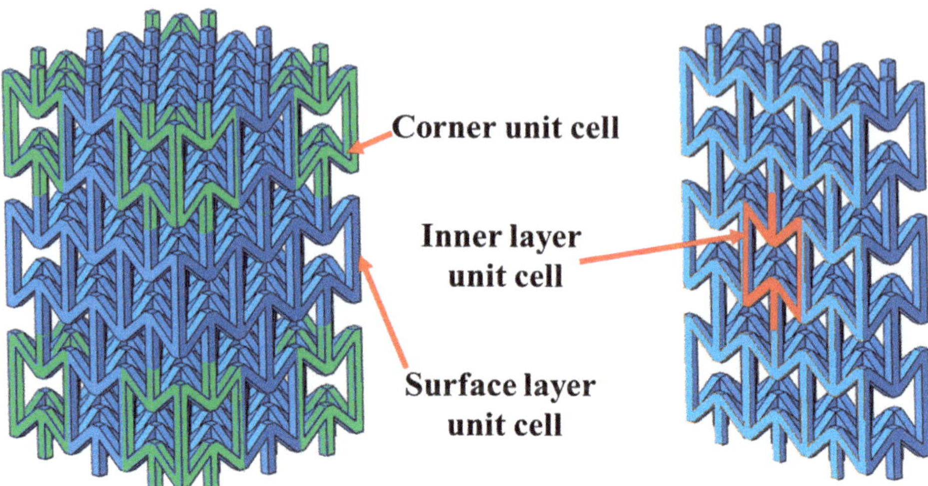

Figure 9. Structure size effect in 3D auxetic structure.

In Wang's previous research [22], the reduction of the effective length Δl of the 3D auxetic structure is consisted of Δl_1 and Δl_2, as shown in Equation (3):

$$\Delta l = \varphi \Delta l_1 + (1 - \varphi)\Delta l_2 \tag{3}$$

where $\Delta l_1 = \frac{t/2}{sin\theta}$ and $\Delta l_2 = \frac{t/2}{tan\frac{\theta}{2}}$. In this equation, the reduction of a single unit cell under compression stress is related with the pillar thickness t and the re-entrant angle. The coefficient φ is a semi-empirical coefficient which is calculated via curve fitting. This model is quite accurate according to the experimental and simulation result, but the stress-strain relationship of the 3D auxetic is not easy to obtain.

In this research, an uncoupled model based on Lemaitre's damage model [27] is presented. According to Lemaitre's model, the damage value D of material is defined as the effective area $\overline{S}$ to the entire area of a unit cell selected from the material S, and can be represented as,

$$D = \frac{S - \overline{S}}{S} \tag{4}$$

And the damage equivalent stress σ^* is similar to the Von Mises stress and can be represented as,

$$\sigma^* = \sqrt{\frac{2}{3}(1+v)\frac{\sigma}{1-D} + 3(1-2v)\sigma_m^2} \tag{5}$$

where v is the Poisson's ratio and σ_m is the mean stress.

For inner unit cells of the 3D auxetic structure, the damage value D can be applied as the volume of the vertical pillar and the oblique structure to the volume of a unit cell with a solid core according to Equation (4), which is:

$$D_{inner} = \frac{V_{total} - V_{pillar}}{V_{total}} = 1 - \frac{4t^2\left[H + \frac{1-cos\theta}{sin\theta}t\right] + 16t^2\left(l - \frac{t}{sin\theta}\right)}{(2lsin\theta)^2(2(H - lcos\theta))} = 1 - \frac{\left[\frac{H}{l} + 4 - \frac{3+cos\theta}{sin\theta}\frac{t}{l}\right]t^2}{2(lsin\theta)^2\left(\left(\frac{H}{l} - cos\theta\right)\right)} \tag{6}$$

And for surface auxetic structure, the damage value D can be distinguished into two parts, which is the surface layer unit cells and the corner unit cells of the 3D auxetic structure. These two kinds of structures can be represented as:

$$D_{corner} = \frac{V_{total} - V_{pillar}}{V_{total}} = 1 - \frac{4t^2\left[H + \frac{1-cos\theta}{sin\theta}t\right] + 16t^2(l - \frac{t}{sin\theta}) + 4t^2 l + 7H\frac{t^2}{2}}{(2lsin\theta)^2(2(H - lcos\theta))} = 1 - \frac{t^2\left[20 - \frac{12+4cos\theta}{sin\theta}\frac{t}{l} + \frac{15H}{2l}\right]}{(2lsin\theta)^2\left(2\left(\frac{H}{l} - cos\theta\right)\right)} \quad (7)$$

$$D_{surface} = \frac{V_{total} - V_{pillar}}{V_{total}} = 1 - \frac{4t^2\left[H + \frac{1-cos\theta}{sin\theta}t\right] + 16t^2(l - \frac{t}{sin\theta}) + 4\frac{t^2}{2}l + 4H\frac{t^2}{2}}{(2lsin\theta)^2(2(H - lcos\theta))} = 1 - \frac{4t^2\left[\frac{3H}{2l} - \frac{3+cos\theta}{sin\theta}\frac{t}{l} + \frac{9}{2}\right]}{(2lsin\theta)^2\left(2\left(\frac{H}{l} - cos\theta\right)\right)} \quad (8)$$

When the auxetic structure contains a large number of unit cells, the proportion of the surface and corner unit cells will be very small. For example, the auxetic structure contains 100 × 100 × 100 unit cells have 8 corner unit cells and 58,800 surface unit cells, and the number of the inner unit cells is 941,192, the size factor η is 5.88% and the structure size effect is not obvious. The auxetic structure contains 10 × 10 × 10 unit cells have 8 corner unit cells and 480 surface unit cells, and the number of the inner unit cells is 512, the size factor η is 48.8%. However, when the structure size or the total number of unit cells is becoming smaller, the size factor becomes larger. In this research, the auxetic structure contains 3 × 3 × 3 unit cells and the size factor of the auxetic structure is 96.3%, which is consisted of 29.6% of corner unit cells and 66.7% of surface layer unit cells. Therefore, the equivalent stress of a 3D auxetic structure with 3 × 3 × 3 unit cells can be represented as:

$$\begin{cases} \sigma = \sqrt{\frac{2}{3}(1+v)\frac{\sigma}{1-D} + 3(1-2v)\sigma_m^2} \\ D_{surface} = 1 - \frac{4t^2\left[\frac{3H}{2l} - \frac{3+cos\theta}{sin\theta}\frac{t}{l} + \frac{9}{2}\right]}{(2lsin\theta)^2\left(2\left(\frac{H}{l} - cos\theta\right)\right)} \\ D_{corner} = 1 - \frac{t^2\left[20 - \frac{12+4cos\theta}{sin\theta}\frac{t}{l} + \frac{15H}{2l}\right]}{(2lsin\theta)^2\left(2\left(\frac{H}{l} - cos\theta\right)\right)} \\ D_{inner} = 1 - \frac{\left[\frac{H}{l} + 4 - \frac{3+cos\theta}{sin\theta}\frac{t}{l}\right]t^2}{2(lsin\theta)^2\left(\left(\frac{H}{l} - cos\theta\right)\right)} \\ \sigma^* = \eta\sigma_{surface} + (1-\eta)\sigma_{inner} \end{cases} \quad (9)$$

Equation (9) is the full version of the structure surface layer model.

3.3. Model Application

In order to examine the accuracy of the model above, the SLM auxetic structure contains 3 × 3 × 3 unit cells with smaller size are manufactured, as shown in Figure 10a. Its overall dimension is 4.8 × 4.8 × 7.8 mm. The compression test is conducted and the result is compared with simulation result.

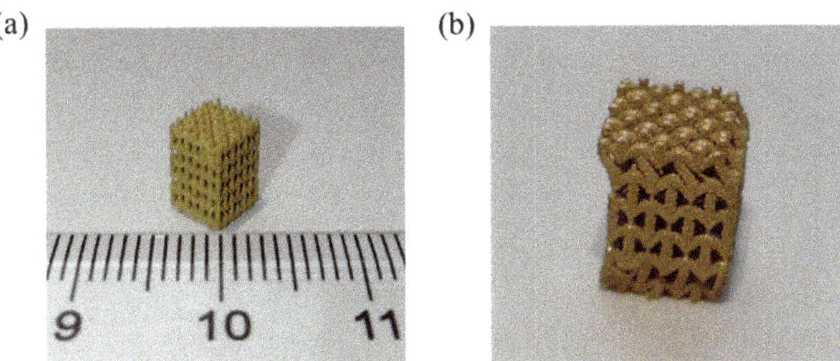

Figure 10. Micro scaled 3D auxetic structure. (**a**) Before compression test (**b**) After compression test.

It is found that the top layer of the auxetic structure is first destabilized during the compression test, and the position of destabilization is similar to macro scaled auxetic structure, as shown in Figure 10b. The Load-stroke curve is obtained, and it is consistent with Li's [23] research, as shown in Figure 11. The main reason of the cyclic load-stroke is due to the structure failure during the compression process. The vertical pillars are always under compression first during the compression test, and stress will concentrate on other pillars if one of them bends. The rest of the pillars will fail instantly and finally all the unit cells in the same layer will collapse, and the next layer of the structure will repeat this process. As each pillar of the structure is close to each other, the corner of the unit cell will contact when layer collapse, and the load will increase. It is obvious that the number of the cyclic equals to the number of the layer of the structure.

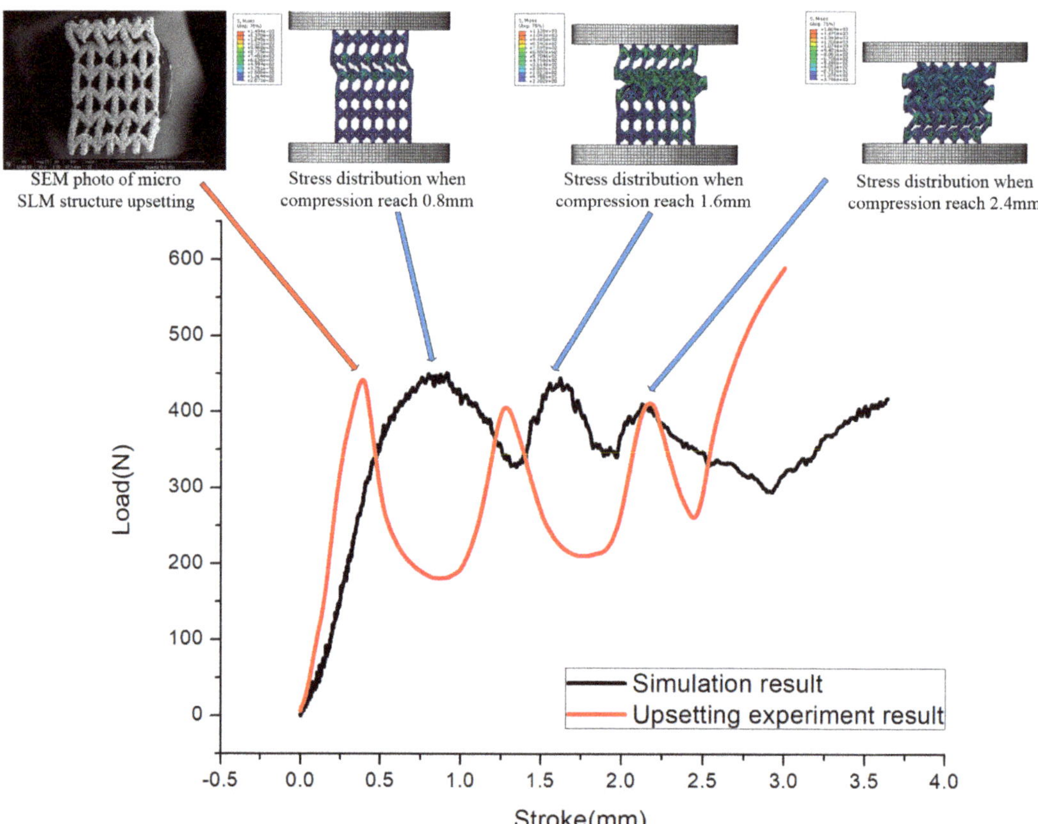

Figure 11. Load-stroke curve of micro auxetic structure obtain via FE simulation and actual experiment.

The simulation result has also been shown in Figure 11. It has the same number of cyclic load-stroke as the experiment result. The maximum load of each cyclic obtained via FE simulation is close to the experiment result, which proves that the structure surface layer model can be used to predict the mechanical behavior of the auxetic structure. The stroke of the simulation result is slightly different from the experiment. This is mainly because some of the supporting structure inside the SLM structure has not been fully removed, thus enhance the mechanical property of the structure.

4. Conclusions

A 3D auxetic structure manufactured by SLM and MA has been conducted in this research. The 3D structures with different pillar thickness have been produced to study the structure size effect of the 3D auxetic structure. By comparing the simulation and experiment result, it is found that the negative Poisson's ratio exists in both the SLM and MA structure during the compression test, and the Poisson's ratio decrease when the pillar thickness increase.

During the elastic stage of the compression test, the auxetic effect of MA structure is more obvious, but the Young's modulus is relatively small. In the plastic stage, both SLM and MA structure has destabilized. Micro cracks exist on SLM structure, and the surface quality of MA structure is good. By establishing the structure surface layer model of 3D auxetic structure, the influence of structure size effect is revealed and quantified. Comparing with the experiment result, it is found that the structure surface layer model is relatively helpful for the prediction of the mechanical behavior of 3D auxetic structure.

Author Contributions: Conceptualization, J.R.; methodology, J.R.; software, G.C.; validation, F.Z. and L.X.; formal analysis, J.R.; investigation, T.X.; resources, F.G.; data curation, J.R.; writing—original draft preparation, J.R.; writing—review and editing, J.R.; visualization, J.R.; supervision, F.G.; project administration, F.G.; funding acquisition, T.X. All authors have read and agreed to the published version of the manuscript.

Funding: This research was funded by [The National Natural Science Foundation of China] grant number [51705333, 52005341], [The Natural Science Foundation of GuangDong Province] grant number [2017A030310352], [The open project of key laboratory of GuangDong Province] grant number [PEM201605], [The Shenzhen Science and Technology Program] grant number [JCYJ20160520175255386].

Institutional Review Board Statement: Not applicable.

Informed Consent Statement: Not applicable.

Data Availability Statement: The data presented in this study are available on request from the corresponding author.

Acknowledgments: The work described in this paper was supported by the grants from the National Natural Science Foundation of China (Grant No. 51705333, 52005341), the Natural Science Foundation of GuangDong Province (Grant No. 2017A030310352), the open project of key laboratory of GuangDong Province (PEM201605), the Shenzhen Science and Technology Program (Grant No. JCYJ20160520175255386).

Conflicts of Interest: The authors declare no conflict of interest.

References

1. Gunton, D.J.; Saunders, G.A. The Young's modulus and Poisson's ratio of arsenic, antimony and bismuth. *J. Mater. Sci.* **1972**, *7*, 1061–1068. [CrossRef]
2. Lakes, R. Foam structures with a negative Poisson's ratio. *Science* **1987**, *23*, 1038–1040. [CrossRef]
3. Evans, K.E.; Caddock, B.D. Microporous materials with negative Poisson's ratios: II. Mechanisms and interpretation. *J. Phys. D: Appl. Phys.* **1989**, *22*, 1883–1887. [CrossRef]
4. Evans, K.E.; Nkansah, M.A.; Hutchinson, I.J.; Rogers, S.C. Molecular network design. *Nature* **1991**, *353*, 124. [CrossRef]
5. Wojciechowski, K.W. Constant thermodynamic tension Monte Carlo studies of elastic properties of a two-dimensional system of hard cyclic hexamers. *Mol. Phys.* **1987**, *61*, 1247–1258. [CrossRef]
6. Wojciechowski, K.W. Two-dimensional isotropic system with a negative Poisson ratio. *Phys. Lett. A* **1989**, *137*, 60–64. [CrossRef]
7. Wojciechowski, K.W.; Brańka, A.C. Negative Poisson ratio in a two-dimensional "isotropic" solid. *Phys. Rev. A Gen. Phys.* **1989**, *40*, 7222. [CrossRef] [PubMed]
8. Kolpakov, A.G. Determination of the average characteristics of elastic frameworks. *J. Appl. Math. Mech.* **1985**, *49*, 739–745. [CrossRef]
9. Hoover, W.G.; Hoover, C.G. Searching for auxetics with DYNA3D and ParaDyn. *Phys. Status Solidi B* **2005**, *242*, 585–594. [CrossRef]
10. Pozniak, A.A.; Wojciechowski, K.W. Poisson's ratio of rectangular anti-chiral structures with size dispersion of circular nodes. *Phys. Status Solidi* **2014**, *251*, 367–374. [CrossRef]

11. Bilski, M.; Pigłowski, P.M.; Wojciechowski, K.W. Extreme Poisson's Ratios of Honeycomb, Re-Entrant, and Zig-Zag Crystals of Binary Hard Discs. *Symmetry* **2021**, *13*, 1127. [CrossRef]
12. Smardzewski, J.; Maslej, M.; Wojciechowski, K.W. Compression and low velocity impact response of wood-based sandwich panels with auxetic lattice core. *Eur. J. Wood Wood Prod.* **2021**, *79*, 797–810. [CrossRef]
13. Baughman, R.H.; Shacklette, J.M.; Zakhidov, A.A.; Stafstrom, S. Negative Poisson's ratios as a common feature of cubic metals. *Nature* **1998**, *392*, 362–365. [CrossRef]
14. Piglowski, P.M.; Narojczyk, J.W.; Wojciechowski, K.W.; Tretiakov, K.V. Auxeticity enhancement due to size polydispersity in fcc crystals of hard-core repulsive Yukawa particles. *Soft Matter.* **2017**, *13*, 7916–7921. [CrossRef] [PubMed]
15. Narojczyk, J.W.; Wojciechowski, K.W. Elastic properties of the fcc crystals of soft spheres with size dispersion at zero temperature. *Phys. Status Solidi B* **2008**, *245*, 606–613. [CrossRef]
16. Narojczyk, J.W.; Wojciechowski, K.W. Poisson's Ratio of the f.c.c. Hard Sphere Crystals with Periodically Stacked (001)-Nanolayers of Hard Spheres of Another Diameter. *Materials* **2019**, *12*, 700. [CrossRef]
17. Almgren, R. An isotropic three-dimensional structure with Poisson's ratio. *J. Elast.* **1985**, *15*, 427–430.
18. Larsen, U.D.; Sigmund, O.; Bouwstra, S. Design and Fabrication of Compliant Micromechanisms and Structures with Negative Poisson's Ratio. *J. Microelectromechanical Syst.* **1997**, *6*, 99–106. [CrossRef]
19. Grima, J.N.; Gatt, R.; Alderson, A.; Evans, K.E. On the potential of connected stars as auxetic systems. *Mol. Simul.* **2005**, *31*, 925–935. [CrossRef]
20. Shokri, R.M.; Prawoto, Y.; Ahmad, Z. Analytical solution and finite element approach to the 3D re-entrant structures of auxetic materials. *Mech. Mater.* **2014**, *74*, 76–87. [CrossRef]
21. Hengsbach, S.; Lantada, A.D. Direct laser writing of auxetic structures: Present capabilities and challenges. *Smart Mater. Struct.* **2014**, *23*, 085033. [CrossRef]
22. Wang, X.; Li, X.; Ma, L. Interlocking assembled 3D auxetic cellular structures. *Mater. Des.* **2016**, *99*, 467–476. [CrossRef]
23. Li, Y.; Cormier, D.; West, H.; Harrysson, O.L.A.; Knowlson, K. Non-stochastic Ti–6Al–4V foam structures with negative Poisson's ratio. *Mater. Sci. Eng. A* **2012**, *558*, 579–585.
24. Evans, K.E.; Nkansah, M.A.; Hutchinson, I.J. Auxetic foams: Modelling negative Poisson's ratios. *Acta Metall. Mater.* **1994**, *42*, 1289–1294. [CrossRef]
25. Schwerdtfeger, J.; Schury, F.; Stingl, M.; Wein, F.; Singer, R.F.; Körner, C. Mechanical characterisation of a periodic auxetic structure produced by SEBM. *Phys. Status Solidi B* **2012**, *249*, 1347–1352. [CrossRef]
26. Ran, J.Q.; Fu, M.W.; Chan, W.L. The influence of size effect on the ductile fracture in micro-scaled plastic deformation. *Int. J. Plast.* **2013**, *41*, 65–81. [CrossRef]
27. Lemaitre, J. A continuous damage mechanics model for ductile fracture. *Trans. Asme J. Eng. Mater. Technol.* **1985**, *107*, 83–89. [CrossRef]

Article

Characterization of Refining the Morphology of Al–Fe–Si in A380 Aluminum Alloy due to Ca Addition

Meng Wang *, Yu Guo *, Hongying Wang and Shengsheng Zhao

School of Mechanical and Electrical Engineering, Shenzhen Polytechnic, Shenzhen 518055, China; wanghy@szpt.edu.cn (H.W.); sszhao@szpt.edu.cn (S.Z.)

* Correspondence: wangmeng04@szpt.edu.cn (M.W.); guoyu_@szpt.edu.cn (Y.G.); Tel.: +86-185-6587-3580 (M.W.); +86-180-0453-7358 (Y.G.)

Abstract: Aluminum–silicon (Al–Si) alloys are the most commonly cast aluminum alloys. Fe is the most deleterious element for Al–Si die casting alloys, as its existence causes the precipitation of substantial intermetallics that result in the unsatisfactory mechanical performance of the alloy, such as its ductility. Hence, controlling the morphology and formation of the AlFeSi phase, particularly the β-AlFeSi phase, is vital for improving the ductility of Al–Si die casting alloys. Herein, Ca was added to the A380 alloy, and the morphological changes resulting from the influence of Ca on the AlFeSi phase were characterized. The outcomes revealed that according to different cooling rates, specific amounts of Ca addition (0.01–0.1 wt.%) were capable of refining α-AlFeSi and β-AlFeSi morphology and transforming the β-AlFeSi phase into α-AlFeSi. Moreover, Ca addition could also modify eutectic silicon. The transformation mechanism and refining role of Ca in AlFeSi and the different morphologies of Al_2CaSi_2 were analyzed.

Keywords: A380 alloy; Ca; AlFeSi phase; refine

Citation: Wang, M.; Guo, Y.; Wang, H.; Zhao, S. Characterization of Refining the Morphology of Al–Fe–Si in A380 Aluminum Alloy due to Ca Addition. *Processes* **2022**, *10*, 672. https://doi.org/10.3390/pr10040672

Academic Editors: Guoqing Zhang and Prashant K Sarswat

Received: 24 January 2022
Accepted: 29 March 2022
Published: 30 March 2022

1. Introduction

A380 aluminum alloy is one of the most extensively utilized Al–Si pressure casting alloys, and it is widely used in electronic and communicational devices, auto components, and engine parts.

For the majority of Al–Si die casting Al alloys, iron is a double-edged sword. In the presence of Fe, die soldering can be suppressed, whereas the fragile AlFeSi phase can affect the mechanical performance of Al–Si alloys in addition to causing a reduction in ductility. The existence of Fe induces the precipitation of Fe-rich intermetallics with platelet-like or fibrous morphologies, causing unsatisfactory mechanical performances of the overall alloy. Those intermetallics remarkably affect the tensile strength and plasticity of Al alloys by avoiding interdendritic feeding during the casting process and by causing the formation of numerous apertures in the eutectic phase. Thus, it is imperative to regulate the levels and morphologies of iron intermetallics in Al–Si compound metals to ensure optimal performance [1–6].

Among the iron-containing phases, the two most significant are the β-AlFeSi and α-AlFeSi phases. The β-AlFeSi phase is harmful due to its platelet or needle-shaped morphology, which is fragile, and therefore, the alloy can be more severely influenced by the stress concentration. The crystal structure of β-AlFeSi is monoclinic, with lattice parameter a = b = 0.611 nm, c = 4.15 nm, β = 91°. The α-AlFeSi phase, which presents a certain compact morphological status, such as Chinese scripts, stars, or polygon shapes, is less deleterious to the material performance [7–9]. The crystal structure of α-AlFeSi is cubic, with parameter a = 1.26 nm. Hence, substantial amounts of research have been committed to reducing the fraction of β-AlFeSi, either directly or through conversion into α-AlFeSi [10–13].

Thus, different approaches have been utilized to decrease the detrimental influence of AlFeSi intermetallic phases, including iron level reduction, element alloying, cooling rate control, and heat treatment. Among these approaches, element alloying is the most commonly used method, and many related studies have been conducted [9]. Mn possesses an atom lattice similar to Fe and can be used as an Fe substitute in β-AlFeSi to generate α-Al(Mn,Fe)Si, which has the same lattice structure as α-AlFeSi. Therefore, it can be used to reliably avoid the deleterious influence of β-AlFeSi. Nevertheless, Mn can facilitate the formation of primary α-Al(Mn,Fe)Si precipitation in Al–Si cast alloys of a large size, resulting in the loss of ductility. Sr is another satisfactory Fe substitute, but its high cost and associated sintering challenges limit its usage. There are also other elements for alloying, while the effect still has limitations. [7,14–17]

Calcium can also act as a valid Fe substitute and functions via the refinement of the AlFeSi phase and the eutectic Si phase, improving the strength and ductility, particularly via elongation. Sreeja Kumari claimed that calcium supplementation could refine both eutectic Si and platelet Fe-intermetallic phases in Al–7Si–0.3Mg–0.6Fe alloy, which also enhanced the tensile properties and impacted the strength of the material [18,19]. Moreover, Ca can be combined with other elements, such as Be, Mn, and Sr, to mitigate the detrimental influence of the AlFeSi phase [20,21]. In a previous study, we found that the addition of Ca and Mn to the A380 alloy resulted in a microstructure consisting of FCC α-Al, modified eutectic Si, α-AlFeSi Chinese script phases, and refined platelet β-AlFeSi phase [21].

Nevertheless, comprehensive studies on the roles of Ca in Fe-rich phases of A380 alloy are insufficient. On the one hand, in previous studies, the content of Fe was comparably low, with low Ca addition (less than 0.01 wt.%). On the other hand, in die cast aluminum A380 alloy, a high level of Fe (about 1 wt.%) needs a high amount of Ca addition (even 0.3 wt.%) as a correction agent. In addition, the function of Al_2CaSi_2 introduced by Ca addition has not been clearly illustrated. Therefore, in the present research, the roles of the Ca level and the cooling rates in the AlFeSi phase of A380 alloy were analyzed, and the findings can elucidate the mechanisms underlying Ca refinement of the AlFeSi phase.

2. Materials and Methods

In this study, three different cooling rates were studied. A copper mold with a V-shaped slope and a straight water quench method were used to emulate the high cooling rates of die casting. As per the secondary dendritic arm space identified, cooling rates around 50 °C/s were studied. A steel mold was employed to realize moderate cooling rates around 5 °C/s. Furnace cooling was utilized for low cooling rates at 0.05 °C/s. Moderate and low cooling rates were documented using National Instruments cDAQ-9171 data collection devices. The association between the secondary dendritic arm spacing (d) and the relevant cooling velocity (t) can be delineated by:

$$d = A\left(\dot{t}\right)^{n} \tag{1}$$

As per the outcomes of Samuel [14], A and n are approximately 31 and −0.366, respectively. Hence, for the fast cooling rate samples, since the values of the secondary dendritic arm spacing were around 14 μm, the cooling rates computed from Equation (1) for high-cooling-rate specimens were approximately 50 °C/s, which qualified our research design.

The sample utilized in this study was the commercially available Alcoa Al A380, which is a hypoeutectic aluminum with a comparatively low liquidus (593 °C) and solidus (527 °C) temperature. The A380 + Ca additive samples were fabricated from the master alloy. The constituents of the master alloy and the initial A380 are presented in Tables 1 and 2, respectively. The Ca additions were 0.01, 0.05, 0.1, and 0.3 wt.%.

Table 1. Composition of the master alloy.

Material	Major	Minor Element Composition (% by weight)					Al
		Si	Fe	Mg	Others Each	Others Total	
Ca10	Ca = 10%	≤0.20	≤0.30	≤0.05	≤0.05	≤0.10	Balance

Table 2. Composition of the A380 alloy.

Element	Si	Fe	Cu	Mn	Mg	Ni	Zn	Ti	Al
wt.%	9.0	1.0	3.5	0.4	0.2	0.3	0.35	0.08	Balance

For the preparation of all the alloy specimens, the alloys were melted in a Kerl automatic standard electrical melt stove (Cincinnati, OH, USA). The cast metal for a single casting process was approximately 180 g. The melted alloys were held under 710 °C for 120 min and subjected to a slow cooling process until reaching the target temperature of 660 °C for casting.

The prepared specimens were sliced, milled, and subjected to careful polishing. The treated specimens were subjected to etching via a 0.5% HF in distilled water. Optical images were captured using a Leica DM LM/P 11888500 light microscope (Berlin, Germany). The SEM analyses were conducted via a Hitachi S-4800 field emission SEM (Tokyo, Japan) or an FEI Philips XL-40 SEM (Hillsboro, OR, USA).

3. Results

3.1. A380–Ca Equilibrium Phase Diagram

The A380–Ca equilibrium phase diagram obtained by using Thermo-Calc is shown in Figure 1. From the phase diagram, it can be seen that if the Ca content is lower than 0.6 wt.%, α-AlFeSi precipitates in advance of β-AlFeSi, and therefore no large primary β-AlFeSi can form. Therefore, the highest desired Ca addition is 0.6 wt.%. Additionally, the introduction of Ca can induce the formation of the Al_2CaSi_2 phase, which can serve as a nucleation center for other secondary AlFeSi phases. Thus, from the calculated phase diagram, Ca addition may improve the ductility of the A380 alloy.

3.2. The Effect of Ca Addition on the Morphology and Phase Fraction of A380 at Different Cooling Rates

3.2.1. Equilibrium Situation (Low Cooling Rate)

In the equilibrium situation, several phenomena were observed (Figure 2a–d): (1) No evident fraction change in the β-AlFeSi phase was observed. Nevertheless, the average size of the β-AlFeSi phase was reduced, and no primary β-AlFeSi phase was found if the Ca addition was higher than 0.05 wt.%, which revealed the refinement effect of Ca on the β-AlFeSi phase. (2) Polygon-shaped Al_2CaSi_2 tended to occur under the low cooling rates when the Ca content was 0.1 wt.%, whereas no obvious Al_2CaSi_2 was formed if the Ca content was lower than 0.1 wt.%. (3) Eutectic silicon became refined with the addition of Ca, with an obvious decrease in particle size observed. From the images, we can conclude that at a low cooling rate, Ca can reduce the size of the β-AlFeSi phase, and the best addition of Ca content is around 0.1 wt.%.

3.2.2. Moderate Cooling Rates

Figure 3 shows the microstructure of the A380 alloy with different Ca additions at 5 °C/s. When the cooling rate was elevated, several other phenomena could be identified: (a) With an increase in Ca, the β-AlFeSi phase was transformed into an α-AlFeSi phase. No evident variations were observed in the A380–0.01 Ca specimen (Figure 3a), whereas a noticeable decrease in the β-AlFeSi phase in the A380–0.05 Ca specimen was identified (Figure 3b). When the Ca level was higher than 0.05 wt.%, the β-AlFeSi phase was entirely transformed into the α-AlFeSi phase. (b) When the Ca level was 0.1 wt.%, a polygon-shaped Al_2CaSi_2 phase formed (Figure 3c), while some needle-shaped Al_2CaSi_2 were observed when the Ca addition level was up to 0.3 wt.%, which was detrimental to the performance of the alloy (Figure 3d). Thus, at a 5 °C/s cooling rate, 0.05 wt.% Ca addition is the optimal content.

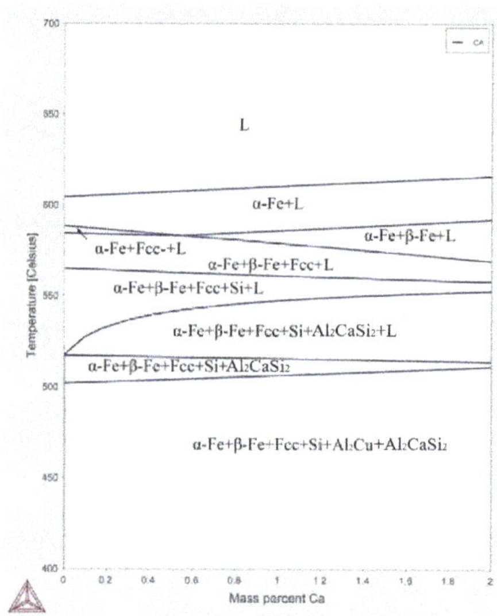

Figure 1. Al-Si-Cu-xCa phase diagram (Si = 9%, Cu = 3.5%, Mn = 0.4%, Fe = 1%, Mg = 0.2%, Zn = 0.35%, and Al = balance).

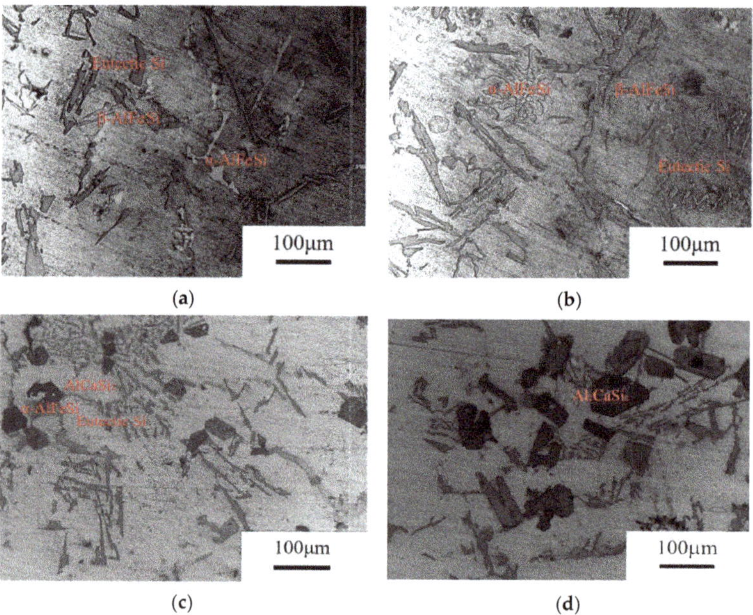

Figure 2. Microstructure of the A380 alloy with different Ca additions at 0.05 °C/s cooling rate: (**a**) 0.01 wt.%; (**b**) 0.05 wt.%; (**c**) 0.1 wt.%; (**d**) 0.3 wt.%.

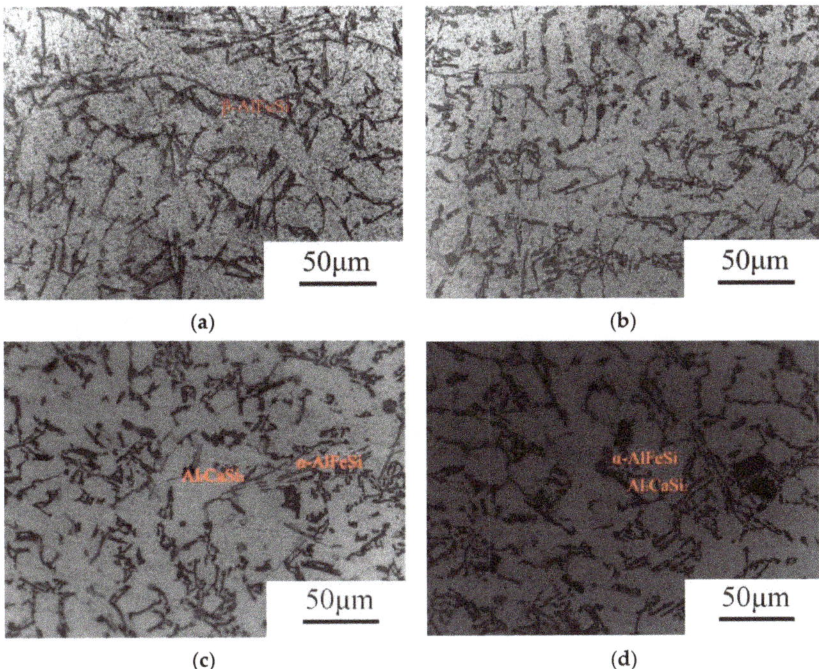

Figure 3. Microstructure of A380 alloy with different Ca additions at 5 °C/s cooling rate: (**a**) 0.01 wt.%; (**b**) 0.05 wt.%; (**c**) 0.1 wt.%; (**d**) 0.3 wt.%.

3.2.3. High Cooling Rates

When the cooling rate was increased even further to 50 °C/s, the AlFeSi phase was further refined under the influence of both the high cooling rates and Ca addition. Figure 4 shows the microstructure of the A380 alloy with diverse Ca levels at 50 °C/s. When Ca additions ranged from 0.01 to 0.05 wt.%, no large blocks of α-AlFeSi and β-AlFeSi phases were found. Ca further modified the eutectic Si, and the morphology of the eutectic Si changed from a platelet morphology to a coral-like morphology. Thus, 0.01–0.05 wt.% Ca addition is enough for A380 alloy at a 50 °C/s cooling rate.

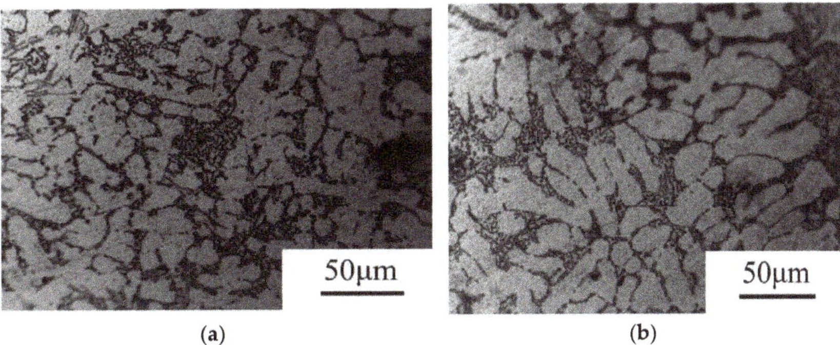

Figure 4. Microstructure of A380 alloy with different Ca additions at 50 °C/s cooling: (**a**) 0.05 wt.%; (**b**) 0.1 wt.%.

4. Discussion

4.1. The Refinement Effect of Ca on the AlFeSi Phase

There have been several studies on the mechanism of Ca in the AlFeSi phase. Kumari stated that the refining role of Ca in the β-AlFeSi phase was induced by the fragmented β-platelets. With the addition of Ca, he argued that the greater diffusion coefficient of Si relative to Fe could cause fragmentation of the β-AlFeSi-phase platelets due to the presence of Ca [18].

Low levels of Ca were associated with reductions in the particle size of β-AlFeSi, which coincided with the fragmentation theory. Nevertheless, when the Ca level reached 0.1 wt.%, more α-AlFeSi was generated, and no evident β-AlFeSi was identified. Such a phenomenon cannot be sufficiently explained by the fragmentation theory but is possible based on the nucleation theory. Our team discovered that the Al_2CaSi_2 phase was related to α-AlFeSi. We identified that polygon-sized α-AlFeSi were nucleated from the polygon-size Al_2CaSi_2, and α-AlFeSi entirely wrapped the Al_2CaSi_2 nucleating center, which was reported in our previous work (Figure 5a) [21]. In addition, the needle-shaped Al_2CaSi_2 phase spread across the dendrites, which acted as the primary phase to generate α-AlFeSi (Figure 5b). Although in the calculated phase diagram of Figure 1, primary α-AlFeSi and β-AlFeSi formed prior to Al_2CaSi_2 during the cooling process, secondary α-AlFeSi grew from the Al_2CaSi_2 phase and consumed the iron during the transformation of primary β-AlFeSi to secondary α-AlFeSi. Thus, the function of Al_2CaSi_2 can reduce the amount of β-AlFeSi and reduce the size of β-AlFeSi. Additionally, the Al_2CaSi_2 phase was capable of nucleating other phases, such as Si and Al_2Cu.

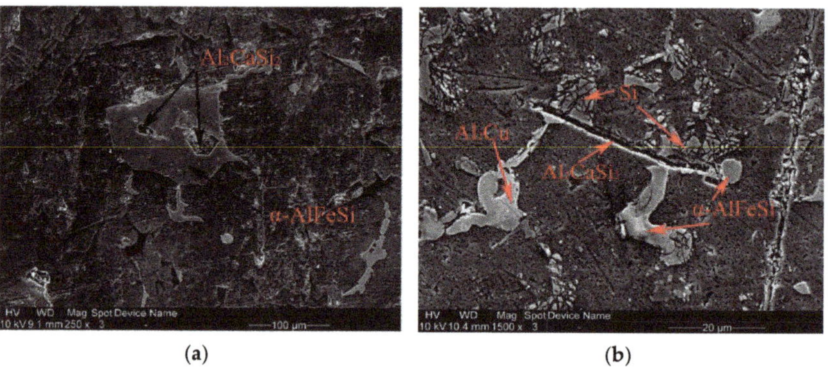

(a) (b)

Figure 5. Relations of Al_2CaSi_2 and α-AlFeSi: (**a**) SEM images showing nucleation of α-AlFeSi from polygon-shaped Al_2CaSi_2, retrieved from [21]; (**b**) SEM images showing nucleation of α-AlFeSi from needle-shaped Al_2CaSi_2.

Intriguingly, we also discovered that β-AlFeSi could be thoroughly refined by Al_2CaSi_2 at a high cooling rate (50 °C/s). The reason for this can be attributed to two factors: First, it is common knowledge that increasing cooling rates can both decrease the size of the AlFeSi phase and prompt the conversion of β-AlFeSi into α-AlFeSi. The β-AlFeSi is a stable phase that can be generated at slow cooling rates, while α-AlFeSi is a metastable phase. Second, in the equilibrium situation, the Al_2CaSi_2 phase was generated subsequent to the formation of the primary AlFeSi phase. The high cooling rate postponed the forming temperature of the AlFeSi phase, whereas it played no role in the formation of the Al_2CaSi_2 phase. Thereby, the Al_2CaSi_2 phase could form first and directly nucleate α-AlFeSi.

Another way to determine the validity of the nucleating AlFeSi phase in the nucleation substrate is to compute the planar disregistry δ. As aforementioned, this is the only way of realizing the calculation in a quantitative way. Hence, it was utilized as the main computation for the possible nucleation potential and was used to assess the experiment outcomes; δ has been described in [22] as:

$$\delta_{(hkl)_n}^{(hkl)_s} = \sum_{i=1}^{i=3} \frac{\frac{\left| d_{[uvw]_s} \cos\theta - d_{(uvw)_n} \right|}{d_{(uvw)_n}}}{3} \times 100 \quad (2)$$

where $(hkl)_s$ is the low-index plane of the substrate, $(uvw)_s$ is the low-index orientation in $(hkl)_s$, $(hkl)_n$ is the low-index plane of the nucleated solid, $(uvw)_n$ is the low-index orientation in $(hkl)_n$, $d_{(uvw)_n}$ is the interatomic spacing along $(uvw)_n$, $d_{(uvw)_s}$ is the interatomic spacing along $(uvw)_s$, and θ is an angle between $(uvw)_s$. The crystalline structure of Al_2CaSi_2 is hexagonal, with lattice parameters of a = 0.4130 nm and c = 0.7145 nm. The crystal structure of α-AlFeSi is cubic, with the parameters of a = 1.26 nm. Therefore, in Equation (2), s is for Al_2CaSi_2 and n is for α-AlFeSi, and all the parameters above can be calculated. Campbell [23] hypothesized that the electronic contribution to the interfacial energy was beneficial when δ was below 12%. Table 3 presents the outcomes computed in this study for δ, which show that Al_2CaSi_2 is a satisfactory nucleation substrate of α-AlFeSi, hence, revealing how Ca is capable of modifying β-AlFeSi.

Table 3. Planar disregistries between α-Al_2CaSi_2 and α-AlFeSi. s is for Al_2CaSi_2 and n is for α-AlFeSi.

Match Planes	[hkl] s	[hkl] n	d[hkl] s (nm)	d[hkl] n (nm)	θ	δ (Pct)
(0001)s//(001)n	$[\bar{1}010]$	[100]	2.146011	2.53	0 deg	7.90
	$[\bar{5}410]$	[110]	3.651899	3.57796	4.10 deg	
	$[\bar{1}2\bar{1}0]$	[010]	2.700338	2.53	0 deg	

4.2. Ca Refinement Effect on Si

Although Al–Si alloy modification is usually realized via adding Na and Sr, Ca is also utilized as a modifier. From the experimental outcomes, we could clearly observe that the morphology of the eutectic structure of silicon changed from a coarse plate-like morphology to a coral or fibrous shape. There are several reports that Al–Si alloy modification could be achieved via adding Ca, and the duration times were longer than those for Sr and Mn addition.

There are several explanations for the refinement mechanism of Ca on eutectic silicon. The reason for the modification effect of Ca on those alloys was the destructive effects of active impurity particulates, which served as the nucleation center of eutectic Si in an unmodified alloy. Thus, greater undercooling was introduced, which induced the formation of a significant number of solidification centers of Si, refining the eutectic silicon. Ca can also be adsorbed on the small crystals of eutectic Si to slow down their growth [24].

Another explanation is impurity-induced twinning. The addition of elements into molten alloys that contain Si can facilitate a transition of the morphology of the Si phase via the formation of growth twins. A modifier of an appropriate atom size (with respect to Si ($r_{modifier}/r_{silicon}$ = 1.646), upon incorporation into the molten alloy, can accumulate adequately via inducing growth twins for Si 3D growth. The radius ratio reported for Ca is 1.68, very close to 1.646 [25].

In our study, Ca introduced greater undercooling, which could spontaneously increase the formation of nucleation centers of Si. In addition, a growth twin was created with the absorbed Ca on the liquid/solid interface of Si, contributing to the modification of the Al–Si alloy upon Ca addition.

Moreover, higher cooling rates facilitated further refinement, whereas it could be clearly seen that the high cooling rate was responsible for refinement rather than modification by merely introducing additional Si nucleation centers.

It has also been previously reported that Ca could facilitate the formation of primary Al_2CaSi_2, which could serve as the nucleation substrate of primary Si [25]. Hence, we can conclude that both Ca and Al_2CaSi_2 can refine eutectic silicon.

4.3. Al_2CaSi_2 Morphology

As shown in the SEM images in Figure 5, two morphologies of Al_2CaSi_2 were formed in our experiment. The first was polygon-shaped, which could be observed at low cooling rates. The second was needle-shaped, which could be observed at high cooling rates. The reason for this is that the crystalline structure of Al_2CaSi_2 is hexagonal, with the lattice parameters of a = 0.4130 nm and c = 0.7145 nm. Therefore, at low cooling rates, growth occurred in a 3D manner, whereas at high cooling rates, there was no time for c-axle growth, and hence, growth proceeded in a 2D manner. Nevertheless, in both cases, they can serve as a nucleation substrate for α-AlFeSi. Hence, it can reduce the fraction of β-AlFeSi.

In summary, the scheme of the relation of nucleation by polygon- and needle-shaped Al_2CaSi_2 is shown in Figure 6. In Figure 6a, polygon-shaped Al_2CaSi_2 can serve as the nucleate center of secondary α-AlFeSi, and the formed α-AlFeSi entirely trapped the polygon-shaped Al_2CaSi_2 nucleate center. In Figure 6b, needle-shaped Al_2CaSi_2 can serve as a dendrite nucleate origin of secondary α-AlFeSi, ball shape Al_2Cu, and particulate silicon. All three phases grow outside of the needle-shaped Al_2CaSi_2.

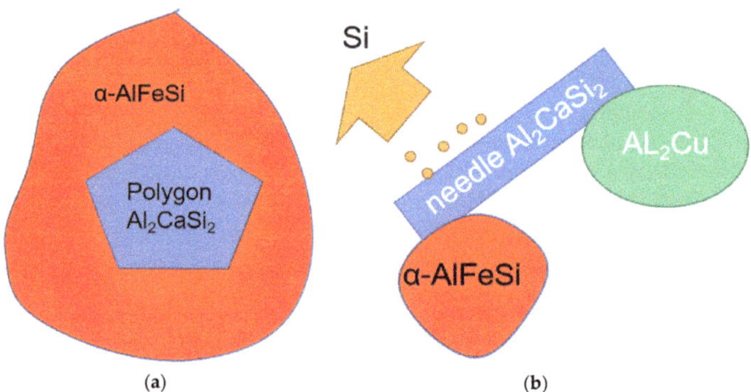

Figure 6. Relation of nucleation by (**a**) polygon- and (**b**) needle-shaped Al_2CaSi_2.

5. Conclusions

(1) Ca can accelerate the fragmentation of the β-AlFeSi phase and introduce Al_2CaSi_2, which acts as the nucleation substrate of α-AlFeSi. At a low cooling rate, Al_2CaSi_2 can transform primary β-AlFeSi to secondary α-AlFeSi. At a high cooling rate, primary Al_2CaSi_2 can directly nucleate secondary α-AlFeSi.

(2) A high cooling rate can facilitate the transformation of β-AlFeSi to α-AlFeSi with lower Ca addition. Therefore, the best Ca addition for A380 alloy at the cooling rates of 0.05, 5, and 50 °C/s are 0.1 wt.%, 0.05 wt.%, and 0.01 wt.%, respectively.

(3) Ca itself and Al_2CaSi_2 can serve as the modification agents of eutectic silicon, and high cooling rates can only refine eutectic Si.

(4) At low cooling rates, polygon-shaped Al_2CaSi_2 form in a 3D growth manner, while at high cooling rates, needle-shaped Al_2CaSi_2 form in a 2D growth manner, and both can serve as a nucleation substrate for α-AlFeSi.

Author Contributions: Conceptualization, M.W. and Y.G.; methodology, M.W.; experiment, M.W.; validation, S.Z. and H.W.; writing—original draft preparation, M.W.; writing—review and editing, M.W.; funding acquisition, M.W., H.W. and S.Z. All authors have read and agreed to the published version of the manuscript.

Funding: This research was funded by the Post-Doctoral Later-Stage Foundation Project of Shenzhen Polytechnic (6020271008K), the Young Innovative Talents Project of General Colleges and Universities in Guangdong Province (2019GKQNCX127), the Shenzhen Science and Technology Project (JCYJ20190809150001747), and the Key Project of Shenzhen Polytechnic (6020310007K).

Institutional Review Board Statement: Not applicable.

Informed Consent Statement: Not applicable.

Data Availability Statement: Not applicable.

Conflicts of Interest: The authors declare no conflict of interest.

References

1. Ren, J.; Fang, X.; Chen, D. The effect of heat treatments on the microstructural evolution of twin-roll-cast Al-Fe-Si alloys. *J. Materi. Eng. Perform.* **2021**, *30*, 4401–4410. [CrossRef]
2. Li, G.; Qu, W.-Y.; Luo, M.; Cheng, L.; Guo, C.; Li, X.-G.; Xu, Z.; Hu, X.-G.; Li, D.-Q.; Lu, H.-X.; et al. Semi-solid processing of aluminum and magnesium alloys: Status, opportunity, and challenge in China. *Trans. Nonferrous Met. Soc. China* **2021**, *31*, 3255–3280. [CrossRef]
3. Liu, X.; Jia, H.-L.; Wang, C.; Wu, X.; Zha, M.; Wang, H.-Y. Enhancing mechanical properties of twin-roll cast Al–Mg–Si–Fe alloys by regulating Fe-bearing phases and macro-segregation. *Mater. Sci. Eng. A* **2021**, *831*, 142256. [CrossRef]
4. Bardziński, P.J. New Al3Si7 phase with tetragonal silicon structure in quasicrystal-forming near-eutectic Al-Cu-Fe-Si alloys. *J. Alloy. Compd.* **2021**, *869*, 159349. [CrossRef]
5. Gan, J.; Du, J.; Wen, C.; Zhang, G.; Shi, M.; Yuan, Z. The Effect of Fe Content on the Solidification Pathway, Microstructure and Thermal Conductivity of Hypoeutectic Al–Si Alloys. *Int. J. Met.* **2021**, *16*, 178–190. [CrossRef]
6. Gao, T.; Li, Z.; Zhang, Y. Phase Evolution of β-Al5FeSi During Recycling of Al–Si–Fe Alloys by Mg Melt. *Int. J. Metalcast.* **2019**, *13*, 473–478. [CrossRef]
7. Lu, L.; Dahle, A.K. Iron-rich intermetallic phases and their role in casting defect formation in hypoeutectic Al-Si alloys. *Metall. Mater. Trans. A* **2005**, *36*, 819–835.
8. Dinnis, C.M.; Taylor, J.A.; Dahle, A.K. As-cast morphology of iron-intermetallics in Al–Si foundry alloys. *Scr. Mater.* **2005**, *53*, 955–958. [CrossRef]
9. Zhang, L.; Gao, J.; Damoah, L.; Robertson, D.G. Removal of Iron From Aluminum: A Review. *Miner. Process. Extr. Met. Rev.* **2012**, *33*, 99–157. [CrossRef]
10. Becker, H.; Bergh, T.; Vullum, P.E.; Leineweber, A.; Li, Y. Effect of Mn and cooling rates on α-, β- and δ-Al–Fe–Si intermetallic phase formation in a secondary Al–Si alloy. *Materialia* **2019**, *5*, 100198. [CrossRef]
11. Becker, H.; Bergh, T.; Vullum, P.; Leineweber, A.; Li, Y. β- and δ-Al-Fe-Si intermetallic phase, their intergrowth and polytype formation. *J. Alloy. Compd.* **2019**, *780*, 917–929. [CrossRef]
12. Zhu, Y.Z.; Peng, H.; Huang, H.; Li, J.C. Abnormal Grain Growth Mechanism in the Twin-Roller Cast Al-Fe-Si Alloy in the Annealing Process. *Adv. Mater. Sci. Eng.* **2020**, *2020*, 1056274. [CrossRef]
13. Gao, T.; Li, Z.-Q.; Liu, X.-F.; Zhang, Y.-X. Evolution Behavior of γ-$Al_{3.5}$FeSi in Mg Melt and a Separation Method of Fe from Al–Si–Fe Alloys. *Acta Met. Sin. Engl. Lett.* **2017**, *31*, 48–54. [CrossRef]
14. Samuel, M.; Samuel, F.H. Effect of alloying elements and dendrite arm spacing on the microstructure and hardness of an Al-Si-Cu-Mg-Mg-Fe-Mn (380). *J. Mater. Sci.* **1995**, *301*, 1698–1708. [CrossRef]
15. Wang, M.; Xu, W.; Han, Q.Y. Effect of heat treatment on controlling the morphology of AlFeSi phase in A380 alloy. *Int. J. Metalcast.* **2016**, *10*, 516–523. [CrossRef]
16. Wang, M.; Xu, W.; Han, Q. The Influence of Sr Addition on the Microstructure of A380 Alloy. *Int. J. Met.* **2017**, *11*, 321–327. [CrossRef]
17. Wang, P.; Deng, Y.; Dai, Q.; Jiang, K.; Chen, J.; Guo, X. Microstructures and strengthening mechanisms of high Fe containing Al–Mg–Si–Mn–Fe alloys with Mg, Si and Mn modified. *Mat. Sci. Eng. A-Struct.* **2021**, *803*, 140477. [CrossRef]
18. Kumari, S.S.; Pillai, R.; Pai, B. A study on the structural, age hardening and mechanical characteristics of Mn and Ca added Al–7Si–0.3Mg–0.6Fe alloy. *J. Alloy. Compd.* **2008**, *453*, 167–173. [CrossRef]
19. Kumari, S.S.; Pillai, R.; Rajan, T.; Pai, B. Effects of individual and combined additions of Be, Mn, Ca and Sr on the solidification behaviour, structure and mechanical properties of Al–7Si–0.3Mg–0.8Fe alloy. *Mater. Sci. Eng. A* **2007**, *460–461*, 561–573. [CrossRef]
20. Belov, N.; Naumova, E.; Akopyan, T. Effect of 0.3% Sc on microstructure, phase composition and hardening of Al-Ca-Si eutectic alloys. *Trans. Nonferrous Met. Soc. China* **2017**, *27*, 741–746. [CrossRef]
21. Wang, M.; Xu, W.; Han, Q. Study of Refinement and Morphology Change of AlFeSi Phase in A380 Alloy due to Addition of Ca, Sr/ Ca, Mn and Mn, Sr. *Mater. Trans.* **2016**, *57*, 1509–1513. [CrossRef]
22. Cao, X.; Campbell, J. Morphology of Beta-Al5FeSi Phase in Al-Si Cast Alloys. *Mater. Trans.* **2006**, *4*, 1303–1312. [CrossRef]
23. Campbell, J. *Casting*; Butterworth-Heinemann: Oxford, UK, 1991.
24. Korotkova, N.O.; Belov, N.A.; Avxentieva, N.N. Effect of Calcium additives on the phase composition and physicomechanical properties of a conductive alloy Al-0.5% Fe-0.2% Si-0.2% Zr-0.1% Sc. *Phys. Met. Metallogr.* **2020**, *121*, 95–101. [CrossRef]
25. Kumari, S.S.S.; Pillai, R.M.; Pai, B.C. Role of calcium in aluminium based alloys and composites. *Int. Mater. Rev.* **2005**, *50*, 216–238. [CrossRef]

Review

Recent Advances in the Equal Channel Angular Pressing of Metallic Materials

Lang Cui [1], Shengmin Shao [2], Haitao Wang [3], Guoqing Zhang [4], Zejia Zhao [1,*] and Chunyang Zhao [4,*]

1 Institute of Semiconductor Manufacturing Research, College of Mechatronics and Control Engineering, Shenzhen University, Shenzhen 518060, China
2 China Electronics Technology Group Corporation 52nd Research Institute, Hangzhou 310061, China
3 School of Mechanical and Electrical Engineering, Shenzhen Polytechnic, Shenzhen 518055, China
4 Shenzhen Key Laboratory of High Performance Nontraditional Manufacturing, Guangdong Provincial Key Laboratory of Electromagnetic Control and Intelligent Robots, College of Mechatronics and Control Engineering, Shenzhen University, Shenzhen 518060, China
* Correspondence: zhaozejia@szu.edu.cn (Z.Z.); zcy724317@163.com (C.Z.)

Abstract: Applications of a metallic material highly depend on its mechanical properties, which greatly depend on the material's grain sizes. Reducing grain sizes by severe plastic deformation is one of the efficient approaches to enhance the mechanical properties of a metallic material. In this paper, severe plastic deformation of equal channel angular pressing (ECAP) will be reviewed to illustrate its effects on the grain refinement of some common metallic materials such as titanium alloys, aluminum alloys, and magnesium alloys. In the ECAP process, the materials can be processed severely and repeatedly in a designed ECAP mold to accumulate a large amount of plastic strain. Ultrafine grains with diameters of submicron meters or even nanometers can be achieved through severe plastic deformation of the ECAP. In detail, this paper will give state-of-the-art details about the influences of ECAP processing parameters such as passes, temperature, and routes on the evolution of the microstructure of metallic materials. The evolution of grain sizes, grain boundaries, and phases of different metallic materials during the ECAP process are also analyzed. Besides, the plastic deformation mechanism during the ECAP process is discussed from the perspectives of dislocation slipping and twinning.

Keywords: grain size; ECAP; metallic materials; processing parameters; deformation mechanism

Citation: Cui, L.; Shao, S.; Wang, H.; Zhang, G.; Zhao, Z.; Zhao, C. Recent Advances in the Equal Channel Angular Pressing of Metallic Materials. *Processes* **2022**, *10*, 2181. https://doi.org/10.3390/pr10112181

Academic Editor: George Z. Kyzas

Received: 17 September 2022
Accepted: 15 October 2022
Published: 25 October 2022

1. Introduction

The mechanical properties of metallic materials depend to a large extent on the grain size, and reducing the grain size has great potential to improve the mechanical properties of metals and their alloys. The yield strength σ_y of the material and the grain diameter d conform to the Hall–Petch relationship ($\sigma_y = \sigma_0 + k_y d^{-\frac{1}{2}}$) [1,2], where σ_0 and k_y are constants about the material. It can be seen from the relationship that the smaller the grain size, the higher the yield strength of the material. For example, Magnesium and its alloys have high biosafety and can be used in the medical field. Reducing the grain size can improve their mechanical properties and deformability, showing great application potential in the preparation of medical devices and medical materials [3]. Furthermore, a number of studies have shown that titanium and its alloys with ultrafine grains have superior strength, plasticity, fatigue resistance, corrosion resistance, and many other excellent properties, which have been widely used in the fields of aerospace, biomedicine and chemical machinery [4–7].

Many methods have been applied to grain refinement nowadays, such as traditional thermomechanical and metal forming processes, ultrasonic vibration, and severe plastic deformation [8,9]. Ultrafine and nanoscale grains are difficult to obtain by conventional thermomechanical and metal forming processes, and ultrasonic vibration also has some limitations due to the corrosion and reactivity of the radiator under high-temperature conditions [10,11]. Among these methods, severe plastic deformation (SPD) was reported to be an efficient and low-cost technique that has several advantages over other approaches to refine the grains to micro or nano meters of metallic materials, such as wide applicability of workpiece shapes, suitable for handling bulk materials [10] and high levels of strain [12]. Commonly used severe plastic deformation methods include high-pressure torsion (HPT), equal channel angular pressing (ECAP), twist channel angular pressing (TCAP), twist channel multi-angular pressing (TCMAP), cyclic extrusion and compression (CEC), accumulative rolling bonding(ARB), etc. [12–16].

The ECAP technique can prepare large-scale ultrafine grain bulk materials without changing the workpiece size and area of the material, which has great application potential in comparison to other SPD methods [14,17]. Many scholars have used this method to alter the microstructure and improve the mechanical properties of non-ferrous metals including Al, Ti, Mg, Cu, and their alloys [17,18]. Previous studies have shown that the improvement of material properties by ECAP is affected by many factors, such as passes, temperature, routes, back pressures, channel angle, etc., and the grain size can be effectively reduced by controlling the processing parameters [14,17,19,20]. The strain distribution of aluminum after different channel angles, curvature angles, and passes was simulated by the finite element method. The strain uniformity was quantitatively studied by using a non-uniformity index and standard deviation, and the best suggestions for die parameters were given. When the channel angle is obtuse, the effective strain decreases with the increase of the angle; When the channel angle is acute, the deformation is more uneven and complex; When the channel angle is a right angle, the effective change is uniform [20,21]. Figure 1 schematically shows the ECAP process and schematic diagram of different routes. In addition, the addition of subsequent different treatments based on the study of ECAP also affects the properties of metal materials. For example, studies have shown that the strength of the material is further improved after ECAP annealing, but the plasticity is almost unchanged. The deformation microstructure and phase structure of the brass before and after heat treatment were also not changed [22,23]. The ECAP technology shows great potential for industrial application in grain refinement, making great contributions to the development of various fields.

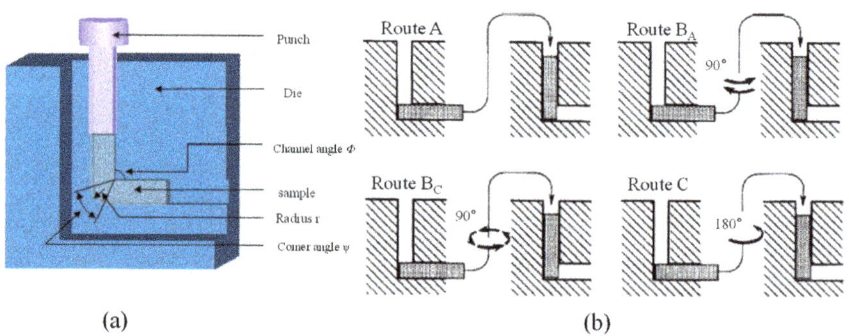

Figure 1. (**a**) Working principle of ECAP [24]. Reproduced with permission from Elsevier B.V. (**b**) Schematic diagram of different routes [17]. Reproduced with permission from Elsevier Ltd.

Even there are several reviews about the ECAP, most of the existing studies focus on summarizing the effects of ECAP parameters on material properties, and little research was conducted on reviewing the microstructure evolution after the ECAP from perspectives of various grain sizes, grain boundaries and phase compositions. Therefore, In this review, the microstructure evolution of the metallic materials induced by the ECAP is to be analyzed and discussed from the views of grain size evolution, grain boundaries, and phase evolution, as summarized in Figure 2.

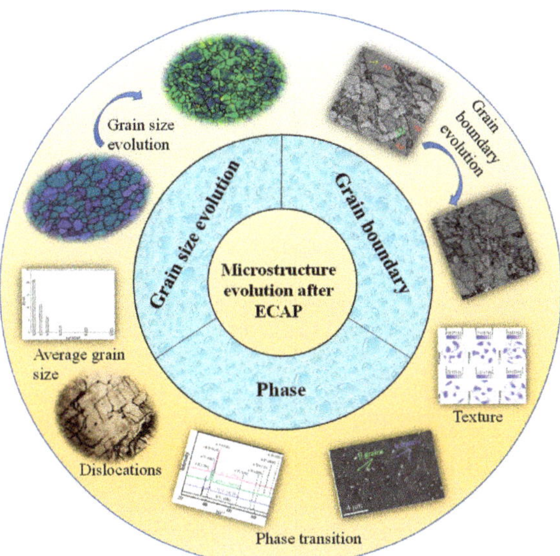

Figure 2. Overview map [25–31]. Reproduced with permission from Elsevier B.V. Copyright 2014. Reproduced with permission from Elsevier Editor Ltd. Reproduced with permission from Springer Nature. Reproduced with permission from MDPI. Reproduced with permission from Elsevier B.V. Reproduced with permission from Elsevier Ltd. Reproduced with permission from Elsevier Ltd.

2. Microstructure Evolution after ECAP

The microstructure can strongly influence the properties of materials, which primarily includes grain sizes, grain boundaries and phases. In this section, the effects of ECAP on the microstructural evolution of metallic materials will be discussed in detail by analyzing the evolution of grain sizes, grain boundaries, and phases. The metallic materials mainly include Ti alloys, Mg alloys and Al alloys.

2.1. Grain Size Evolution

To demonstrate the grain refinement induced by the ECAP, the grain size is divided into three scales, i.e., from 1 to 10 μm, from 0.1 to 1 μm, and below 0.1 μm. The processing parameters in the ECAP process can directly affect the microstructure of the material and its grain size. Different processing parameters have different effects on the microstructure of the material, which contributes to various mechanical properties. The contributory parameters affecting the grain size evolution primarily include ECAP passes, routes, and temperature. Next, the grains of different size ranges will be discussed mainly from the above factors.

2.1.1. Grain Size from 1 to 10 μm

ECAP pass is one of the primary factors that affect the evolution of grain sizes. Zhao et al. preheated the sample to 923 K for 5 min before extrusion and carried out non-isothermal ECAP with the extrusion rate of 20 mm/s and B_C route. They found that the average grain size of titanium alloys decreased from 15–20 μm to 5–10 μm and from 30 μm to 10 μm after four passes of ECAP, respectively [32,33]. Jahadi et al. reported that the grain size of magnesium alloys was reduced from the initial 20.4 μm to 3.9 μm after four passes of ECAP at 275 °C [34]. Matsubara et al. found that the size of magnesium alloys decreased by almost the same proportion after four passes at 300 °C, from the initial 50–10 μm [35]. It should be noted that the initial grain size refers to the grain size of the metallic materials before the ECAP processing. In comparison, it is found that the grain size of the material decreases to 1–10 μm under similar conditions. Figure 3 shows the microstructures of titanium and magnesium alloys after four passes of ECAP.

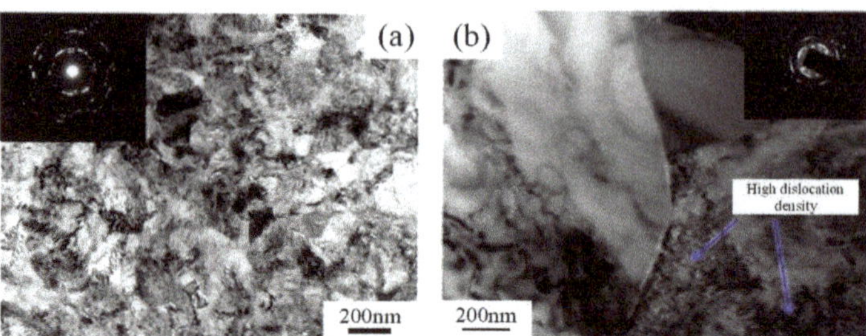

Figure 3. TEM image of the material after four passes: (**a**) titanium alloy [32]. Reproduced with permission from MDPI. (**b**) magnesium alloy [36]. Reproduced with permission from Springer Nature.

Figure 4 shows the EBSD images of the commercially pure aluminum after different passes, and the microstructure variation with ECAP passes could be clearly seen from the images. Kawasaki et al. studied the microstructure of commercially pure Al by selecting the Bc route and pressing rate of 7 mm/s at room temperature. When other conditions are the same, the grain size will be refined to different degrees after different passes. The grain size was reduced to 1.8 μm after two passes and 1.4 μm after three passes. With the increase of pass number, the average grain size was measured to be 1.3 μm after four passes of ECAP and 1.1 μm after eight passes of ECAP [25]. Terhune et al. obtained similar data at room temperature and B_C routes, with a grain size of 1.2 μm in four passes and 1.0 μm in twelve passes [37]. Generally speaking, the degree of grain size reduction becomes smaller and smaller as the pass increases. However, when the number of passes increased to twelve, the grain size was measured to be 1.2 μm, which increased somewhat for the commercially pure Al [25]. A similar situation is also found in the pure Ti and Al alloy, which proves that the grain size of the material will hardly change or even become coarse after a certain number of passes [38,39].

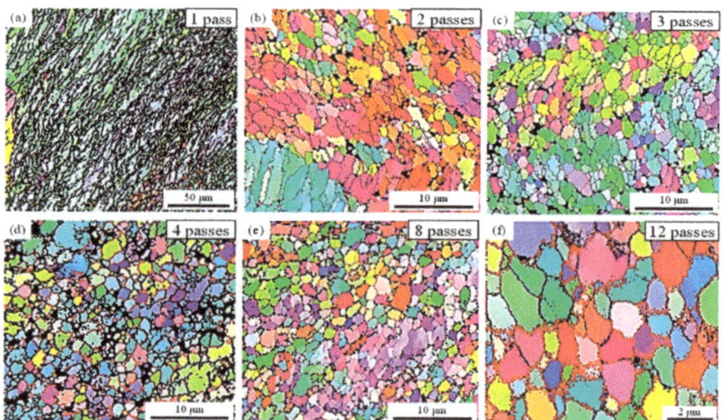

Figure 4. OIM images under different passes [25]: (**a**) one pass; (**b**) two passes; (**c**) three passes; (**d**) four passes; (**e**) eight passes; (**f**) twelve passes. Reproduced with permission from Elsevier B.V.

The grain size of the metal matrix composites (MMCs) can also be reduced by the ECAP process. Ramu and Bauri prepared the aluminum matrix composites with SiCp volume fractions of five and ten by the stirring casting method. After two passes of ECAP, the grain size of the Al/5SiCp composite was reduced to 8 μm, while the Al/10SiCp could only undergo one ECAP (up to 16 μm), and then cracks appeared. This is because higher SiC content affects the strain rate sensitivity and the degree of strain hardening [40].

The ECAP route also affects the change in grain sizes. Kocich et al. showed that the grain size of commercially pure aluminum decreased to 8 μm and 7.4 μm after two-pass ECAP under the A and B_C routes, respectively [41].Gottstein et al. reported that the average grain size of pure Mg reached 6–8 μm after four passes under the A, B_C, and C routes [42]. Tong et al. found that the average grain size of Mg alloys decreased from 4.3 μm to 1.3 μm under the A route, and the grain size decreased to less than 1 μm under both the B_C and C routes [43]. El-Danaf et al. performed eight passes of ECAP on pure Al under the B_C route and the C route, and the measured average grain sizes were 1.1 μm and 2.9 μm, respectively [44]. Tong et al. also found that the ECAP process of the A and B_C routes increased the material yield strength due to the grain refinement strengthening effect and the texture softening competition, while the C route induced a reduction of the yield strength [43]. The microstructures of Mg under different routes are shown in Figure 5. Based on previous studies, the B_C route is more likely to contribute to the best grain refinement effect.

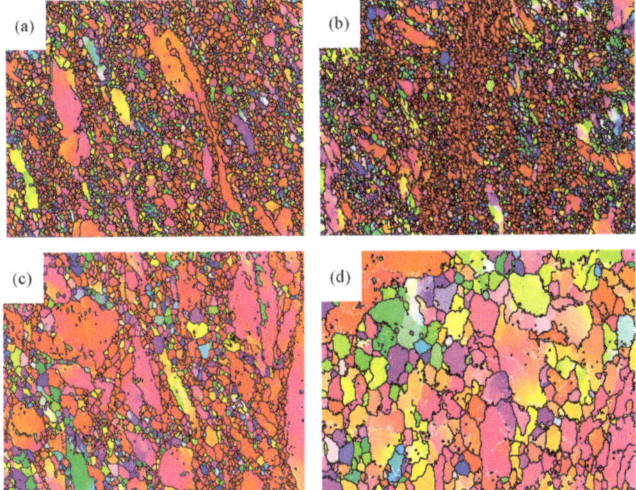

Figure 5. Microstructure of magnesium under different routes [45]: (**a**) A route; (**b**) C route; (**c**) B_C route after 1 pass; (**d**) B_C route after four passes. Reproduced with permission from MDPI.

Some special routes were also proposed to refine the grain sizes. Gajanan et al. compared the microstructure and mechanical properties of Mg alloys after passing through the B_C and R (inversions per pass) routes at 390 isothermal conditions. The research shows that the R route has a better microstructure refinement effect; grain refinement and secondary phase distribution are more uniform, and the mechanical properties are better than those processed by the B_C route [46]. Li et.al. developed a new route B_{135} by defining 135° clockwise rotations of the samples per pass and ECAP treatment at a temperature of 573 K. After four passes, the grain sizes under the B_{135} route and the B_C route are 2.36 μm and 2.82 μm, respectively. Compared with the B_C route, it is found that the new route B_{135} has a better grain refinement effect [47]. Figure 6 shows the evolution of grain sizes after different ECAP routes.

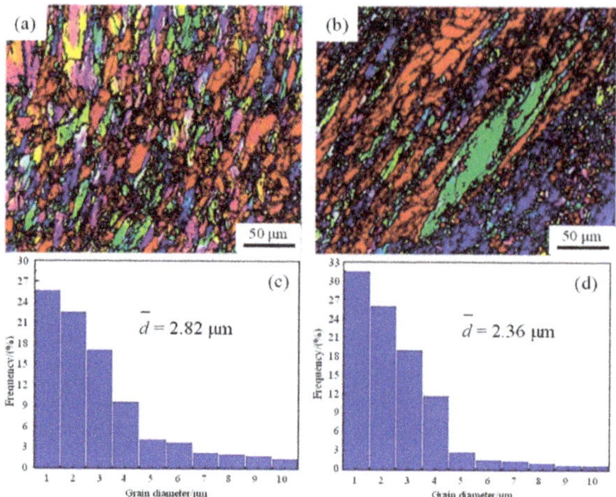

Figure 6. EBSD images under different routes [47]: (**a**) B_C route; (**b**) B_{135} route; (**c**,**d**) are the corresponding grain size histograms, respectively. Reproduced with permission from IOP Publishing Ltd.

Processing temperature also has some effect on grain size during ECAP. Lin et al. found that the average grain size of AZ31 Mg alloy reached about 2.5 μm after one pass of ECAP at 300 °C [48]. Kim and Jeong reported that the average grain size of AZ31 alloy was reduced from 24 μm to 4.8 μm, 3.2 μm and 2.2 μm after six passes of ECAP at three different temperatures (553 K, 523 K, 493 K), respectively [49]. The average grain size of pure Mg reaches 2.6 μm and 1.4 μm after eight passes of ECAP at 200 °C and 150 °C [50]. Figure 7 shows the grain distribution at different temperatures. The lower the temperature, the smaller the grain size and the more uniform the microstructure. This is because the low temperature can suppress dynamic grain growth and dynamic recrystallization [49]. Due to the high temperature and long pass time of Mg alloys treated at high temperatures, recrystallization will occur and the grains will grow rapidly [51,52]. Minárik et al. showed that temperatures between 180 °C and 250 °C generally result in average grain sizes greater than 1μm for the Mg alloys [53].

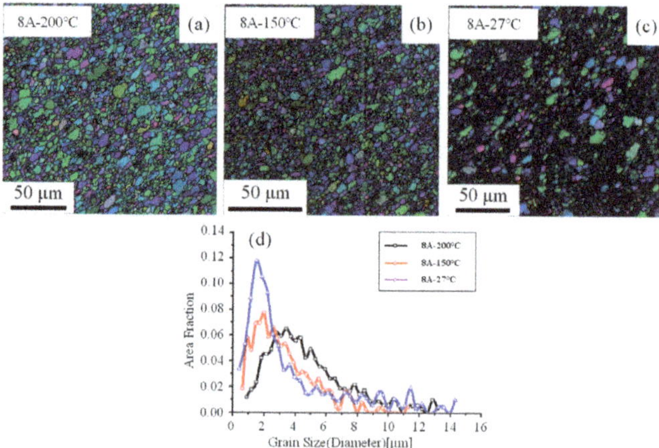

Figure 7. Inverse pole plot at different temperatures [50]: (**a**) 200 °C; (**b**) 150 °C; (**c**) 27 °C; (**d**) is the area fraction particle size distribution at different temperatures. Reproduced with permission from Elsevier B.V.

Consequently, grain sizes ranging from 1 to 10 μm can usually be achieved through four passes of ECAP for titanium and magnesium alloys, while two passes are required for pure aluminum. Furthermore, the B_C route has the best grain refinement effect among the four traditional routes, and low processing temperatures usually result in small grain sizes.

2.1.2. Grain Size from 0.1–1 μm

Further increase in the ECAP passes normally gives rise to smaller grain sizes. Hyun et al. found that the average grain size of pure Ti reached 1 μm after two passes of ECAP, and the grain size decreased to 0.4 μm and 0.3 μm after four passes and six passes at 683 K and B_C route conditions [1]. Liu et al. Obtained a more uniform equiaxed grain size in four passes using route C at room temperature (0.17 μm) [54]. On the B_C route and at temperatures around 400 °C, Zhilyaev et.al. and Fan et.al performed eight passes of ECAP on pure Ti to reduce the grain size to 0.4 μm and 0.5 μm, respectively [55,56]. Figure 8 shows the microstructure changes of pure titanium at different passes. Hajizadeh et al. performed ten passes of ECAP, and the average grain size was reduced to 183 nm, which is more significant than in previous studies [57]. For Ti and its alloys, under different conditions, the grain size can reach a sub-micron level after four passes, and the grain size decreases correspondingly with the increase of passes.

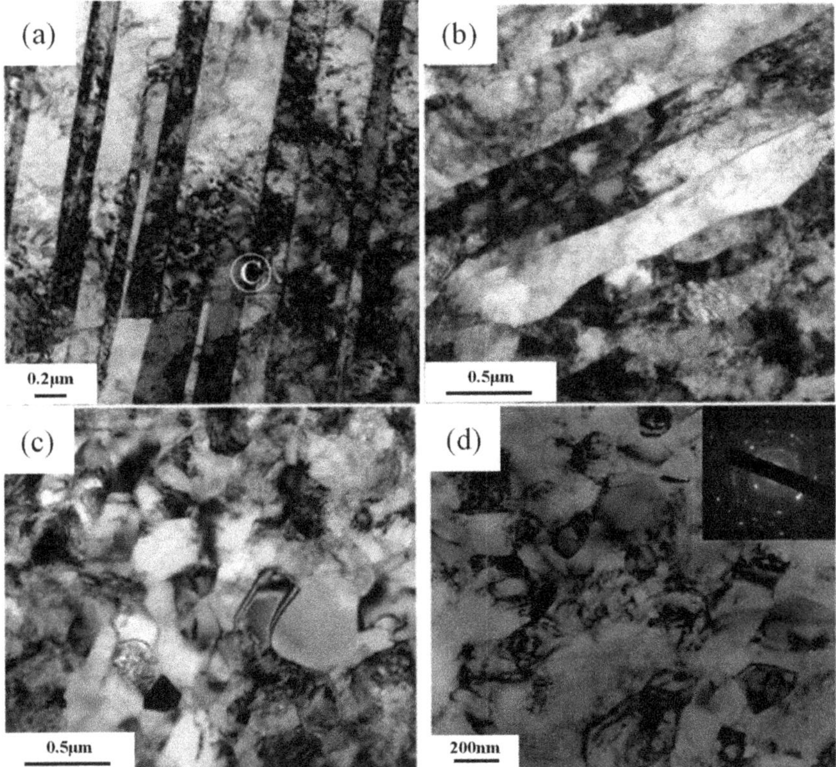

Figure 8. Microstructure of pure titanium under different passes [55]: (**a**) one pass; (**b**) two passes; (**c**) four passes; (**d**) eight passes. Reproduced with permission from Elsevier B.V.

Similar to titanium, the grain size of Al can be reduced to a sub-micron scale with the increasing pass. Aal et al. showed that the average grain size of pure Al was 1.7 μm after two passes of ECAP and decreased to 0.43 μm and 0.23 μm after four and ten passes at room temperature through route A, respectively [58]. The same conditions are applied in his other articles paper, it was also found that the average grain size was reduced from 390 μm to 1.8 μm, 0.4 μm and 0.3 μm after two, four and ten passes, respectively [59]. Song et al. observed that the grain size of pure Al decreased from 200 μm to 0.5 μm after sixteen passes [60].

However, for Mg alloys, it is difficult to refine the grain size to submicron grains with the increase of the pass. Wang et al. showed that the grain size decreased from 160–180 μm to 6–8 μm after twelve passes at 380 °C for Mg alloys with long-period stacking structure [61], Jiang et al. reported that ultrafine equiaxed grains of 2.5 μm were obtained after sixteen passes via route A at 603 K [62]. Ma et al. found that the grain size of Mg alloys decreased from 80 μm to 1.5 μm after thirty-two passes at 603 K [63]. Figure 9 shows the microstructure and grain size histograms for different passes.

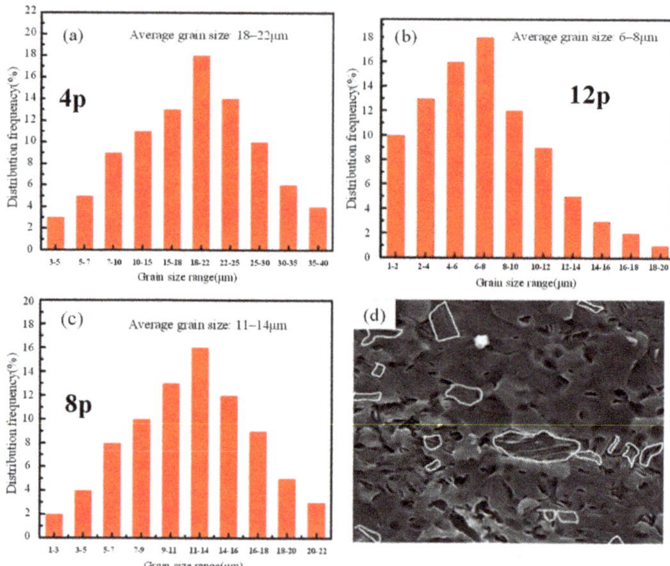

Figure 9. Grain size histogram [61]: (**a**) four passes; (**c**) eight passes; (**b**,**d**) is the grain size distribution map and SEM image of twelve passes. Reproduced with permission from Springer Nature.

The same tendency to transmit the effect of grain size is found in nickel and its alloys as in Al and Ti. Zhilyaev et al. reported ultrafine grains with an average size of about 0.35 μm after eight passes during ECAP pure nickel [64]. This is similar to the grain size (0.3 μm) reported by Neishi et al. for eight passes of pure Ni at room temperature [65]. The grain size was further reduced to about 0.24 μm when the passes increased to twelve passes [66]. Moreover, the grain size of Fe–Cr–Ni alloy was reduced to 0.4 μm after eight passes of ECAP processing, achieving a submicron-scale microstructure [67]. It appears that Ni and its alloys are more likely to reach submicron structures after eight passes.

The grain size of the copper and its alloys also decreases with the increase of passes. Jayakumar et al. refined the Cu-Cr-Zr alloy to 0.15~0.2 μm after eight passes of pressing, and it has high thermal stability [68]. Wongsa-Ngam et al. showed that the Cu-0.1% Zr alloy also formed a submicron structure after eight passes [69]. Khereddine et al. evaluated the grain size of Cu–Ni–Si alloy after high-pressure torsion (HPT) and ECAP treatment. The grain size after twelve passes of ECAP treatment was 0.2 μm [70]. Abib et al. found that the deformed microstructure evolved from fibrous cast grains to almost equiaxial microstructure after sixteen passes of pressing, and the high angle grain boundary fraction increased with the increase of ECAP passes [18]. The grain size of oxygen-free pure Cu decreased to 0.6 μm after twenty-four passes of equal-pass angular pressing at room temperature, but an increase in grain size was observed with increasing strain after sixteen passes because of dynamic recovery [71]. Figure 10 shows the microstructure and orientation distribution of the Cu alloy after different passes. For copper and its alloys, the grain size can also be reduced to the submicron level after eight passes.

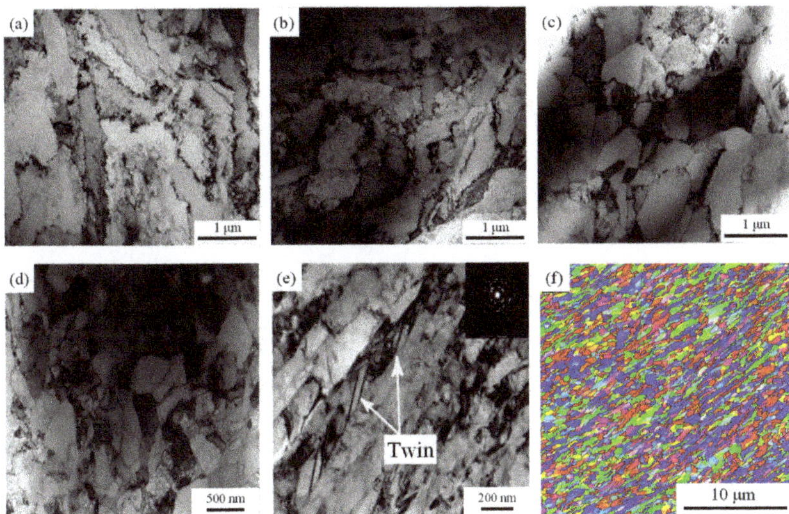

Figure 10. TEM and EBSD images of the microstructure of Copper Alloys after ECAP for (**a**) one pass; (**b**) two passes; (**c**) four passes; (**d**) eight passes [69]. Reproduced with permission from Elsevier B.V. (**e**) twelve passes [70]. Reproduced with permission from Elsevier B.V. (**f**) sixteen passes [18]. Reproduced with permission from Elsevier Inc.

Schafler et al. studied the evolution of Cu lattice defects under different routes. With the increase of deformation, the corresponding dislocation densities of different routes are different and show an increasing trend. The comparison shows that the B_C route has the largest dislocation density under the same strain [72]. Purcek et al. carried out eight passes of ECAP under different routes, and all three routes finally refined the Cu-Cr-Zr alloy to the submicron structure. The grain size is reduced to about 0.2–0.3 μm under three routes, but the B_C route obtained the smallest grain size and the best mechanical properties as well as good thermal stability [73]. The same conclusion was found in pure Ti [74] and pure Al [75]. Figure 11 shows the microstructures under different route treatments.

Bulutsuz et al. obtained the average grain size of pure Ti after two passes of ECAP at the composite temperature, which were 1.5 μm and 1.7 μm, respectively [76]. Attarilar et al. found that the average grain size of pure Ti reached 1.09 μm after four passes of ECAP at 400 °C [77], Ebrahimi and Attarilar measured an average grain size below 0.5 μm after four passes of ECAP at 450 °C [78]. However, Zhao et al. reported that the average grain size reached 0.25 μm after four passes of ECAP at room temperature, and even decreased to 0.2 μm after six to eight passes, resulting in a better microstructure than pure Ti treated at high temperature [79]. Similar conclusions have been obtained in Ti-Ni alloys, where lower temperatures lead to smaller grain sizes [80]. Under the low-temperature condition of liquid nitrogen treatment, the grain size of pure Ti is reduced to 0.56 μm after ECAP processing [81]. Psaruk and Lapovok studied the microstructure of pure Ti at different temperatures. As the temperature increases, more high-angle grain boundaries appear. However, the decrease in ECAP temperature leads to an increase in strength properties, indicating that the material undergoes thermally activated plastic deformation [82]. It can be seen that the temperature has a certain influence on the ECAP processing technology.

Figure 12 shows the microstructure and grain size distribution of pure Ti at different temperatures.

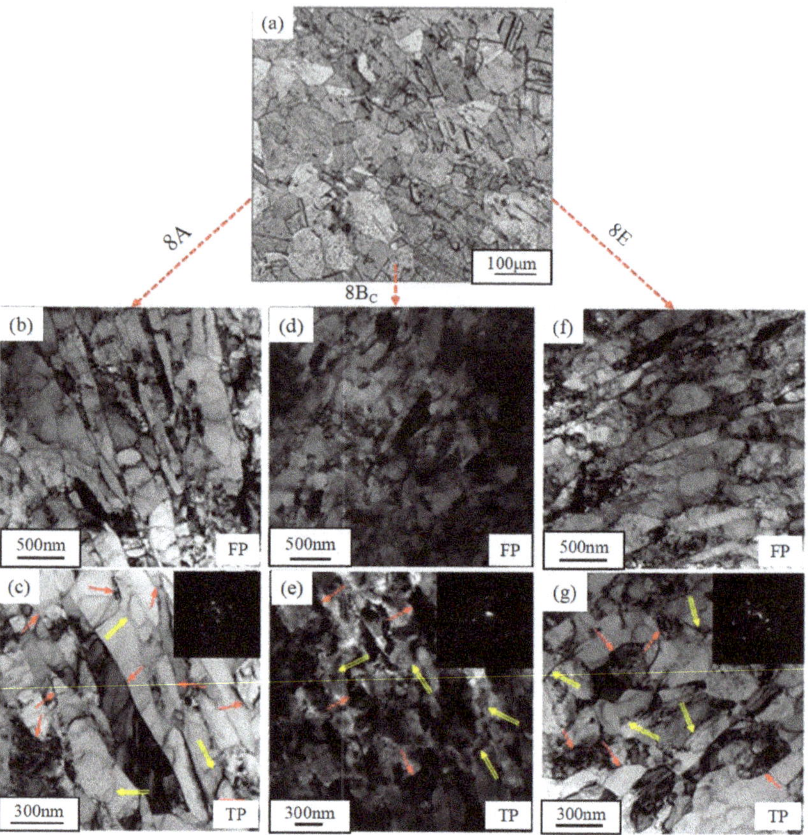

Figure 11. TEM images of the microstructure of Cu Alloys after eight passes of ECAP [73]: (**a**) is the pre-ECAP microstructure; (**b**,**c**) is the lateral and flow surface of the A route; (**d**,**e**) is the lateral and flow surface of the B_C route; (**f**,**g**) are the lateral and flow surface of the E route. Reproduced with permission from Elsevier B.V.

Applying back pressure during ECAP can improve the microstructure. The grain size of no back pressure and after applying 100 MPa back pressure is 0.24 μm and 0.2 μm, respectively. Applying a certain back pressure to electrolytic hard-pitch pure Cu can reduce the grain size and increase the proportion of high-angle grain boundaries [83]. Similar conclusions were obtained for ECAP with back pressure applied in Al alloys [84,85] and Ti alloys [86]. The grain size of AA5083 alloy was even reduced to 0.25 μm after three passes of ECAP under 200 MPa back pressure [87]. Increasing back pressure will reduce the thermal stability of the material, but the effect of back pressure on the improvement of the mechanical properties of the material makes this defect worthless [83]. Figure 13 shows the microstructural evolution of Al alloys with and without back pressure.

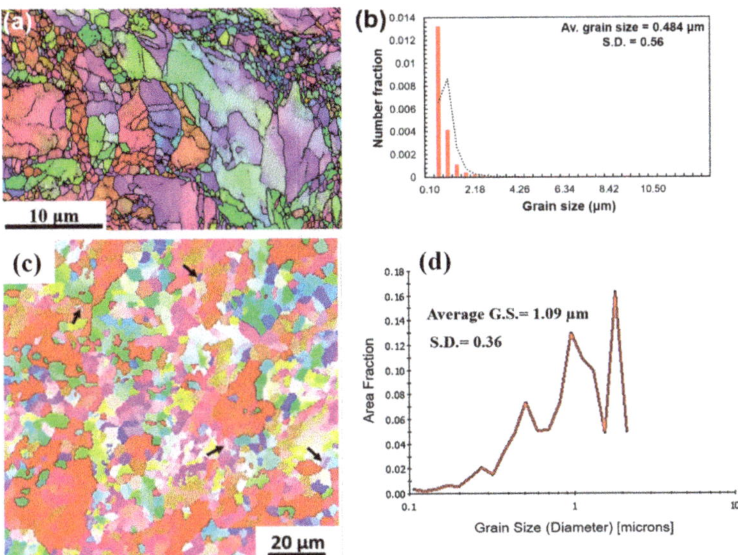

Figure 12. Microstructure of 4-pass pure titanium at different temperatures and grain size distribution:(**a**,**b**) at 450 °C [78]. Reproduced with permission from Springer Nature. (**c**,**d**) at 400 °C [77]. Reproduced with permission from Elsevier B.V.

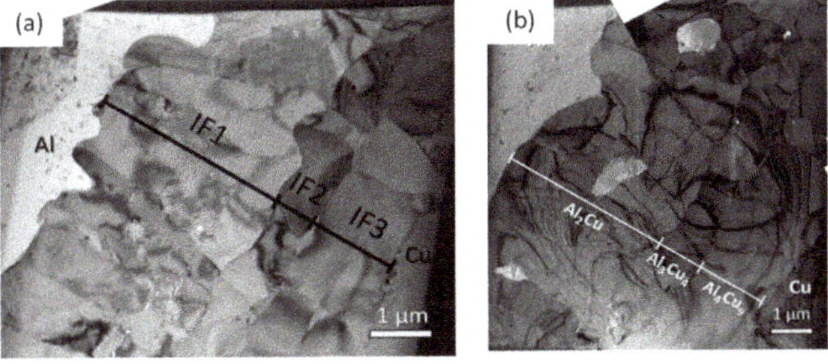

Figure 13. TEM images and SAED patterns of the substructure evolution after ECAP [88]: (**a**) no back pressure; (**b**) with back pressure. Reproduced with permission from Elsevier Ltd.

ECAP technology has a high strain rate during processing, which will improve production efficiency [89]. Demirtas et al. achieved maximum superplasticity at high strain rates and obtained the smallest grain size (0.2μm) of the Zn-22 Al alloy using two-step ECAP [90]. Afifi et al. refined the grain to 0.8μm and 0.3μm after one pass and four passes at different strain rates, respectively [91]. At different strain rates, it is found that the increase of strength is due to grain refinement, high dislocation density and more precipitates. With the increase of passage and temperature, the strain rate sensitivity will be enhanced, and the strength after dynamic ECAP is higher than that of quasi-static flow [92,93].

Consequently, most metal materials can obtain submicron grain after four to eight passes of ECAP. At the same time, it is found that the B_C route still has the best grain refinement effect. Under the condition of not exceeding the recrystallization temperature, choosing the appropriate temperature is beneficial to grain refinement. Moreover, the increase of back pressure is beneficial to the decrease of grain size. Compared with quasi-static ECAP, dynamic ECAP will obtain higher strength and improve machining efficiency.

2.1.3. Grain Size below 0.1 μm

Aal et al. found in their study of Al-Cu alloys that the pass and Cu content as well as the homogenization process can affect the average grain size. For heterogeneous Cu-Al alloys, the average grain size obtained gradually decreases but is larger than 0.1 μm with increasing pass and Cu content. The copper-aluminum alloy obtained by the homogenization process, after two passes, four passes and nine passes of ECAP processing, the average grain size decreased from 261 μm before ECAP to 177 nm, 165 nm and 55 nm, respectively. Homogeneous alloys have better structure and better deformation uniformity than heterogeneous alloys. It can be seen that the increase of the homogenization treatment and the number of passes can obtain smaller grain sizes, and even obtain nano-scale grains [94,95]. Figure 14 shows the variation trends of grain size and high-angle grain boundaries under different conditions.

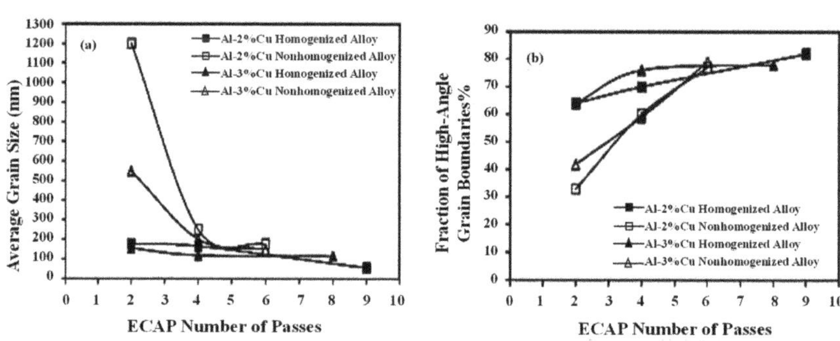

Figure 14. (**a**) shows the average grain size of the alloy as a function of the pass, and (**b**) shows the high angle grain boundary ratio of the alloy as a function of the pass [95]. Reproduced with permission from Elsevier B.V.

Stacking fault energy (SFE) plays a role in the fabrication of ultrafine and nanocrystallites. Lower stacking fault energies allow for the formation of finer microstructures. Qu et al. studied the microstructure of Cu-Al alloys at different stacking fault energies. The value of SFE will also be different for different alloys with Al content. They found that the grain size of Cu-5% Al alloy reached 107 nm after four passes of ECAP, and the average grain size of Cu-8% Al alloy reached 82 nm under the same conditions. Figure 15 shows the grain size distribution of Cu alloys with different Al contents after four passes. With the increase of Al content, the average grain size of the alloy decreased to the nanometer scale, and the microstructure was more uniform [96,97].

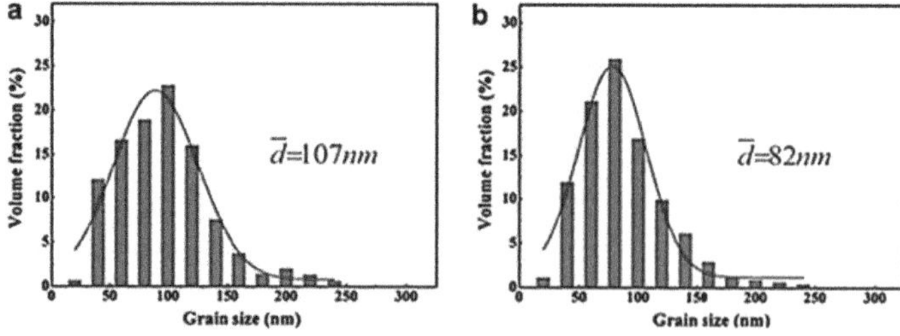

Figure 15. The statistic grain size of alloys after four-pass pressing measured by TEM: (**a**) Cu–5 at.% Al; (**b**) Cu–8 at.% Al [96]. Reproduced with permission from Elsevier Ltd.

Homogenizing the Al alloy will improve the machinability of the material, and a smaller grain size will be obtained after the ECAP process. For Cu alloys, the grain size can be reduced to the nanometer level by changing the stacking fault energy when other processing parameters reach the extreme. Smaller average grain size and uniform microstructure contribute to better metal properties.

It is difficult for the ECAP process to refine grains down to the nanometer level. The effect of processing factors such as pass, route and temperature to refine the grain will reach a limit, and too much change will make the grain coarser. Obtaining nanoscale grain size through the ECAP process requires not only starting from external conditions (changing processing factors, adding other processes, etc.) but also understanding the limitations of the nature of the material itself in refining grains. Table 1 summarizes the changes of grain size of different materials after ECAP process with different processing parameters.

Table 1. Effects of ECAP parameters on the grain size evolution.

Material	Initial D/μm	Passes	Routes	Temperature	D after ECAP/μm	Refs.
Ti alloy	30	4	B_C	723 K	10	[32]
Ti alloy	15–20	4	B_C	723 K	5–10	[33]
Ti alloy	50	6–8	\	500 °C	0.3–0.5	[80]
Pure Ti	2000	4	B_C	573 K	2.82	[47]
Pure Ti	26	4	C	Room T	0.17	[54]
Pure Ti	40–120	8	B_C	390–400 °C	<0.5	[55]
Pure Ti	10	8	B_C	450 °C	<0.4	[56]
Pure Ti	20	10	\	250 °C	0.183	[57]
Pure Ti	\	8	B_C	400–450 °C	0.26	[74]
Pure Ti	20	4	B_C	400 °C	1.09	[77]
Pure Ti	24.96	6	B_C	450 °C	0.29	[78]
Pure Ti	23	8	B_C	Room T	0.2	[79]
Pure Ti	196	4	B_C	Liquid nitrogen T	0.56	[81]
Mg alloy	20.4	4	B_C	275 °C	3.9	[34]
Mg alloy	50	4	B_C	573 K	10	[35]
Mg alloy	45	4	B_C	275 °C	1	[36]
Mg alloy	4.3	4	A	250 °C	1.3	[43]
Pure Mg	200	4	B_C	250 °C	6–8	[42]
Pure Mg	\	4	B_C	225 °C	1.896	[45]
Pure Mg	12	8	A	27 °C	0.75	[50]
AZ80	50.2	4	B_C	390 °C	12.82	[46]
AZ31	75	8	B_C	200 °C	0.7	[48]
AZ31	24	6	B_C	553 K	4.8	[49]
ZK60	37	4	B_C	250 °C	10.9	[51]
ZM21	5–60	4	B_C	200 °C	0.7	[52]
ZE41A	\	16	A	603 K	2.5	[62]
ZE41A	80	32	\	603 K	1.5	[63]
Pure Al	1000	4	B_C	Room T	1.3	[25]
Pure Al	\	4	B_C	Room T	1.2	[37]
Pure Al	39	2	B_C	Room T	7.4	[41]
Pure Al	300	8	C	\	2.9	[44]
Pure Al	\	10	A	Room T	0.23	[58]
Pure Al	390	10	A	Room T	0.3	[59]
Pure Al	200	16	\	Room T	0.5	[60]
Al-SiCp	55	2	B_C	Room T	8	[40]
AA5083	\	3	B_C	Room T	0.25	[87]
Al alloy	1.3	4	B_C	423 K	0.3	[91]
Al-Cu	261	9	A	Room T	0.055	[95]
Cu alloy	40–80	8	B_C	Room T	0.15–0.2	[68]
Cu alloy	30	8	B_C	Room T	0.33	[69]
Cu alloy	\	12	A	423 K	0.2	[70]
Cu alloy	24	24	B_C	Room T	0.6	[71]
Cu alloy	55	8	B_A + C	Room T	0.3	[73]
Pure Cu	28	12	B_C	Room T	0.24	[83]
Cu-Al	\	4	B_C	Room T	0.082	[96]

2.2. High/Low Angle Grain Boundary Evolution

High angle grain boundaries are when the misorientation angle is greater than 15°, while low angle grain boundaries are when the misorientation angle is between 3° and 15°. It is well known that the formation of high/low angle grain boundaries contributes to grain refinement, thereby improving the mechanical properties of materials.

For the FCC metals represented by Al and Cu, grain boundary angle has a certain tendency of evolution with increasing pass times. Khelfa et al. studied the grain boundary angle change of Al alloy (AA6061) [98]. Figure 16 shows the grain boundary angle distribution of AA6061 alloy under different passes. It can be seen from the figure that with the increase in the number of passes, the proportion of high-angle grain boundaries increases, but the proportion of low angle grain boundaries decreases. Alateyah et al. believed that there is a certain relationship between dislocations and grain boundary angle in the ECAP process, i.e., the formation and development of dislocations promote the formation of high and low angle grain boundaries, and dislocation slip will be inhibited when the high/low angle grain boundary reaches a certain proportion [13]. And Reihanian et al. found the same trend of grain boundary angle in pure Al [99]. Zhang et al. gave the change curve of grain boundary angle and dislocation of Cu and its alloys with the pass as shown in Figure 17 [97]. Therefore, the evolution of the grain boundary angle can be seen more intuitively.

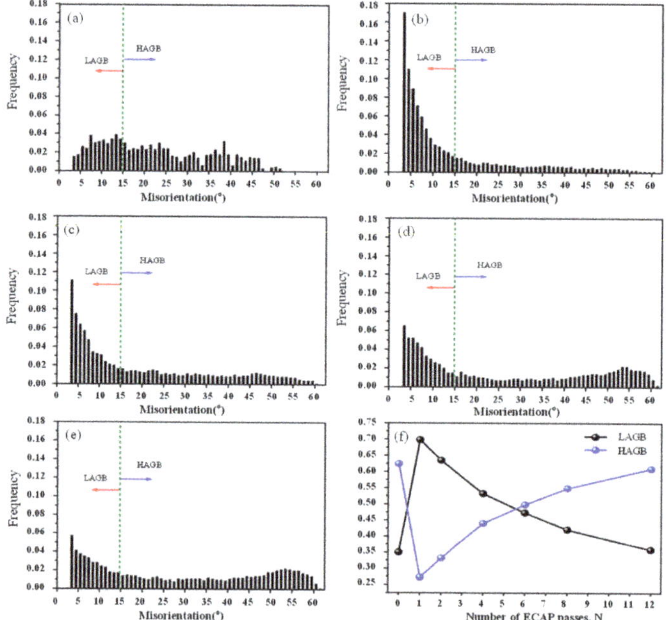

Figure 16. Variation of angular grain boundary of aluminum alloy after different passes [98]: (**a**) zero pass; (**b**) one pass; (**c**) four passes; (**d**) eight passes; (**e**) twelve passes; (**f**) is the angular grain boundary variation curve of different passes. Reproduced with permission from Springer Nature.

For HCP metals represented by magnesium and titanium, studies have shown that the changing trend of grain boundary angle in Mg alloys [100] and pure Ti [101] is similar to that of FCC metals. Li et al. found that the phase change in Mg alloys with the pass also affects the transition between high and low angle grain boundaries. The twisting and fragmentation of the internal phase of the grains not only facilitates the transformation of low-angle grain boundaries to high-angle grain boundaries but also refines the grains [102].

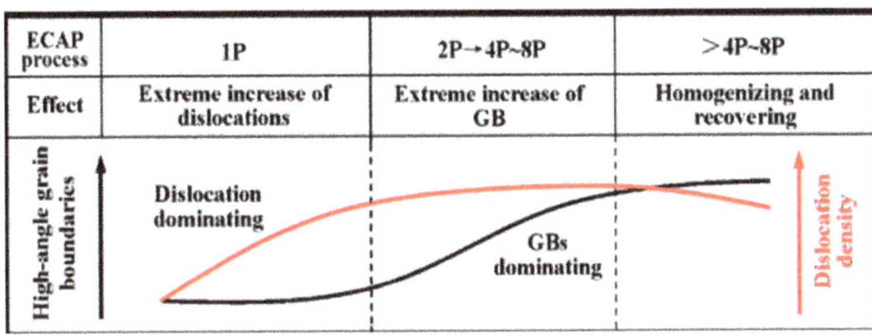

Figure 17. Variation trends of angular grain boundaries and dislocations in copper and its alloys [97]. Reproduced with permission from Elsevier B.V.

Low-angle grain boundaries have been reported to play an important role in the strengthening of alloys [98]. Luo et al. quantitatively studied the strength contribution of high and low-angle grain boundaries to pure Ti and found that the Hall-Petch law largely follows the strength contribution of low-angle grain boundaries [103]. However, Balasubramanian and Langdon believed that high-angle grain boundaries play a role in ameliorating the microstructure, and low-angle grain boundaries have little effect on the strengthening of materials because low-angle grain boundaries are more likely to produce slip and have a lower barrier effect [104]. Furthermore, Qarni et al. believe that existing high-angle grain boundaries provide nucleation sites, while low-angle grain boundaries transform around them. And they think that grain refinement starts from the outside and gradually spreads to the inside [105].

In different alloys, after a certain strain deformation, the high-angle grain boundary and the low-angle grain boundary are no longer transformed, and an equilibrium state is reached [106]. Due to dynamic recrystallization at the grain boundaries, a stable microstructure will be formed [107]. The grain refinement also reaches the extreme when the deformation of the alloy reaches saturation [108].

2.3. Phase

2.3.1. Titanium Alloys

For (α + β) two-phase titanium alloys, Suresh et al. found that the evolution of the two phases is different for different passes [109]. By observing the texture evolution of α phase for the first four passes, they think it is the result of dynamic recrystallization. After four passes, the α-phase texture weakens and the β-phase texture appears, which is due to the local lattice rotation of the β-phase around the α-phase. After eight passes, the β phase formed an ideal texture structure, indicating that the delayed dynamic recrystallization of the β phase occurred [109]. Figure 18 shows the evolution of the two phases after different passes. Kocich et al. also showed that the phase evolution would be affected by the increase of pass number and the decrease of deformation temperature in special titanium alloys [110]. In the study of the α phase, it was found that the high dislocation density and a large number of non-equilibrium grain boundaries/sub-grain boundaries generated during the ECAP process, which provide a good nucleation site for α-phase precipitation, and the ultrafine equiaxed α grains were formed in the β matrix. Moreover, vacancies and non-equilibrium grain boundaries in the microstructure enhance atomic diffusivity and accelerate the formation of the α-phase [111]. Dyakonov et al. argued that the evolution of the primary α phase was driven by continuous dynamic recrystallization [112]. Polyakova et al. also found that the fragmentation of the primary α-phase grains was achieved due to the slip and accumulation of dislocations [113].

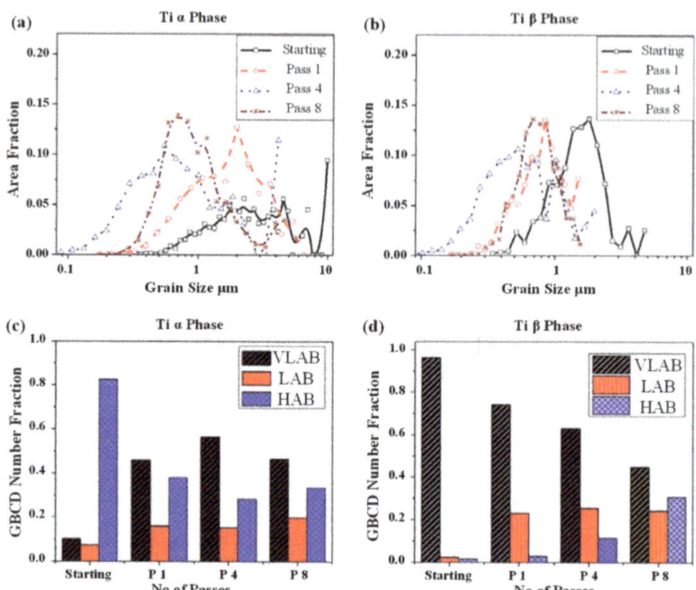

Figure 18. Grain distribution and angular grain boundary distribution of α and β phases under different passes [109]: (**a**,**c**) is the α phase and (**b**,**d**) is the β phase. Reproduced with permission from Elsevier Inc.

The slender primary α phase was formed in hot extruded Ti alloy after ECAP treatment, and the primary α slender grains play an important role in the failure behavior of Ti alloy. Semenova et al. found that the lowest fracture toughness was obtained in the sample with elongated grains parallel to the loading direction [114]. Bartha et al. showed that the formation of the alpha phase is accelerated in regions with a higher concentration of lattice defects. The equiaxed α-phase is mainly formed in the area with a dense grain boundary network and high dislocation density, and the flaky α-phase particles precipitate along the coarse β-grain boundary, which inhibits the precipitation of the α-phase inside the β-grain [115].

Sun and Xie reported that the β phase transformed into an acicular martensite α′ phase after ECAP processing, which was helpful for further grain refinement in other phase regions [116]. Li et al. also found that after ECAP processing, the preferred orientations vary. Figure 19 shows the XRD images of the phases under different conditions. A shift from the preferred orientation of the β(200) peak to a combination of the β(200) peak and the β(211) peak as the preferred orientation. The different intensities of the β peaks in different regions are considered to be caused by the changes in the temperature field and stress field during the ECAP process. They also showed that the deformation mechanism of hot-extruded Ti alloys includes stress-induced transformation of α″ phase to β phase, and it is believed that the deformation mechanism of Ti alloys depends to a large extent on the grain size of the β phase [117].

2.3.2. Mg Alloys

Phase evolution affects the microstructure and properties of metallic materials during ECAP. Liu et al. found that ECAP could refine the grain size of each phase but could not change the dispersion of the phases. And they found that the refining rate of the β phase is higher than that of the α phase in Mg alloys [118]. Both phases are more uniform as the number of passes increases [119]. Moreover, Xu et al. also found that the formation of the β phase can inhibit grain growth [120]. Figure 20 shows the microstructure and

(0002) pole figures of Mg alloys at different passes. For Mg alloys containing second-phase particles, ECAP can promote the refinement of the matrix α-phase, but after too many passes, the α-Mg grains will grow and form a block, which is caused by the decline of the mechanical properties of Mg alloys due to particle dissolution [121]. However, Eyram Klu et al. demonstrated that the strong basal texture formed by the α phase of the Mg matrix can improve the strength of the material [122]. Mostaed et al. studied the grain size of the second phase in Mg alloys after ECAP. They found that the reduction in the grain size of the second phase also made the material microstructure more uniform [123]. Yang et al. show that the corrosion properties of Mg alloys are not only related to grain size but also the morphology and distribution of the β phase [124].

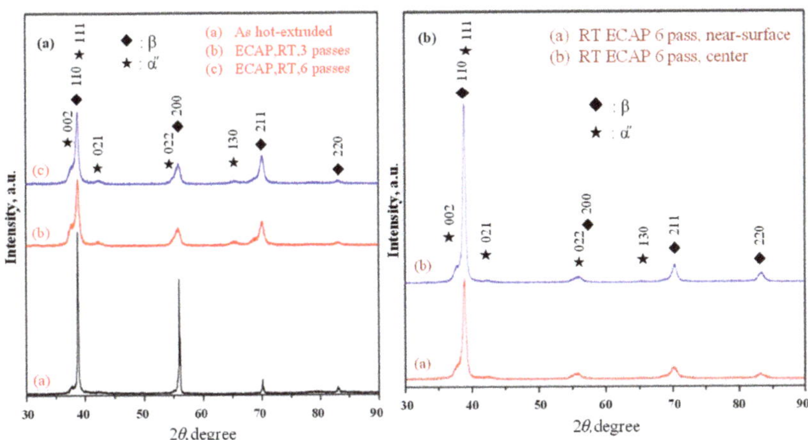

Figure 19. XRD patterns of different phases [117]: (**a**) Different processing passes; (**b**) different regions. Reproduced with permission from Springer Nature.

In addition, Furui et al. found that temperature also affects phase evolution [125]. Several phases in Mg alloys continue to grow with increasing temperatures at different temperatures. Figure 21 shows the evolution of the phases at different temperatures. As the temperature increases, the particles of different phases will grow to different degrees. Grain growth is also different at different temperature stages. When the processing temperature is at the lowest conditions, the material will obtain the most uniform and fine microstructure [126].

The combination of ECAP with other treatments can affect the phase evolution and thus change the mechanical properties of magnesium alloys. Mg alloys undergo aging treatment after ECAP. Although the grain size will grow slightly and reduce the material properties, the precipitation hardening caused by the precipitation particles will compensate for the loss and further increase the strength [127]. Yan et al. produced a homogeneous near-single-phase microstructure during the deformation process by ECAP and hot stretching techniques, thereby obtaining Mg alloys with better mechanical properties [3]. By adding the Zr element in the ECAP process, the Mg alloy can produce smaller second-phase particles, inhibit grain growth, reduce the final grain size, and improve the properties of the material [128].

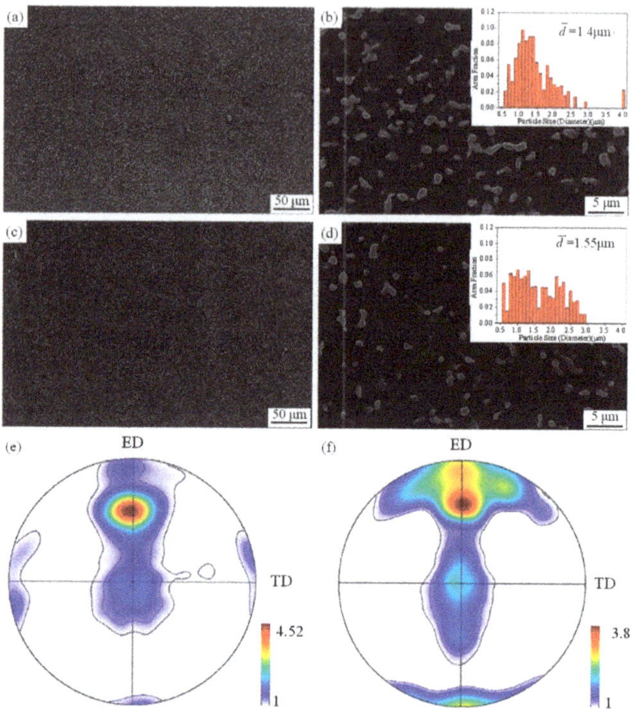

Figure 20. SEM images after different passes of ECAP and (0002) pole figures [120]: (**a**,**b**,**e**) four passes; (**c**,**d**,**f**) twelve passes. Reproduced with permission from Elsevier B.V.

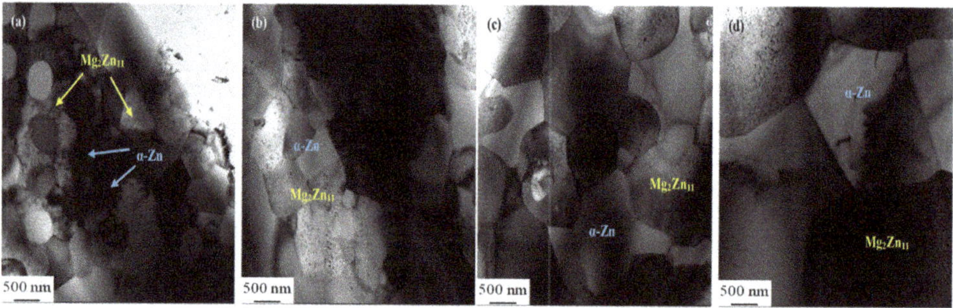

Figure 21. TEM images at different temperatures [126]: (**a**) 150 °C, 4 passes; (**b**) 150 °C, 12 passes; (**c**) 200 °C, 12 passes; (**d**) 250 °C, 12 passes. Reproduced with permission from Elsevier B.V.

2.3.3. Al Alloys

The formation and transformation of η and η′ phases in Al alloys can affect the properties of the material. It is found that the η phase is refined with increasing density. The η′ phase is precipitated after one pass, and many nano-scale precipitates are formed with the increase of the pass to improve the performance of the material [129,130]. And after ECAP, the size of the η′ phase in the matrix is smaller and the distribution is more uniform [131]. Temperature also affects the precipitation of the η phase and η′ phase [132]. Shaeri et al. reported that when the temperature was raised from room temperature to 120 °C, small η′ particles were formed to enhance the mechanical properties of the material. Between 120 °C and 180 °C, the mechanical properties of Al alloys decrease with the increase in temperature, which is related to the increase in the size of grains and precipitates and the transformation of the η′ phase into the η phase [133]. Xu et al. found that at 200 °C the η′ phase would partially dissolve, resulting in a reduction in the strength of the alloy, which could be recovered by aging treatment [134]. Figure 22 shows the XRD patterns of Al alloys at different temperatures.

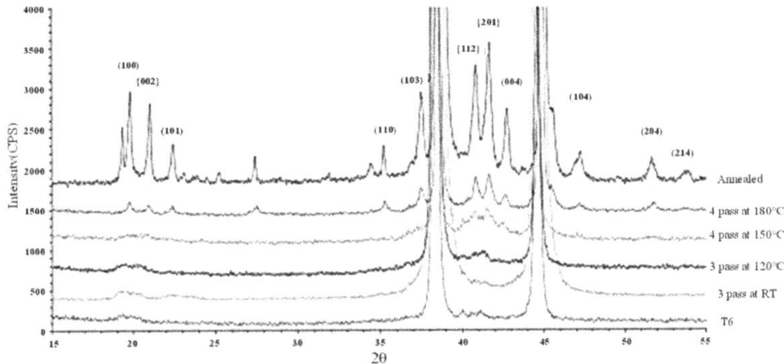

Figure 22. XRD patterns after ECAP treatment at different temperatures [133]. Reproduced with permission from Elsevier B.V.

Different heat treatments after ECAP also affect the phase change and alter the properties of the alloy. Wang et al. An annealing treatment after one pass of ECAP increases the hardness of the material due to the generation of very fine η′ phase particles. However, after eight passes of ECAP, the hardness of the alloy decreases due to recrystallization and the transformation of the η′ phase into the η phase. And they found that annealing at higher temperatures did not further improve the alloy properties [135]. Afifi et al. also reported that with the increase of the pass, more fine precipitates will be generated, and the increase of the annealing temperature after ECAP does not further refine the grains, but because the increase of the heat treatment temperature will cause grain coarsening [136]. Figure 23 shows the microstructures annealed at different temperatures after eight passes of ECAP. Combining the ECAP process and heat treatment optimization will produce metal materials with better properties.

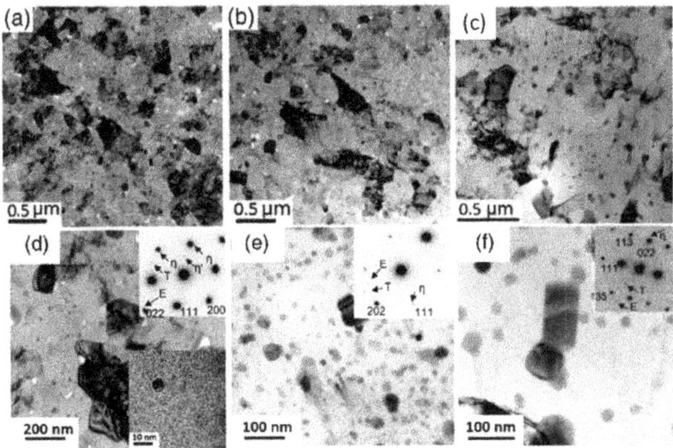

Figure 23. TEM images of annealing at different temperatures after 8 passes of ECAP [135]: (**a**,**d**) 393 K; (**b**,**e**) 423 K; (**c**,**f**) 473 K. Reproduced with permission from John Wiley & Sons, Inc.

3. Deformation Mechanism

In metallic materials, dislocations and twinning are the main deformation mechanisms [137]. At different temperatures, the deformation mechanism remains the same, which is dominated by slip [50]. Figure 24 shows the slip system for HCP metal, FCC metal and BCC metal. Chen et al. found that when the temperature is between 250 °C and 300 °C, the deformation of the alloy is controlled by grain boundary diffusion; when the temperature is between 300 °C and 350 °C, the deformation is controlled by lattice diffusion. When the temperature is further increased, dislocation creep, which requires larger activation energies, plays a role in the deformation. During the deformation process of the alloy, not all grains have a good sliding direction, and only the grains with the most favorable orientation will undergo grain boundary sliding [138]. Chen et al. studied the twinning behavior at different temperatures, as shown in Figure 25, and found that there is a certain relationship between the grain orientation and the evolution of twinning. Under the condition of variable temperature, the formation of twins causes lattice rotation and changes in grain orientation. In addition, the twin formation has different tendencies in grains with different orientations during deformation. The formation of twins changes the grain orientation, which affects the evolution of twins [27]. In the ECAP process at low temperatures, twins are the prevalent microstructure and the deformation mechanism is supposed to be induced by the twining instead of dislocation slipping. With the increase of temperature, both twinning and dislocation slipping contribute to the material deformation behavior. Moreover, the fracture of twins after increasing the temperature is conducive to forming ultrafine grains [27].

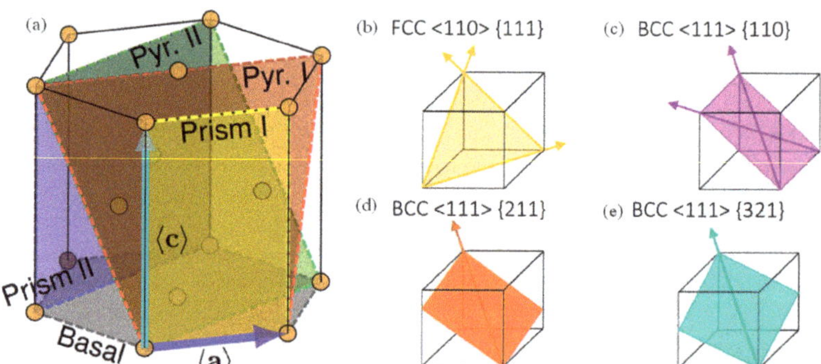

Figure 24. Slip system: (**a**) HCP metals [139]. Reproduced with permission from Elsevier Ltd. (**b**) FCC slip systems; (**c–e**) BCC slip systems [140]. Reproduced with permission from Elsevier Ltd.

It is found that the c/a ratio in HCP metal affects the critical shear stress (CRSS) between substrate slip and non-substrate slip, and thus affects the deformation mechanism of the material [141]. For example, the c/a of Mg is 1.624, which leads to the lower CRSS of the base plane and is conducive to the slip of the base plane. However, due to the low c/a ratio of Ti, the prism CRSS is lower, which is conducive to the prism slip [142]. Minarik also found that different c/a ratios lead to different texture evolution during ECAP processing due to the activation of different slip systems [143]. Bednarczyk et al. showed that an increase in dislocation density would also increase the CRSS of slip systems in HCP metals. The decrease of grain size will cause the deformation mechanism to change from non-basal slip to basal slip [144].

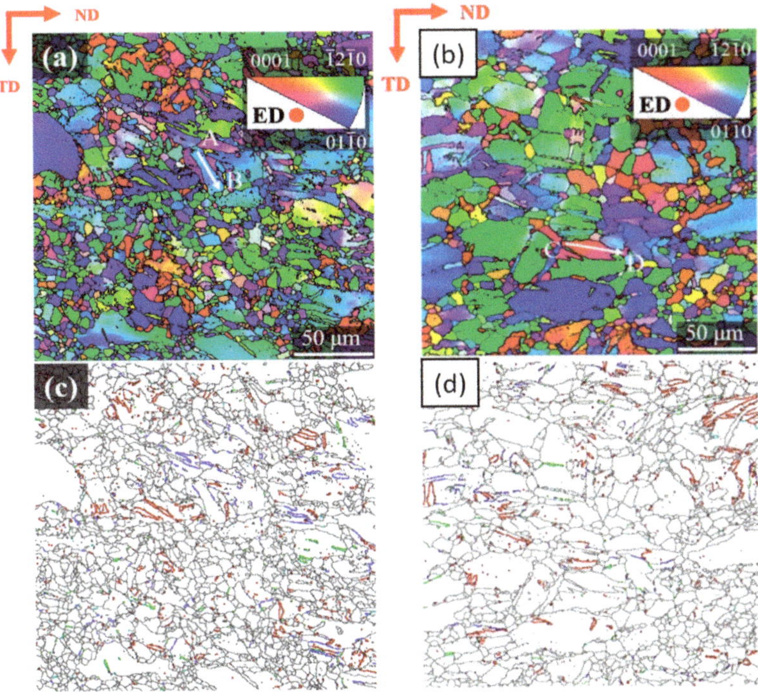

Figure 25. EBSD map and twin distribution map at different temperatures [27]: (**a**,**c**) 250 °C; (**b**,**d**) 300 °C. Reproduced with permission from Springer Nature.

The effect of the ECAP route on the microstructure is due to the differences in the shear paths of the different routes concerning the grain extension texture plane and the direction of the crystal structure [19]. Roodposhti et al. investigated the effect of different machining routes on the ECAP processing of commercially pure titanium. They showed that the processing routes directly affect the microstructure of the material [17]. Different routes activate different slip regimes, resulting in different microstructures. Figure 26 shows the clipping paths generated by different routes. Among them, X, Y, and Z represent three different orthogonal planes, and the numbers represent the shearing directions of different passes. From the figure, we can see that the strain of route C recovers after the even number of passes, and also the third and fourth slips of route B_C offset the first two slips, so route C and route B_C are called redundant strain channels. The non-redundant strain passages Route A and Route Ba have separate shear plane intersections, and additional shear strain accumulates with each pass. The shear directions of the C route are all switched in the orthogonal direction, resulting in equiaxed grains. Other routes deform differently in different directions because of additional shear strain and crystal anisotropy. This also explains the reason why the B_C route obtains a more uniform microstructure in the experiment [14,17,19]. However, Ravisankar and Park found that route A was more effective in grain refinement for BCC alloys. This is related to the redundant strain in the machining process. The crystal structure, slip system and fault energy can affect the route efficiency. For example, Ag, which is also an FCC metal, is more prone to twinning deformation, while Cu and Al are prone to slip deformation because of the low fault energy of Ag. HCP metals prefer c/a-based twinning due to the lack of slip systems [20].

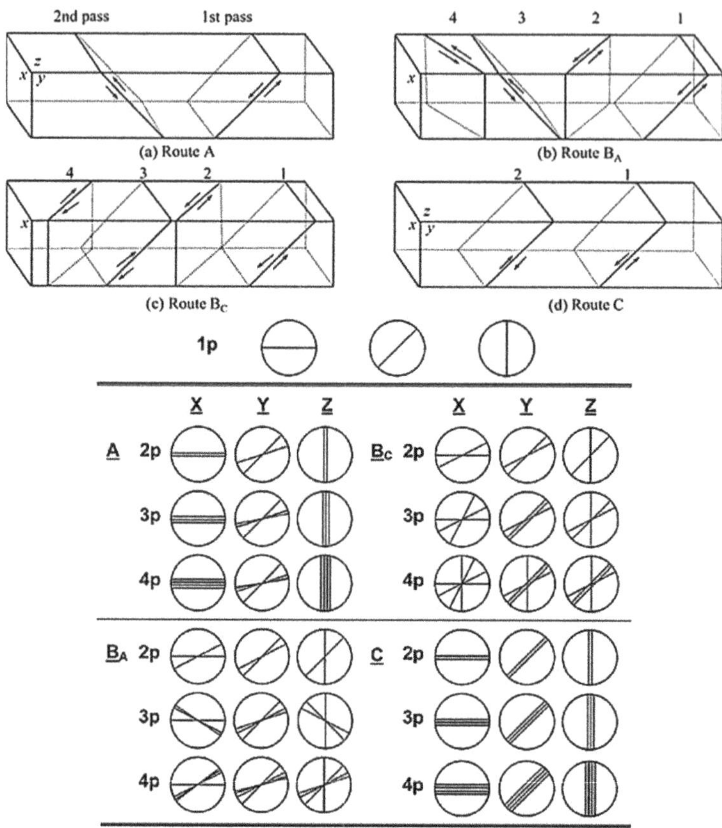

Figure 26. Slip system for different routes [14,20]:(**a**) Route A; (**b**) Route B_A; (**c**) Route B_C; (**d**) Route C. Reproduced with permission from Springer Nature. Reproduced with permission from Elsevier Ltd.

In the initial stage of compression deformation, twinning dominates the deformation mechanism due to the larger grain size, and as the strain increases, the grain size is refined and the twinning decreases [145]. Moreover, Xiang et al. also showed that the formation of twins can improve the deformability of HCP alloys by increasing the slip system [146]. Figure 27 shows the microstructure of the material after different passes.

Li et al. reported that fine grains (3 μm) can suppress the formation of twins and that the twins in the coarse grains are mostly tensile twinning, and the tensile twinning is less than the twinning and compression twinning during deformation. With grain refinement, the deformation mechanism changes from tensile twinning and <a> dislocation slip to <a> dislocation and <c + a> dislocation slip. Figure 28 shows the distribution of different dislocations [147]. Zhu and Langdon also showed that the deformation mechanism under low stress and strain is mainly grain boundary sliding and diffusion creep, while the deformation mechanism under medium and high stress and strain is mainly the slip and climbing of intragranular dislocations. When the grain size is reduced to the nanometer scale, the deformation of the material depends on its grain size. When the grain size is between 50 and 100 nm, the dislocation emission and annihilation at the grain boundary are the main deformation mechanisms, which can also explain the creep mechanism of ultrafine-grained titanium. Between 10 and 50 nm, partial dislocations and deformation twinning are the main deformation mechanisms. For grains smaller than 10 nm, grain boundary slip is still the main deformation mechanism [148,149]. Gautam and Biswas found that with decreasing grain size, the compressive mechanical behavior changed from S-shaped to parabolic, further demonstrating that the deformation mechanism changed from twinning to slipping with decreasing grain size [150].

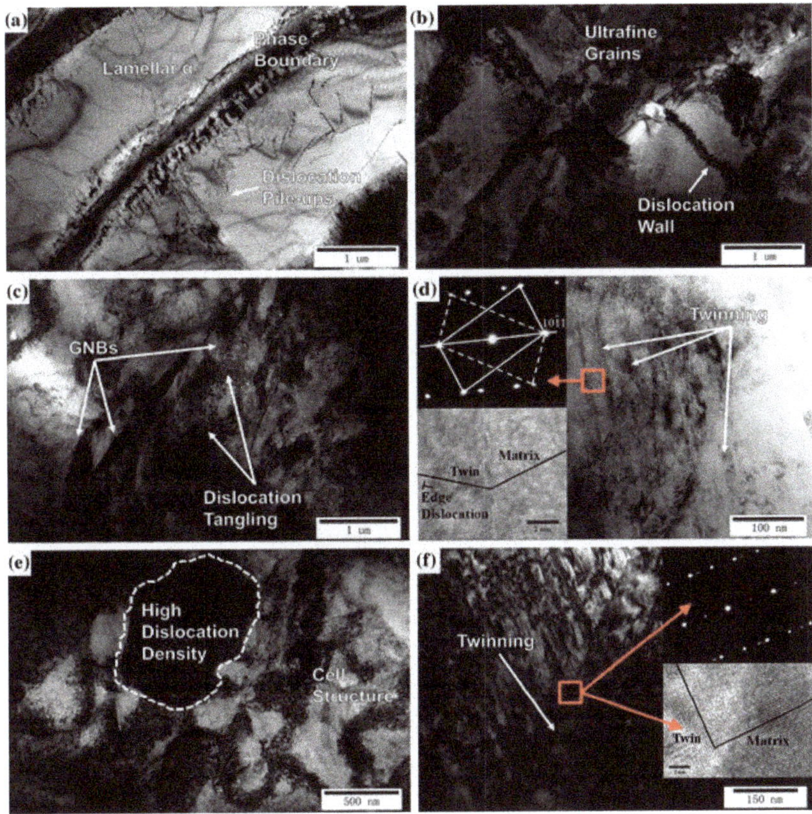

Figure 27. TEM images after different passes [146]: (**a**) as-received; (**b**) one pass; (**c**,**d**) two passes; (**e**,**f**) three passes. Reproduced with permission from Springer Nature.

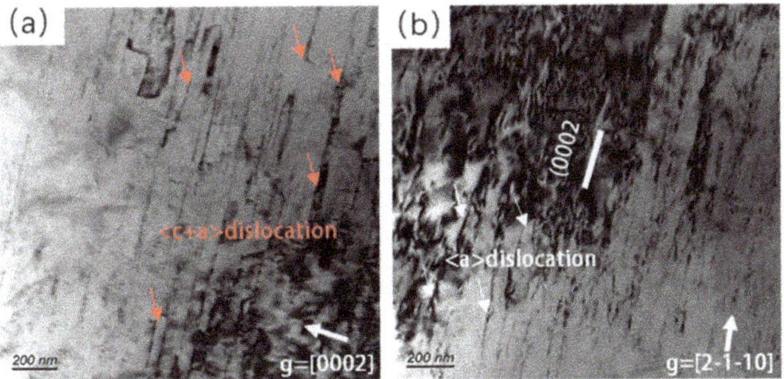

Figure 28. *Cont.*

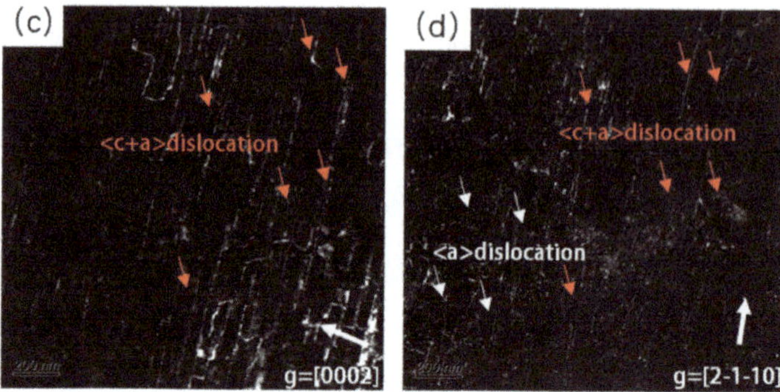

Figure 28. Distribution of dislocations in bright and dark fields [147]: (**a**,**b**) bright fields;(**c**,**d**) dark fields. Reproduced with permission from Elsevier B.V.

4. Conclusions and Perspectives

(1) Starting from the grain size, this paper discusses the processing parameters required to achieve submicron and nanoscale particles for different metal materials. The grain size required for different metal materials is obtained by analyzing the influencing factors such as processing route, pass, and temperature. The research on processing temperature and processing speed is still not comprehensive enough, and no quantitative and continuous trend of influence has been obtained. However, performing the ECAP process at a relatively low temperature is beneficial to obtain finer grains, because the low temperature can inhibit the dynamic recrystallization and grain growth of metal materials.

(2) The grain refinement of the ECAP process is via plastic deformation of the workpiece, during which grain boundaries also change simultaneously. Even though some efforts have been devoted to investigating the evolution of the high/low angle grain boundaries, the shear deformation and its interaction during grain refinement are still under discussion due to the evolution of grain boundaries in the material. Some advanced characterization techniques such as in-situ TEM and transmission Kikuchi diffraction (TKD) can provide more detail information, which are suitable candidates for investigating the grain boundary evolution in future studies.

(3) The quantitative relationship between the effective strain and the pass, channel angle, and curvature angle during ECAP processing is discussed based on previous research under ideal conditions. However, the friction in the actual machining process is not negligible, and the influence of the friction between the sample part and the mold on the ECAP process still needs to be further studied.

(4) During thermal deformation, subgrain boundaries transform from low-angle grain boundaries to high-angle grain boundaries by absorbing dislocations, resulting in grain refinement. Therefore, the dynamic recrystallization that occurs during the ECAP process is considered to be one of the main mechanisms of grain refinement. Existing studies have been able to prove that the evolution of dislocations, grain boundary angle, phase and other factors will reduce the grain size, but no clear refinement mechanism has been given. Therefore, there is no unified consensus on the mechanism of grain refinement in the ECAP process, which still needs further exploration.

(5) In the study of ECAP, the deformation of metallic materials is controlled by twinning and slip. In general, twinning is the dominant deformation mechanism in the low-strain stage. As the strain increases, dislocation slip gradually becomes the dominant deformation mechanism of metallic materials. Under different conditions (temperature, route, pass), severe plastic deformation usually causes different changes in the dominant deformation mechanism of metal materials. Grain refinement further inhibits the formation of twins. When the grains are reduced to a certain

extent, the deformation mechanism of metallic materials will change from twinning to dislocation slip.

Author Contributions: L.C.: Conceptualization, Writing—original draft preparation. S.S.: Methodology, Software. H.W.: Investigation, Resources. G.Z.: Investigation, Supervision. Z.Z.: Supervision, Writing—review and editing, Funding acquisition. C.Z.: Data Curation, Funding acquisition. All authors have read and agreed to the published version of the manuscript.

Funding: This research was funded by the Research Committee of the Shenzhen University and Shenzhen Natural Science Foundation University Stable Support Project (Grant No. 20200826160002001) as well as the National Natural Science Foundation of China (Grant No. 62003216).

Data Availability Statement: Not applicable.

Conflicts of Interest: No conflict of interest exists in this submitted manuscript, and the manuscript is approved by all authors for publication. I would like to declare on behalf of my co-authors that the work described was original research that has not been published previously, and not under consideration for publication elsewhere, in whole, or in part.

References

1. Hyun, C.-Y.; Lee, J.-H.; Kim, H.-K. Microstructures and mechanical properties of ultrafine grained pure Ti produced by severe plastic deformation. *Res. Chem. Intermed.* **2010**, *36*, 629–638. [CrossRef]
2. Naseri, R.; Kadkhodayan, M.; Shariati, M. Static mechanical properties and ductility of biomedical ultrafine-grained commercially pure titanium produced by ECAP process. *Trans. Nonferrous Met. Soc. China* **2017**, *27*, 1964–1975. [CrossRef]
3. Yan, K.; Sun, J.; Bai, J.; Liu, H.; Huang, X.; Jin, Z.; Wu, Y. Preparation of a high strength and high ductility Mg-6Zn alloy wire by combination of ECAP and hot drawing. *Mater. Sci. Eng. A* **2019**, *739*, 513–518. [CrossRef]
4. Sheremetyev, V.; Churakova, A.; Derkach, M.; Gunderov, D.; Raab, G.; Prokoshkin, S. Effect of ECAP and annealing on structure and mechanical properties of metastable beta Ti-18Zr-15Nb (at.%) alloy. *Mater. Lett.* **2021**, *305*, 130760. [CrossRef]
5. Franz, M.; Mingler, B.; Krystian, M.; Sajti, L.; Pohl, D.; Rellinghaus, B.; Wolf-Brandstetter, C.; Scharnweber, D. Strengthening of Titanium by Equal Channel Angular Pressing—Impact on Oxide Layer Properties of Pure Titanium and Ti6Al4V. *Adv. Mater. Interfaces* **2020**, *7*, 2000552. [CrossRef]
6. An, B.; Li, Z.; Diao, X.; Xin, H.; Zhang, Q.; Jia, X.; Wu, Y.; Li, K.; Guo, Y. In vitro and in vivo studies of ultrafine-grain Ti as dental implant material processed by ECAP. *Mater. Sci. Eng. C* **2016**, *67*, 34–41. [CrossRef]
7. Semenova, I.P.; Polyakov, A.V.; Polyakova, V.V.; Huang, Y.; Valiev, R.Z.; Langdon, T.G. High-Cycle Fatigue Behavior of an Ultrafine-Grained Ti-6Al-4V Alloy Processed by ECAP and Extrusion. *Adv. Eng. Mater.* **2016**, *18*, 2057–2062. [CrossRef]
8. Zhou, W.; Yu, J.; Lin, J.; Dean, T.A. Manufacturing a curved profile with fine grains and high strength by differential velocity sideways extrusion. *Int. J. Mach. Tools Manuf.* **2019**, *140*, 77–88. [CrossRef]
9. Zhou, W.; Yu, J.; Lu, X.; Lin, J.; Dean, T.A. A comparative study on deformation mechanisms, microstructures and mechanical properties of wide thin-ribbed sections formed by sideways and forward extrusion. *Int. J. Mach. Tools Manuf.* **2021**, *168*, 103771. [CrossRef]
10. Faraji, G.; Kim, H.S. Review of principles and methods of severe plastic deformation for producing ultrafine-grained tubes. *Mater. Sci. Technol.* **2016**, *33*, 905–923. [CrossRef]
11. Ruirun, C.; Deshuang, Z.; Jingjie, G.; Tengfei, M.; Hongsheng, D.; Yanqing, S.; Hengzhi, F. A novel method for grain refinement and microstructure modification in TiAl alloy by ultrasonic vibration. *Mater. Sci. Eng. A* **2016**, *653*, 23–26. [CrossRef]
12. Azushima, A.; Kopp, R.; Korhonen, A.; Yang, D.Y.; Micari, F.; Lahoti, G.D.; Groche, P.; Yanagimoto, J.; Tsuji, N.; Rosochowski, A.; et al. Severe plastic deformation (SPD) processes for metals. *CIRP Ann.* **2008**, *57*, 716–735. [CrossRef]
13. Alateyah, A.I.; Ahmed, M.M.Z.; Zedan, Y.; El-Hafez, H.A.; Alawad, M.O.; El-Garaihy, W.H. Experimental and Numerical Investigation of the ECAP Processed Copper: Microstructural Evolution, Crystallographic Texture and Hardness Homogeneity. *Metals* **2021**, *11*, 607. [CrossRef]
14. Koujalagi, M.B.; Siddesha, H.S. ECAP of titanium alloy by sever plastic deformation: A review. *Mater. Today Proc.* **2021**, *45*, 71–77. [CrossRef]
15. Kunčická, L.; Kocich, R.; Král, P.; Pohludka, M.; Marek, M. Effect of strain path on severely deformed aluminium. *Mater. Lett.* **2016**, *180*, 280–283. [CrossRef]
16. Kocich, R.; Kunčická, L.; Macháčková, A. Twist Channel Multi-Angular Pressing (TCMAP) as a method for increasing the efficiency of SPD. In *Proceedings of IOP Conference Series: Materials Science and Engineering*; IOP Publishing: Bristol, UK, 2014; p. 012006.
17. Roodposhti, P.S.; Farahbakhsh, N.; Sarkar, A.; Murty, K.L. Microstructural approach to equal channel angular processing of commercially pure titanium—A review. *Trans. Nonferrous Met. Soc. China* **2015**, *25*, 1353–1366. [CrossRef]
18. Abib, K.; Balanos, J.A.M.; Alili, B.; Bradai, D. On the microstructure and texture of Cu-Cr-Zr alloy after severe plastic deformation by ECAP. *Mater. Charact.* **2016**, *112*, 252–258. [CrossRef]

19. Shaat, M. Effects of processing conditions on microstructure and mechanical properties of equal-channel-angular-pressed titanium. *Mater. Sci. Technol.* **2018**, *34*, 1149–1167. [CrossRef]
20. Ravisankar, B.; Park, J.K. ECAP of commercially pure titanium: A review. *Trans. Indian Inst. Met.* **2008**, *61*, 51–62. [CrossRef]
21. Djavanroodi, F.; Omranpour, B.; Ebrahimi, M.; Sedighi, M. Designing of ECAP parameters based on strain distribution uniformity. *Prog. Nat. Sci. Mater. Int.* **2012**, *22*, 452–460. [CrossRef]
22. Kim, H.S.; Kim, W.Y.; Song, K.H. Effect of post-heat-treatment in ECAP processed Cu–40%Zn brass. *J. Alloys Compd.* **2012**, *536*, S200–S203. [CrossRef]
23. Chen, J.; Su, Y.; Zhang, Q.; Sun, J.; Yang, D.; Jiang, J.; Song, D.; Ma, A. Enhancement of strength-ductility synergy in ultrafine-grained Cu-Zn alloy prepared by ECAP and subsequent annealing. *J. Mater. Res. Technol.* **2022**, *17*, 433–440. [CrossRef]
24. Zhao, G.; Xu, S.; Luan, Y.; Guan, Y.; Lun, N.; Ren, X. Grain refinement mechanism analysis and experimental investigation of equal channel angular pressing for producing pure aluminum ultra-fine grained materials. *Mater. Sci. Eng. A* **2006**, *437*, 281–292. [CrossRef]
25. Kawasaki, M.; Horita, Z.; Langdon, T.G. Microstructural evolution in high purity aluminum processed by ECAP. *Mater. Sci. Eng. A* **2009**, *524*, 143–150. [CrossRef]
26. Minárik, P.; Král, R.; Pešička, J.; Chmelík, F. Evolution of mechanical properties of LAE442 magnesium alloy processed by extrusion and ECAP. *J. Mater. Res. Technol.* **2015**, *4*, 75–78. [CrossRef]
27. Chen, J.; Sun, R.; Li, G.; Fang, M.; Xu, G.; Zhang, M.; Li, J. Dynamic Recrystallization Behaviors and the Texture Evolution in Mg–9Al–1Zn Alloy Produced by ECAP at Different Temperatures. *Met. Mater. Int.* **2022**, *2022*, 1–14. [CrossRef]
28. Baig, M.; Seikh, A.H.; Rehman, A.U.; Mohammed, J.A.; Hashmi, F.H.; Ragab, S.M. Microstructure Evaluation Study of Al5083 Alloy Using EBSD Technique after Processing with Different ECAP Processes and Temperatures. *Crystals* **2021**, *11*, 862. [CrossRef]
29. Damavandi, E.; Nourouzi, S.; Rabiee, S.M.; Jamaati, R.; Szpunar, J.A. EBSD study of the microstructure and texture evolution in an Al–Si–Cu alloy processed by route A ECAP. *J. Alloys Compd.* **2021**, *858*, 157651. [CrossRef]
30. Zhao, Z.; To, S.; Sun, Z.; Ji, R.; Yu, K.M. Microstructural effects of Ti6Al4V alloys modified by electropulsing treatment on ultraprecision diamond turning. *J. Manuf. Process.* **2019**, *39*, 58–68. [CrossRef]
31. Zhao, Z.; To, S. An investigation of resolved shear stress on activation of slip systems during ultraprecision rotary cutting of local anisotropic Ti-6Al-4V alloy: Models and experiments. *Int. J. Mach. Tools Manuf.* **2018**, *134*, 69–78. [CrossRef]
32. Zhao, Z.; Gao, J.; Wang, Y.; Zhang, Y.; Hou, H. Effect of Equal Channel Angular Pressing on the Dynamic Softening Behavior of Ti-6Al-4V Alloy in the Hot Deformation Process. *Materials* **2021**, *14*, 232. [CrossRef] [PubMed]
33. Zhao, Z.; Wang, G.; Zhang, Y.; Gao, J.; Hou, H. Microstructure Evolution and Mechanical Properties of Ti-6Al-4V Alloy Prepared by Multipass Equal Channel Angular Pressing. *J. Mater. Eng. Perform.* **2020**, *29*, 905–913. [CrossRef]
34. Jahadi, R.; Sedighi, M.; Jahed, H. ECAP effect on the micro-structure and mechanical properties of AM30 magnesium alloy. *Mater. Sci. Eng. A* **2014**, *593*, 178–184. [CrossRef]
35. Matsubara, K.; Miyahara, Y.; Horita, Z.; Langdon, T.G. Developing superplasticity in a magnesium alloy through a combination of extrusion and ECAP. *Acta Mater.* **2003**, *51*, 3073–3084. [CrossRef]
36. Gopi, K.; Nayaka, H.S.; Sahu, S. Investigation of microstructure and mechanical properties of ECAP-processed AM series magnesium alloy. *J. Mater. Eng. Perform.* **2016**, *25*, 3737–3745. [CrossRef]
37. Terhune, S.D.; Swisher, D.L.; Oh-Ishi, K.; Horita, Z.; Langdon, T.G.; McNelley, T.R. An investigation of microstructure and grain-boundary evolution during ECA pressing of pure aluminum. *Metall. Mater. Trans. A* **2002**, *33*, 2173–2184. [CrossRef]
38. Wroński, M.; Wierzbanowski, K.; Wojtas, D.; Szyfner, E.; Valiev, R.Z.; Kawałko, J.; Berent, K.; Sztwiertnia, K. Microstructure, Texture and Mechanical Properties of Titanium Grade 2 Processed by ECAP (Route C). *Met. Mater. Int.* **2018**, *24*, 802–814. [CrossRef]
39. Najafi, S.; Eivani, A.R.; Samaee, M.; Jafarian, H.R.; Zhou, J. A comprehensive investigation of the strengthening effects of dislocations, texture and low and high angle grain boundaries in ultrafine grained AA6063 aluminum alloy. *Mater. Charact.* **2018**, *136*, 60–68. [CrossRef]
40. Ramu, G.; Bauri, R. Effect of equal channel angular pressing (ECAP) on microstructure and properties of Al–SiCp composites. *Mater. Des.* **2009**, *30*, 3554–3559. [CrossRef]
41. Kocich, R.; Kunčická, L.; Král, P.; Macháčková, A. Sub-structure and mechanical properties of twist channel angular pressed aluminium. *Mater. Charact.* **2016**, *119*, 75–83. [CrossRef]
42. Suwas, S.; Gottstein, G.; Kumar, R. Evolution of crystallographic texture during equal channel angular extrusion (ECAE) and its effects on secondary processing of magnesium. *Mater. Sci. Eng. A* **2007**, *471*, 1–14. [CrossRef]
43. Tong, L.B.; Zheng, M.Y.; Hu, X.S.; Wu, K.; Xu, S.W.; Kamado, S.; Kojima, Y. Influence of ECAP routes on microstructure and mechanical properties of Mg–Zn–Ca alloy. *Mater. Sci. Eng. A* **2010**, *527*, 4250–4256. [CrossRef]
44. El-Danaf, E.A.; Soliman, M.S.; Almajid, A.A.; El-Rayes, M.M. Enhancement of mechanical properties and grain size refinement of commercial purity aluminum 1050 processed by ECAP. *Mater. Sci. Eng. A* **2007**, *458*, 226–234. [CrossRef]
45. Alateyah, A.I.; El-Garaihy, W.H.; Alawad, M.O.; Sanabary, S.E.; Elkatatny, S.; Dahish, H.A.; Kouta, H. The Effect of ECAP Processing Conditions on Microstructural Evolution and Mechanical Properties of Pure Magnesium—Experimental, Mathematical Empirical and Response Surface Approach. *Materials* **2022**, *15*, 5312. [CrossRef]
46. Gajanan, M.N.; Narendranath, S.; Kumar, S.S.S. Influence of ECAP processing routes on microstructure mechanical properties and corrosion behavior of AZ80 Mg alloy. In *AIP Conference Proceedings*; AIP Publishing LLC: Melville, NY, USA, 2019; p. 030016.

47. Li, X.; Du, M.; Mao, Z.; Huang, T.; Ban, C. Study of microstructure and mechanical properties of pure titanium processed by new route of equal channel angular pressing. *IOP Conf. Series. Mater. Sci. Eng.* **2020**, *2*, 768. [CrossRef]
48. Lin, H.K.; Huang, J.C.; Langdon, T.G. Relationship between texture and low temperature superplasticity in an extruded AZ31 Mg alloy processed by ECAP. *Mater. Sci. Eng. A* **2005**, *402*, 250–257. [CrossRef]
49. Kim, W.J.; Jeong, H.T. Grain-Size Strengthening in Equal-Channel-Angular-Pressing Processed AZ31 Mg Alloys with a Constant Texture. *Mater. Trans.* **2005**, *46*, 251–258. [CrossRef]
50. Gautam, P.C.; Biswas, S. On the possibility to reduce ECAP deformation temperature in magnesium: Deformation behaviour, dynamic recrystallization and mechanical properties. *Mater. Sci. Eng. A* **2021**, *812*, 141103. [CrossRef]
51. Dumitru, F.D.; Higuera-Cobos, O.F.; Cabrera, J.M. ZK60 alloy processed by ECAP: Microstructural, physical and mechanical characterization. *Mater. Sci. Eng. A* **2014**, *594*, 32–39. [CrossRef]
52. Mostaed, E.; Fabrizi, A.; Dellasega, D.; Bonollo, F.; Vedani, M. Microstructure, mechanical behavior and low temperature superplasticity of ECAP processed ZM21 Mg alloy. *J. Alloys Compd.* **2015**, *638*, 267–276. [CrossRef]
53. Minárik, P.; Veselý, J.; Král, R.; Bohlen, J.; Kubásek, J.; Janeček, M.; Stráská, J. Exceptional mechanical properties of ultra-fine grain Mg-4Y-3RE alloy processed by ECAP. *Mater. Sci. Eng. A* **2017**, *708*, 193–198. [CrossRef]
54. Liu, X.; Zhao, X.; Yang, X.; Jia, J.; Qi, B. The evolution of hardness homogeneity in commercially pure Ti processed by ECAP. *J. Wuhan Univ. Technol. Mater. Sci. Ed.* **2014**, *29*, 578–584. [CrossRef]
55. Fan, Z.; Jiang, H.; Sun, X.; Song, J.; Zhang, X.; Xie, C. Microstructures and mechanical deformation behaviors of ultrafine-grained commercial pure (grade 3) Ti processed by two-step severe plastic deformation. *Mater. Sci. Eng. A* **2009**, *527*, 45–51. [CrossRef]
56. Zhilyaev, A.; Parkhimovich, N.; Raab, G.; Popov, V.; Danilenko, V. Microstructure and texture homogeneity of ECAP titanium. *Rev. Adv. Mater. Sci.* **2015**, *43*, 61–66.
57. Hajizadeh, K.; Eghbali, B.; Topolski, K.; Kurzydlowski, K.J. Ultra-fine grained bulk CP-Ti processed by multi-pass ECAP at warm deformation region. *Mater. Chem. Phys.* **2014**, *143*, 1032–1038. [CrossRef]
58. El Aal, M.I.A.; El Mahallawy, N.; Shehata, F.A.; El Hameed, M.A.; Yoon, E.Y.; Lee, J.H.; Kim, H.S. Tensile properties and fracture characteristics of ECAP-processed Al and Al-Cu alloys. *Met. Mater. Int.* **2010**, *16*, 709–716. [CrossRef]
59. Abd El Aal, M.I.; Sadawy, M.M. Influence of ECAP as grain refinement technique on microstructure evolution, mechanical properties and corrosion behavior of pure aluminum. *Trans. Nonferrous Met. Soc. China* **2015**, *25*, 3865–3876. [CrossRef]
60. Song, D.; Ma, A.-b.; Jiang, J.-h.; Lin, P.-h.; Yang, D.-h. Corrosion behavior of ultra-fine grained industrial pure Al fabricated by ECAP. *Trans. Nonferrous Met. Soc. China* **2009**, *19*, 1065–1070. [CrossRef]
61. Wang, L.-S.; Jiang, J.-H.; Saleh, B.; Xie, Q.-Y.; Xu, Q.; Liu, H.; Ma, A.-B. Controlling Corrosion Resistance of a Biodegradable Mg–Y–Zn Alloy with LPSO Phases via Multi-pass ECAP Process. *Acta Metall. Sin.* **2020**, *33*, 1180–1190. [CrossRef]
62. Jiang, J.; Aibin, M.; Saito, N.; Zhixin, S.; Dan, S.; Fumin, L.; Nishida, Y.; Donghui, Y.; Pinghua, L. Improving corrosion resistance of RE-containing magnesium alloy ZE41A through ECAP. *J. Rare Earths* **2009**, *27*, 848–852. [CrossRef]
63. Ma, A.; Jiang, J.; Saito, N.; Shigematsu, I.; Yuan, Y.; Yang, D.; Nishida, Y. Improving both strength and ductility of a Mg alloy through a large number of ECAP passes. *Mater. Sci. Eng. A* **2009**, *513–514*, 122–127. [CrossRef]
64. Zhilyaev, A.; Kim, B.-K.; Nurislamova, G.; Baró, M.; Szpunar, J.; Langdon, T. Orientation imaging microscopy of ultrafine-grained nickel. *Scr. Mater.* **2002**, *46*, 575–580. [CrossRef]
65. Neishi, K.; Horita, Z.; Langdon, T.G. Grain refinement of pure nickel using equal-channel angular pressing. *Mater. Sci. Eng. A* **2002**, *325*, 54–58. [CrossRef]
66. Raju, K.S.; Krishna, M.G.; Padmanabhan, K.A.; Muraleedharan, K.; Gurao, N.P.; Wilde, G. Grain size and grain boundary character distribution in ultra-fine grained (ECAP) nickel. *Mater. Sci. Eng. A* **2008**, *491*, 1–7. [CrossRef]
67. Sun, C.; Yu, K.Y.; Lee, J.H.; Liu, Y.; Wang, H.; Shao, L.; Maloy, S.A.; Hartwig, K.T.; Zhang, X. Enhanced radiation tolerance of ultrafine grained Fe–Cr–Ni alloy. *J. Nucl. Mater.* **2012**, *420*, 235–240. [CrossRef]
68. Jayakumar, P.K.; Balasubramanian, K.; Rabindranath Tagore, G. Recrystallisation and bonding behaviour of ultra fine grained copper and Cu–Cr–Zr alloy using ECAP. *Mater. Sci. Eng. A* **2012**, *538*, 7–13. [CrossRef]
69. Wongsa-Ngam, J.; Wen, H.; Langdon, T.G. Microstructural evolution in a Cu–Zr alloy processed by a combination of ECAP and HPT. *Mater. Sci. Eng. A* **2013**, *579*, 126–135. [CrossRef]
70. Khereddine, A.Y.; Larbi, F.H.; Kawasaki, M.; Baudin, T.; Bradai, D.; Langdon, T.G. An examination of microstructural evolution in a Cu–Ni–Si alloy processed by HPT and ECAP. *Mater. Sci. Eng. A* **2013**, *576*, 149–155. [CrossRef]
71. Alawadhi, M.Y.; Sabbaghianrad, S.; Huang, Y.; Langdon, T.G. Evaluating the paradox of strength and ductility in ultrafine-grained oxygen-free copper processed by ECAP at room temperature. *Mater. Sci. Eng. A* **2021**, *802*, 140546. [CrossRef]
72. Schafler, E.; Steiner, G.; Korznikova, E.; Kerber, M.; Zehetbauer, M.J. Lattice defect investigation of ECAP-Cu by means of X-ray line profile analysis, calorimetry and electrical resistometry. *Mater. Sci. Eng. A* **2005**, *410–411*, 169–173. [CrossRef]
73. Purcek, G.; Yanar, H.; Demirtas, M.; Alemdag, Y.; Shangina, D.V.; Dobatkin, S.V. Optimization of strength, ductility and electrical conductivity of Cu–Cr–Zr alloy by combining multi-route ECAP and aging. *Mater. Sci. Eng. A* **2016**, *649*, 114–122. [CrossRef]
74. Stolyarov, V.V.; Zhu, Y.T.; Alexandrov, I.V.; Lowe, T.C.; Valiev, R.Z. Influence of ECAP routes on the microstructure and properties of pure Ti. *Mater. Sci. Eng. A* **2001**, *299*, 59–67. [CrossRef]
75. Valder, J.; Rijesh, M.; Surendranathan, A.O. Forming of Tubular Commercial Purity Aluminum by ECAP. *Mater. Manuf. Process.* **2012**, *27*, 986–989. [CrossRef]

76. Bulutsuz, A.G.; Chrominski, W.; Huang, Y.; Kral, P.; Yurci, M.E.; Lewandowska, M.; Langdon, T.G. A Comparison of Warm and Combined Warm and Low-Temperature Processing Routes for the Equal-Channel Angular Pressing of Pure Titanium. *Adv. Eng. Mater.* **2019**, *22*, 1900698. [CrossRef]
77. Attarilar, S.; Djavanroodi, F.; Irfan, O.M.; Al-Mufadi, F.A.; Ebrahimi, M.; Wang, Q.D. Strain uniformity footprint on mechanical performance and erosion-corrosion behavior of equal channel angular pressed pure titanium. *Results Phys.* **2020**, *17*, 103141. [CrossRef]
78. Ebrahimi, M.; Attarilar, S. Grain Refinement Affected Machinability in Commercial Pure Titanium. *Metall. Mater. Trans. A* **2021**, *52*, 1282–1292. [CrossRef]
79. Zhao, X.; Yang, X.; Liu, X.; Wang, X.; Langdon, T.G. The processing of pure titanium through multiple passes of ECAP at room temperature. *Mater. Sci. Eng. A* **2010**, *527*, 6335–6339. [CrossRef]
80. Khmelevskaya, I.Y.; Prokoshkin, S.D.; Trubitsyna, I.B.; Belousov, M.N.; Dobatkin, S.V.; Tatyanin, E.V.; Korotitskiy, A.V.; Brailovski, V.; Stolyarov, V.V.; Prokofiev, E.A. Structure and properties of Ti–Ni-based alloys after equal-channel angular pressing and high-pressure torsion. *Mater. Sci. Eng. A* **2008**, *481–482*, 119–122. [CrossRef]
81. Wang, H.; Ban, C.; Zhao, N.; Li, L.; Zhu, Q.; Cui, J. Cryogenic temperature equal channel angular pressing of pure titanium: Microstructure and homogeneity. *J. Mater. Res. Technol.* **2021**, *14*, 1167–1179. [CrossRef]
82. Podolskiy, A.V.; Ng, H.P.; Psaruk, I.A.; Tabachnikova, E.D.; Lapovok, R. Cryogenic equal channel angular pressing of commercially pure titanium: Microstructure and properties. *J. Mater. Sci.* **2014**, *49*, 6803–6812. [CrossRef]
83. Wang, Y.L.; Lapovok, R.; Wang, J.T.; Qi, Y.S.; Estrin, Y. Thermal behavior of copper processed by ECAP with and without back pressure. *Mater. Sci. Eng. A* **2015**, *628*, 21–29. [CrossRef]
84. McKenzie, P.W.J.; Lapovok, R. ECAP with back pressure for optimum strength and ductility in aluminium alloy 6016. Part 1: Microstructure. *Acta Mater.* **2010**, *58*, 3198–3211. [CrossRef]
85. McKenzie, P.W.J.; Lapovok, R.; Estrin, Y. The influence of back pressure on ECAP processed AA 6016: Modeling and experiment. *Acta Mater.* **2007**, *55*, 2985–2993. [CrossRef]
86. Lapovok, R.; Tomus, D.; Mang, J.; Estrin, Y.; Lowe, T.C. Evolution of nanoscale porosity during equal-channel angular pressing of titanium. *Acta Mater.* **2009**, *57*, 2909–2918. [CrossRef]
87. Stolyarov, V.V.; Lapovok, R. Effect of backpressure on structure and properties of AA5083 alloy processed by ECAP. *J. Alloys Compd.* **2004**, *378*, 233–236. [CrossRef]
88. Medvedev, A.E.; Lapovok, R.; Koch, E.; Höppel, H.W.; Göken, M. Optimisation of interface formation by shear inclination: Example of aluminium-copper hybrid produced by ECAP with back-pressure. *Mater. Des.* **2018**, *146*, 142–151. [CrossRef]
89. Wu, X.; Pu, L.; Xu, Y.; Shi, J.; Liu, X.; Zhong, Z.; Luo, S.N. Deformation of high density polyethylene by dynamic equal-channel-angular pressing. *RSC Adv.* **2018**, *8*, 22583–22591. [CrossRef]
90. Demirtas, M.; Purcek, G.; Yanar, H.; Zhang, Z.J.; Zhang, Z.F. Improvement of high strain rate and room temperature superplasticity in Zn–22Al alloy by two-step equal-channel angular pressing. *Mater. Sci. Eng. A* **2015**, *620*, 233–240. [CrossRef]
91. Afifi, M.A.; Wang, Y.C.; Cheng, X.; Li, S.; Langdon, T.G. Strain rate dependence of compressive behavior in an Al-Zn-Mg alloy processed by ECAP. *J. Alloys Compd.* **2019**, *791*, 1079–1087. [CrossRef]
92. Khan, A.S.; Meredith, C.S. Thermo-mechanical response of Al 6061 with and without equal channel angular pressing (ECAP). *Int. J. Plast.* **2010**, *26*, 189–203. [CrossRef]
93. Chatterjee, A.; Sharma, G.; Sarkar, A.; Singh, J.B.; Chakravartty, J.K. A study on cryogenic temperature ECAP on the microstructure and mechanical properties of Al–Mg alloy. *Mater. Sci. Eng. A* **2012**, *556*, 653–657. [CrossRef]
94. Abd El Aal, M.I.; El Mahallawy, N.; Shehata, F.A.; Abd El Hameed, M.; Yoon, E.Y.; Kim, H.S. Wear properties of ECAP-processed ultrafine grained Al–Cu alloys. *Mater. Sci. Eng. A* **2010**, *527*, 3726–3732. [CrossRef]
95. Aal, M.I.A.E. Influence of the pre-homogenization treatment on the microstructure evolution and the mechanical properties of Al–Cu alloys processed by ECAP. *Mater. Sci. Eng. A* **2011**, *528*, 6946–6957. [CrossRef]
96. Qu, S.; An, X.H.; Yang, H.J.; Huang, C.X.; Yang, G.; Zang, Q.S.; Wang, Z.G.; Wu, S.D.; Zhang, Z.F. Microstructural evolution and mechanical properties of Cu–Al alloys subjected to equal channel angular pressing. *Acta Mater.* **2009**, *57*, 1586–1601. [CrossRef]
97. Zhang, Z.J.; Duan, Q.Q.; An, X.H.; Wu, S.D.; Yang, G.; Zhang, Z.F. Microstructure and mechanical properties of Cu and Cu–Zn alloys produced by equal channel angular pressing. *Mater. Sci. Eng. A* **2011**, *528*, 4259–4267. [CrossRef]
98. Khelfa, T.; Rekik, M.A.; Muñoz-Bolaños, J.A.; Cabrera-Marrero, J.M.; Khitouni, M. Microstructure and strengthening mechanisms in an Al-Mg-Si alloy processed by equal channel angular pressing (ECAP). *Int. J. Adv. Manuf. Technol.* **2017**, *95*, 1165–1177. [CrossRef]
99. Reihanian, M.; Ebrahimi, R.; Moshksar, M.M.; Terada, D.; Tsuji, N. Microstructure quantification and correlation with flow stress of ultrafine grained commercially pure Al fabricated by equal channel angular pressing (ECAP). *Mater. Charact.* **2008**, *59*, 1312–1323. [CrossRef]
100. Xu, C.; Horita, Z.; Langdon, T.G. Microstructural evolution in an aluminum solid solution alloy processed by ECAP. *Mater. Sci. Eng. A* **2011**, *528*, 6059–6065. [CrossRef]
101. Dyakonov, G.S.; Mironov, S.; Semenova, I.P.; Valiev, R.Z.; Semiatin, S.L. Microstructure evolution and strengthening mechanisms in commercial-purity titanium subjected to equal-channel angular pressing. *Mater. Sci. Eng. A* **2017**, *701*, 289–301. [CrossRef]
102. Li, B.; Teng, B.; Chen, G. Microstructure evolution and mechanical properties of Mg-Gd-Y-Zn-Zr alloy during equal channel angular pressing. *Mater. Sci. Eng. A* **2019**, *744*, 396–405. [CrossRef]

103. Luo, P.; Hu, Q.; Wu, X. Quantitatively Analyzing Strength Contribution vs Grain Boundary Scale Relation in Pure Titanium Subjected to Severe Plastic Deformation. *Metall. Mater. Trans. A* **2016**, *47*, 1922–1928. [CrossRef]
104. Balasubramanian, N.; Langdon, T.G. The strength–grain size relationship in ultrafine-grained metals. *Metall. Mater. Trans. A* **2016**, *47*, 5827–5838. [CrossRef]
105. Qarni, M.J.; Sivaswamy, G.; Rosochowski, A.; Boczkal, S. Effect of incremental equal channel angular pressing (I-ECAP) on the microstructural characteristics and mechanical behaviour of commercially pure titanium. *Mater. Des.* **2017**, *122*, 385–402. [CrossRef]
106. Muñoz, J.A.; Bolmaro, R.E.; Jorge, A.M.; Zhilyaev, A.; Cabrera, J.M. Prediction of Generation of High- and Low-Angle Grain Boundaries (HAGB and LAGB) During Severe Plastic Deformation. *Metall. Mater. Trans. A* **2020**, *51*, 4674–4684. [CrossRef]
107. Gopi, K.R.; Shivananda Nayaka, H.; Sahu, S. Microstructural Evolution and Strengthening of AM90 Magnesium Alloy Processed by ECAP. *Arab. J. Sci. Eng.* **2017**, *42*, 4635–4647. [CrossRef]
108. Gunderov, D.V.; Polyakov, A.V.; Semenova, I.P.; Raab, G.I.; Churakova, A.A.; Gimaltdinova, E.I.; Sabirov, I.; Segurado, J.; Sitdikov, V.D.; Alexandrov, I.V.; et al. Evolution of microstructure, macrotexture and mechanical properties of commercially pure Ti during ECAP-conform processing and drawing. *Mater. Sci. Eng. A* **2013**, *562*, 128–136. [CrossRef]
109. Suresh, K.S.; Gurao, N.P.; Singh, D.S.; Suwas, S.; Chattopadhyay, K.; Zherebtsov, S.V.; Salishchev, G.A. Effect of equal channel angular pressing on grain refinement and texture evolution in a biomedical alloy Ti13Nb13Zr. *Mater. Charact.* **2013**, *82*, 73–85. [CrossRef]
110. Kocich, R.; Kursa, M.; Szurman, I.; Dlouhý, A. The influence of imposed strain on the development of microstructure and transformation characteristics of Ni–Ti shape memory alloys. *J. Alloys Compd.* **2011**, *509*, 2716–2722. [CrossRef]
111. Shi, Q.; Tse, Y.Y.; Higginson, R.L. Microstructure Evolution and Microhardness Analysis of Metastable Beta Titanium Alloy Ti-15V-3Cr-3Al-3Sn Consolidated Using Equal-Channel Angular Pressing from Machining Chips. *J. Mater. Eng. Perform.* **2020**, *29*, 4142–4153. [CrossRef]
112. Dyakonov, G.S.; Yakovleva, T.V.; Stotskiy, A.G.; Mironov, S. EBSD study of advanced Ti-5.7Al-3.8Mo-1.2Zr-1.3Sn alloy subjected to equal-channel angular pressing. *Mater. Lett.* **2021**, *298*, 130003. [CrossRef]
113. Polyakova, V.V.; Semenova, I.P.; Polyakov, A.V.; Magomedova, D.K.; Huang, Y.; Langdon, T.G. Influence of grain boundary misorientations on the mechanical behavior of a near-α Ti-6Al-7Nb alloy processed by ECAP. *Mater. Lett.* **2017**, *190*, 256–259. [CrossRef]
114. Semenova, I.P.; Modina, J.M.; Polyakov, A.V.; Klevtsov, G.V.; Klevtsova, N.A.; Pigaleva, I.N.; Valiev, R.Z.; Langdon, T.G. Fracture toughness at cryogenic temperatures of ultrafine-grained Ti-6Al-4V alloy processed by ECAP. *Mater. Sci. Eng. A* **2018**, *716*, 260–267. [CrossRef]
115. Bartha, K.; Terynková, A.; Stráský, J.; Minárik, P.; Veselý, J.; Polyakova, V.; Semenova, I.; Janeček, M. Inhomogeneous Precipitation of the α-Phase in Ti15Mo Alloy Deformed by ECAP. *Mater. Sci. Forum* **2018**, *941*, 1183–1188.
116. Sun, Q.J.; Xie, X. Microstructure and mechanical properties of TA15 alloy after thermo-mechanical processing. *Mater. Sci. Eng. A* **2018**, *724*, 493–501. [CrossRef]
117. Li, Z.; Zheng, B.; Wang, Y.; Topping, T.; Zhou, Y.; Valiev, R.Z.; Shan, A.; Lavernia, E.J. Ultrafine-grained Ti–Nb–Ta–Zr alloy produced by ECAP at room temperature. *J. Mater. Sci.* **2014**, *49*, 6656–6666. [CrossRef]
118. Liu, T.; Zhang, W.; Wu, S.D.; Jiang, C.B.; Li, S.X.; Xu, Y.B. Mechanical properties of a two-phase alloy Mg–8%Li–1%Al processed by equal channel angular pressing. *Mater. Sci. Eng. A* **2003**, *360*, 345–349. [CrossRef]
119. Liu, T.; Wu, S.D.; Li, S.X.; Li, P.J. Microstructure evolution of Mg–14% Li–1% Al alloy during the process of equal channel angular pressing. *Mater. Sci. Eng. A* **2007**, *460–461*, 499–503. [CrossRef]
120. Xu, B.; Sun, J.; Yang, Z.; Xiao, L.; Zhou, H.; Han, J.; Liu, H.; Wu, Y.; Yuan, Y.; Zhuo, X.; et al. Microstructure and anisotropic mechanical behavior of the high-strength and ductility AZ91 Mg alloy processed by hot extrusion and multi-pass RD-ECAP. *Mater. Sci. Eng. A* **2020**, *780*, 139191. [CrossRef]
121. Liu, X.; Bian, L.; Tian, F.; Han, S.; Wang, T.; Liang, W. Microstructural evolution and mechanical response of duplex Mg-Li alloy containing particles during ECAP processing. *Mater. Charact.* **2022**, *188*, 111910. [CrossRef]
122. Klu, E.E.; Song, D.; Li, C.; Wang, G.; Gao, B.; Ma, A.; Jiang, J. Achieving ultra-fine grains and high strength of Mg–9Li alloy via room-temperature ECAP and post rolling. *Mater. Sci. Eng. A* **2022**, *833*, 142371. [CrossRef]
123. Mostaed, E.; Vedani, M.; Hashempour, M.; Bestetti, M. Influence of ECAP process on mechanical and corrosion properties of pure Mg and ZK60 magnesium alloy for biodegradable stent applications. *Biomatter* **2014**, *4*, e28283. [CrossRef] [PubMed]
124. Yang, Z.; Ma, A.; Xu, B.; Jiang, J.; Sun, J. Corrosion behavior of AZ91 Mg alloy with a heterogeneous structure produced by ECAP. *Corros. Sci.* **2021**, *187*, 109517. [CrossRef]
125. Furui, M.; Kitamura, H.; Anada, H.; Langdon, T.G. Influence of preliminary extrusion conditions on the superplastic properties of a magnesium alloy processed by ECAP. *Acta Mater.* **2007**, *55*, 1083–1091. [CrossRef]
126. Liu, H.; Huang, H.; Zhang, Y.; Xu, Y.; Wang, C.; Sun, J.; Jiang, J.; Ma, A.; Xue, F.; Bai, J. Evolution of Mg–Zn second phases during ECAP at different processing temperatures and its impact on mechanical properties of Zn-1.6Mg (wt.%) alloys. *J. Alloys Compd.* **2019**, *811*, 151987. [CrossRef]
127. Yuan, Y.; Ma, A.; Jiang, J.; Lu, F.; Jian, W.; Song, D.; Zhu, Y.T. Optimizing the strength and ductility of AZ91 Mg alloy by ECAP and subsequent aging. *Mater. Sci. Eng. A* **2013**, *588*, 329–334. [CrossRef]

128. Minárik, P.; Zemková, M.; Veselý, J.; Bohlen, J.; Knapek, M.; Král, R. The effect of Zr on dynamic recrystallization during ECAP processing of Mg-Y-RE alloys. *Mater. Charact.* **2021**, *174*, 111033. [CrossRef]
129. Elhefnawey, M.; Shuai, G.L.; Li, Z.; Nemat-Alla, M.; Zhang, D.T.; Li, L. On achieving superior strength for Al–Mg–Zn alloy adopting cold ECAP. *Vacuum* **2020**, *174*, 109191. [CrossRef]
130. Afifi, M.A.; Pereira, P.H.R.; Wang, Y.C.; Wang, Y.; Li, S.; Langdon, T.G. Effect of ECAP processing on microstructure evolution and dynamic compressive behavior at different temperatures in an Al-Zn-Mg alloy. *Mater. Sci. Eng. A* **2017**, *684*, 617–625. [CrossRef]
131. Afifi, M.A.; Wang, Y.C.; Pereira, P.H.R.; Huang, Y.; Wang, Y.; Cheng, X.; Li, S.; Langdon, T.G. Mechanical properties of an Al-Zn-Mg alloy processed by ECAP and heat treatments. *J. Alloys Compd.* **2018**, *769*, 631–639. [CrossRef]
132. Cardoso, K.R.; Travessa, D.N.; Botta, W.J.; Jorge, A.M. High Strength AA7050 Al alloy processed by ECAP: Microstructure and mechanical properties. *Mater. Sci. Eng. A* **2011**, *528*, 5804–5811. [CrossRef]
133. Shaeri, M.H.; Shaeri, M.; Ebrahimi, M.; Salehi, M.T.; Seyyedein, S.H. Effect of ECAP temperature on microstructure and mechanical properties of Al–Zn–Mg–Cu alloy. *Prog. Nat. Sci. Mater. Int.* **2016**, *26*, 182–191. [CrossRef]
134. Xu, C.; Furukawa, M.; Horita, Z.; Langdon, T.G. Influence of ECAP on precipitate distributions in a spray-cast aluminum alloy. *Acta Mater.* **2005**, *53*, 749–758. [CrossRef]
135. Wang, Y.C.; Afifi, M.A.; Cheng, X.; Li, S.; Langdon, T.G. An Evaluation of the Microstructure and Microhardness in an Al–Zn–Mg Alloy Processed by ECAP and Post-ECAP Heat Treatments. *Adv. Eng. Mater.* **2019**, *22*, 1901040. [CrossRef]
136. Afifi, M.A.; Wang, Y.C.; Pereira, P.H.R.; Wang, Y.; Li, S.; Huang, Y.; Langdon, T.G. Characterization of precipitates in an Al-Zn-Mg alloy processed by ECAP and subsequent annealing. *Mater. Sci. Eng. A* **2018**, *712*, 146–156. [CrossRef]
137. Zhou, T.; Zhang, Q.; Li, Q.; Wang, L.; Li, Q.; Liu, D. A simultaneous enhancement of both strength and ductility by a novel differential-thermal ECAP process in Mg-Sn-Zn-Zr alloy. *J. Alloys Compd.* **2021**, *889*, 161653. [CrossRef]
138. Chen, D.; Kong, J.; Gui, Z.; Li, W.; Long, Y.; Kang, Z. High-temperature superplastic behavior and ECAP deformation mechanism of two-phase Mg-Li alloy. *Mater. Lett.* **2021**, *301*, 130358. [CrossRef]
139. Yin, B.; Wu, Z.; Curtin, W.A. Comprehensive first-principles study of stable stacking faults in hcp metals. *Acta Mater.* **2017**, *123*, 223–234. [CrossRef]
140. Wilson, D.; Wan, W.; Dunne, F.P.E. Microstructurally-sensitive fatigue crack growth in HCP, BCC and FCC polycrystals. *J. Mech. Phys. Solids* **2019**, *126*, 204–225. [CrossRef]
141. Li, R.; Pan, F.; Jiang, B.; Dong, H.; Yang, Q. Effect of Li addition on the mechanical behavior and texture of the as-extruded AZ31 magnesium alloy. *Mater. Sci. Eng. A* **2013**, *562*, 33–38. [CrossRef]
142. Kim, H.L.; Park, J.S.; Chang, Y.W. Effects of lattice parameter changes on critical resolved shear stress and mechanical properties of magnesium binary single crystals. *Mater. Sci. Eng. A* **2012**, *540*, 198–206. [CrossRef]
143. Minárik, P.; Král, R.; Čížek, J.; Chmelík, F. Effect of different c/a ratio on the microstructure and mechanical properties in magnesium alloys processed by ECAP. *Acta Mater.* **2016**, *107*, 83–95. [CrossRef]
144. Bednarczyk, W.; Wątroba, M.; Kawałko, J.; Bała, P. Can zinc alloys be strengthened by grain refinement? A critical evaluation of the processing of low-alloyed binary zinc alloys using ECAP. *Mater. Sci. Eng. A* **2019**, *748*, 357–366. [CrossRef]
145. Yu, X.; Li, Y.; Wei, Q.; Guo, Y.; Suo, T.; Zhao, F. Microstructure and mechanical behavior of ECAP processed AZ31B over a wide range of loading rates under compression and tension. *Mech. Mater.* **2015**, *86*, 55–70. [CrossRef]
146. Xiang, J.; Han, Y.; Huang, G.; Le, J.; Chen, Y.; Xiao, L.; Lu, W. Microstructural evolution in titanium matrix composites processed by multi-pass equal-channel angular pressing. *J. Mater. Sci.* **2019**, *54*, 7931–7942. [CrossRef]
147. Li, A.; Li, W.; Luo, M.; Yu, H.; Sun, Y.; Liang, Y. Effect of grain size on the microstructure and deformation mechanism of Mg-2Y-0.6Nd-0.6Zr alloy at a high strain rate. *Mater. Sci. Eng. A* **2021**, *824*, 141774. [CrossRef]
148. Zhu, Y.T.; Langdon, T.G. Influence of grain size on deformation mechanisms: An extension to nanocrystalline materials. *Mater. Sci. Eng. A* **2005**, *409*, 234–242. [CrossRef]
149. Liu, X.; Zhang, Q.; Zhao, X.; Yang, X.; Luo, L. Ambient-temperature nanoindentation creep in ultrafine-grained titanium processed by ECAP. *Mater. Sci. Eng. A* **2016**, *676*, 73–79. [CrossRef]
150. Gautam, P.C.; Biswas, S. Effect of ECAP temperature on the microstructure, texture evolution and mechanical properties of pure magnesium. *Mater. Today Proc.* **2021**, *44*, 2914–2918. [CrossRef]

 *processes* 

Article

Numerical Prediction of the Performance of Chamfered and Sharp Cutting Tools during Orthogonal Cutting of AISI 1045 Steel

Zakaria Ahmed M. Tagiuri, Thien-My Dao, Agnes Marie Samuel and Victor Songmene *

Department of Mechanical Engineering, École de Technologie Supérieure (ÉTS), 1100 Notre-Dame Street West, Montréal, QC H3C 1K3, Canada

* Correspondence: victor.songmene@etsmtl.ca; Tel.: +1-514-396-8869; Fax: +1-514-396-8530

Abstract: This paper presents a numerical investigation of the effects of chamfered and sharp cemented carbide tools using finite element method-based DEFORM-2D software and cutting parameters on different machining characteristics during the orthogonal cutting of AISI 1045 steel. The objective is to study the interactions between chamfer width, chamfer angle, sharp angle and the cutting speed and feed rate on the cutting temperature, effective stress and wear depth. These effects were investigated statistically using the analysis of variance (ANOVA) test. The obtained numerical results showed that for the chamfer tool, high values of temperature, stress and wear depth were obtained for chamfer widths of 0.35 mm and 0.45 mm. In terms of combined influences, for the cutting temperature and stress, a strong interaction between the cutting speed and chamfer width was obtained. For the sharp tool design, and in terms of temperature, strong interactions are mostly observed between cutting speeds and feed rates. The ANOVA showed that for both chamfer and sharp tools, the feed rate, the cutting speed and their interactions are the most significant parameters that influence temperature and stress.

Keywords: tool edge preparation; orthogonal cutting; numerical simulation; ANOVA; temperature; stress; tool wear

Citation: Tagiuri, Z.A.M.; Dao, T.-M.; Samuel, A.M.; Songmene, V. Numerical Prediction of the Performance of Chamfered and Sharp Cutting Tools during Orthogonal Cutting of AISI 1045 Steel. *Processes* **2022**, *10*, 2171. https://doi.org/10.3390/pr10112171

Academic Editors: Guoqing Zhang, Zejia Zhao and Wai Sze YIP

Received: 17 August 2022
Accepted: 18 October 2022
Published: 23 October 2022

1. Introduction

In different machining companies, the production demands for machined metal structural components with good surface finish and close dimensional tolerances continue to increase. However, manufacturing such parts requires a good knowledge of cutting technology as, during the machining process, there is much waste of material in the form of chips, resulting in long cycle times, particularly in the case of complex parts that are difficult to cut. Such long cycle times can lead to the eventual presence of defects and irregularities due to excessive heat generation; this can lead to the breakage and damage of cutting tools, reducing the machining performance and increasing manufacturing costs. Thus, in order to circumvent these problems, realize these demands and meet the market requirements, it is necessary to develop new methods for the cutting tool edge preparation, which is one of the important aspects in the development of cutting tools and improving the machining performance.

Cutting tool-manufacturing engineers use the tool edge preparation process to design the cutting-edge geometry and in order to remove edge defects and prepare the tool surface for coating, especially for the machining of difficult-to-cut materials. It has been demonstrated that cutting tool parameters, such as cutting speed, feed rate and the selection of tool edge geometry, e.g., edge radius or rake angle, have an impact on machining operations. There are several research studies in the literature, for example, the work of Rodriguez [1] and the investigations of Shfnir et al. [2], that show the effects of cutting tool edge preparation on tool life and the thermomechanical aspects of the cutting process, such as temperature distribution, cutting forces and effective stresses, chip formation, surface roughness and tool wear resistance.

Denkena and Biermann [3] reviewed the development of different tool edge preparation technologies and the interactions between cutting edge microgeometry and their effects on machining processes. They concluded that cutting processes with high performance are based on the good performance of the cutting tools in order to realize good surface finish quality and thus reduce manufacturing costs. They stated that the adoption of cutting-edge preparation methods results in improving tool wear and increasing the cutting temperature and cutting forces. The majority of such research studies have been conducted on the interactive effects of cutting tool geometry and machining conditions were limited to the round tool geometry.

Yen et al. [4] carried out a numerical investigation of the effect of round/honed edge and T-land/chamfered edge tools on the orthogonal machining performance of AISI 1020 steels. Their results showed that there are no significant variations in terms of maximum temperature and chip thickness. Cheng et al. [5] investigated the influence of honed tool geometry and the rake angle on the machining temperature and stress during the orthogonal machining of stainless steel using numerical methods. They showed that with increasing tool edge radius, the temperature increases slightly and the stress decreases, but increases, instead, with an increase in the rake angle. Emamian [6] performed finite element analysis on the effects of tool edge radius and feed rate during orthogonal turning of AISI 1045 steel. He found that the feed rate increases the cutting force for all studied edge radii and the maximum temperature increases with increasing edge radius only for higher feed rates. Daoud et al. [7] studied the cutting force behavior with variation of rake angle during the orthogonal machining of Al 2024-T3 alloy. They found that cutting forces decrease with increasing rake angle. Davoudinejad and Noordin [8] investigated the effects of honed and chamfered edge tools prepared with ceramic materials on the tool life, cutting forces and surface finish during the hard turning of DF-3 tool steel under various cutting experimental conditions using ANOVA. Their results showed that longer machining lengths were obtained in all cutting conditions (i.e., longer tool life), lower roughness for chamfered edge geometry and higher forces for honed tools design. Gao et al. [9] studied the influences of different cutting tool chamfer lengths on cutting stress, tool wear and surface roughness using a series of slot milling experiments and 3D finite element numerical simulations on aluminum alloy 7075. They found that higher chamfer lengths reduced tool wear (i.e., resulted in longer tool life) but higher flank wear width was obtained for higher stress. In addition, the contact stress with the workpiece increased with the increase of cutting-edge chamfer length.

Wan et al. [10] numerically analyzed the effects of chamfered geometry cutting tool on machining force in orthogonal cutting of P20 material under different cutting speeds. Their findings showed that as the chamfer angle increased the cutting and thrust forces increased but when the cutting speed increased, these forces decreased.

Despite the large volume of research realized on using the concept of cutting-edge preparation for developing different machining processes, machining problems still remain, due to a lack of understanding of the mechanical behavior and mechanisms that occur during cutting, especially in regard to the impact of interactions between microgeometry and cutting conditions, which have not been extensively studied and analyzed thus far [3]. The topic is still of actuality. Gregório et al. [11] have depicted the impact on tool edge preparation (sharp-edge and rounded edge) and its interaction with tool rake angle on chip formation, friction on tool, material flow and pressure on tool. Their work clearly shows the importance that the edge preparation has on the cutting process. With the exception of some recent works, such as the experimental study realized by Javidikia et al. [12] on the interactive impact between the cutting tool parameters of cutting-edge radius and rake angle, the machining parameters of cutting speed, feed rate and rake angle on the cutting and feed forces, chip thickness, maximum and average cutting temperature during orthogonal turning of 6061-T6 aluminum alloy, and that of Zhuang et al. [13] on Ti6Al4V and Inconel 718, few studies report on the interaction between the cutting speed, feed rate and cutting tool geometries (i.e., chamfer and sharp tool designs), and their effects on the machining characteristics of steel, in particular for carbon steels.

While studying the impact of tool geometry on low and high-speed turning of AA6061-T6, Javidikia et al. [12] found that the machining forces increased in both conventional and high-speed cutting regimes when using tool with large nose radius; additionally, the location of the maximum temperature on the tool depends on tool geometry and on cutting parameters. With high-speed machining, however, the average temperatures increased in the tool tip with increasing cutting speed.

Among the few studies on the machining process of metallic materials using chamfered and sharp tools for investigating the effects of the cutting parameters and the geometrical parameters of these tool shapes on the machining performance, there is the finite numerical model developed by Zhuang et al. [13] who examined the impacts of chamfer length, chamfer angle and feed rate on the cutting forces in the orthogonal cutting of Ti6Al4V and Inconel 718. They found that the cutting forces are significantly affected by chamfer length and angle. Altintas and Ren [14] predicted the impacts of chamfer angle and cutting parameters on the cutting forces and temperature. The obtained results showed that with increasing chamfer angle the total forces increase and the temperatures remain nearly constant. Tagiuri et al. [15] studied and numerically predicted the interactive effects between tool nose geometries (honed tools) and the cutting parameters on cutting temperature, effective stress, machining forces and tool wear during the orthogonal cutting of AISI1045 steel. The obtained results showed strong interactions between tool nose radius and cutting speed/feed rate on cutting stress and tool wear rate but not on the cutting temperature. Choudhury et al. [16] investigated the impacts of chamfer tool geometries (chamfer width and chamfer angle) on machining performance in terms of cutting force, chip thickness and tool life in the turning of medium carbon low alloy steel. They observed that the cutting and feed forces increased as the chamfer width and chamfer angle increased but at large values of these parameters, the cutting forces were low. With increase in chamfer width, the chip thickness decreased but presented a non-significant variation with chamfer angle. Khalili and Safaei [17] carried out a numerical study of the effects of the chamfer width and chamfer angle on cutting force, effective stress, tool temperature and tool stress during the two-dimensional (2D) orthogonal cutting of AISI 1045 steel by developing two finite element models. They found that there is almost no variation of cutting forces with increase in cutting speed, and that the thrust force is significantly influenced by the chamfer width and the chamfer angle. As cutting speed increases the maximum temperature increases at the tool tip and presents the optimum chamfer angle.

Despite the investigations presented above on the effects of tool edge geometries on machining process performance in cutting carbon steed materials, the data on different interactive configurations between tool edge geometries, such as chamfer width, chamfer angle, sharp angle and cutting parameters, e.g., cutting speed and feed rate, is still scarce. For various reasons, most of these studies were carried out for few and limited designed tests. This would imply a lack of understanding of the actual mechanisms responsible for the effects involved around the cutting edge during the cutting process, such as tribological and heat transfer aspects, for example. Thus, the ultimate goal would be to comprehend these mechanisms, rendering possible the means for the future preparations of design cutting tools.

In view of the above, the present work focuses on the influence of chamfered tool edge geometries and sharp edge on the machining performance. The objective of this article is to investigate the interactions in terms of the effects of cutting speed, feed rate, and chamfer and sharp tool geometries on the cutting temperature, cutting stress and tool wear depth. The main cutting parameters that influence these performance characteristics will be assessed using statistical analysis employing the analysis of variance (ANOVA) technique.

2. Materials and Methods

The numerical study consists of conducting different simulations of the two-dimensional (2D) orthogonal cutting of AISI 1045 steel using the commercial DEFORM software based on the finite element method (FEM) in order to predict the effects of the specified tool edge geometries and their interactions with machining parameters on the cutting temperature, cutting stress and tool wear. The two tool geometries examined are the chamfer and sharp tool geometry. Figures 1 and 2 show the 2D model geometry of the workpiece and the tool, respectively. These interactive effects will be evaluated based on a design plan which comprises different parameter combinations, namely cutting speed, feed rate, chamfer angle and chamfer width for the chamfer tool geometry and cutting speed, feed rate and cutting angle for the sharp tool geometry.

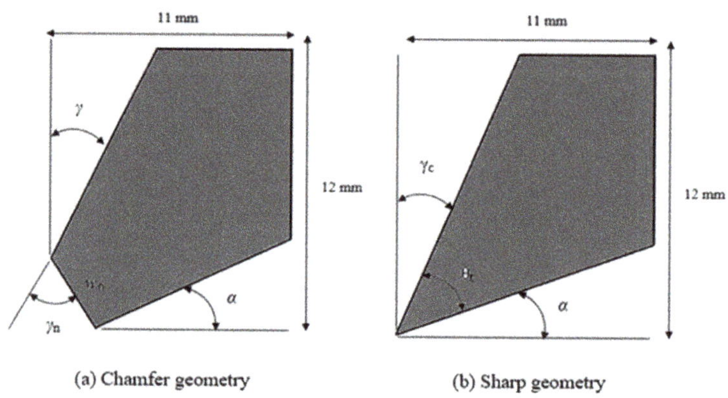

Figure 1. Two-dimensional (2D) models of the cutting tools studied: (**a**) Chamfer tool; (**b**) Sharp tool.

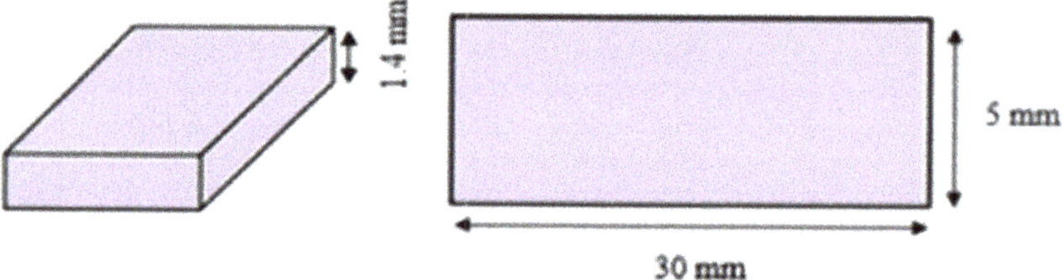

Figure 2. The workpiece model geometry.

2.1. Material and Machining Process Characterization

The machining process specified for the present numerical simulation tests is orthogonal cutting. The studied workpiece material is high carbon steel AISI 1045, which is often selected for different product applications that necessitate higher strength and resistance than other materials of the same category. The material selected for the cutting tool was uncoated cemented carbide, which it is also widely used in many industrial applications due to its hardness, as it gives a better surface finish to the machined part and provides higher productivity than high-speed steel.

2.2. Finite Element Simulation Model

In order to simulate the orthogonal milling process of high carbon steel AISI 1045, a 2D finite element model using DEFORM software was developed. The model includes the cutting tool, the workpiece and the chip. The meshing that was used in this cutting geometry model was created systematically using a default algorithm for solid modeling. The mesh consists of 2D free quadrilateral elements. The DEFORM software uses nodal finite elements for linear approximation. Each element has four nodes, and each node can have different degrees of freedom, such as temperature, force, stress, tool wear, chip thickness, etc.

Tables 1 and 2 present the thermal and mechanical properties of these materials, respectively. It is assumed that the thermal properties of the cutting tool and workpiece materials are constant by neglecting the thermal gradient. Due to the shortened time, the steady state is attained very fast (about 0.3 ms), causing little variation in thermal conductivity and specific heat with temperature. This statement has already been referred to by Zakaria et al. [15].

Table 1. Thermal properties of workpiece and tool materials.

Properties	**Tool: Uncoated Carbide**		**Workpiece: AISI 1045**	
Material	**Used**	**Deform Software (Default)**	**Used**	**Deform Software (Default)**
Density (kg/m^3)	11,900	10,850	7870	7850
Thermal conductivity (W/m °C)	50	59	45	55
Specific heat (J/kg °C)	375	364	590	570

Table 2. Mechanical properties of workpiece and tool materials.

Properties	**Tool: Uncoated Carbide**		**Workpiece: AISI 1045**	
Material	**Used**	**Deform Software (Default)**	**Used**	**Deform Software (Default)**
Young's modulus	620 GPa	-	200 GPa	-
Poisson ratio	0.26	0.22	0.29	0.30
Hardness	93 HRB	93 HRB	163 HB	-

During cutting simulation, the software used two numerical modules: heat transfer and motion analysis. In the DEFORM program, the transient mode was selected and the cutting temperature, force and stress were calculated by solving heat transfer and motion equations as follows [12]:

$$[\mathrm{C_T}]\{\dot{\mathrm{T}}\} + [\mathrm{K_T}]\{\mathrm{T}\} = \{\dot{\mathrm{Q}}\mathrm{g}\} \tag{1}$$

$$[\mathrm{M}]\{\ddot{\mathrm{U}}\} + \{\mathrm{Rint}\} = \{\mathrm{Rext}\} \tag{2}$$

where $[\mathrm{C_T}]$, $[\mathrm{K_T}]$, $\{\dot{\mathrm{Q}}\mathrm{g}\}$, $\{\ddot{\mathrm{U}}\}$, {U}, {Rint} and {Rext} are the volume heat capacitance, the thermal conduction matrices, the total heat generated, the acceleration vector, the displacement, the vector of internal force and the vector of external force, respectively.

During the cutting process, two heat transfer mechanisms occur: conduction and convection. Conduction occurs between cutting tool, workpiece and chip, whereas convection occurs between workpiece, tool surfaces and the ambient air. In order to simulate these mechanisms in 2D, the mentioned components were considered as superficial geometries in contact without internal heat generation.

$$k_n \Delta T = h(T_{air} - T) \tag{3}$$

where T_{air} is the ambient temperature, T is the temperature, k_n is the thermal conductivity of the studied materials, h is the convection heat transfer coefficient (h = 20 W/(m^2 °C)) and ΔT is temperature difference between cutting tool and workpiece at the tool-chip contact.

The considered value of the convection heat transfer coefficient h for air flow was determined by approximation according to the following equation [18]:

$$h = 10.45 - v + 10\, v^{1/2} \tag{4}$$

where v is the relative velocity between object surface and surrounding air (m/s).

The boundary conditions were stated as that (i) there was no motion for the cutting tool and that (ii) the workpiece was cut at the specified machining speed (V).

For material modeling, the Johnson–Cook constitutive model was adopted, because it gives accurate results for machining process simulations. The flow stress model can be formulated as follows [12]:

$$\overline{\sigma} = [A + B(\varepsilon^n)]\left[1 + C\cdot \ln\left(\frac{\dot{\varepsilon}}{\varepsilon_0}\right)\right]\left[1 - \frac{(T - T_{room})}{(T_{melt} - T_{room})^m}\right] \quad (5)$$

where $\overline{\sigma}$ is the cutting stress, ε is the plastic strain, $\dot{\varepsilon}$ (s^{-1}) is the plastic strain rate, $\ddot{\varepsilon}_0$ ($^{-1}$) is the reference plastic strain rate, T (°C) is the workpiece temperature, T_{melt} (°C) is the melt temperature, T_{room} (°C) is the room temperature and A (MPa), B (MPa), C, n and m are the initial yield strength, the hardening modulus, strain rate sensitivity coefficient, hardening coefficient and thermal softening coefficient, respectively. The constants of AISI 1045 steel that were used for Johnson–Cook model are displayed in Table 3.

Table 3. The constants of Johnson–Cook material model for AISI 1045 steel [19]: (Jaspers et al., 1998).

Material	A (MPa)	B (MPa)	n	C	m	T_m (°C)
AISI 1045	553.1	600.8	0.234	0.013	1	1733

For modeling tool wear, the Usui wear model was used by the DEFORM software, based on nodal displacement which is characterized by the calculated wear rate at each node. For the tool wear simulation, the Usui model was expressed as follows [20]:

$$W = \int A\cdot P\cdot V\cdot e^{-\frac{B}{T}}\cdot dt \quad (6)$$

where W is the tool wear, P is the interface pressure, V is the sliding velocity, T is the temperature and dt is the time increment. A and B are constants that are determined experimentally.

2.3. Design of Experiments

The orthogonal cutting numerical tests were realized to investigate the interactive effects of tool chamfer angle, chamfer width and sharp cutting angle on the temperature, stress and tool wear depth under different cutting process parameters. Tables 4 and 5, respectively, present the design plan for the chamfer and sharp tool geometry numerical simulations.

Table 4. Parametric design plan for the chamfer tool geometry numerical simulations.

Test ID	Chamfer Angle γ_n (°)	Cutting Speed (m/min)	Feed Rate (mm/rev)	Chamfer Width w_n (mm)
1	10	150	0.2	0.1–0.75
2	10	350	0.2	0.1–0.75
3	10	500	0.2	0.1–0.75
4	10	600	0.2	0.1–0.75
5	10	150	0.1	0.1–0.75
6	10	150	0.2	0.1–0.75
7	10	150	0.3	0.1–0.75
8	15–45	150	0.1	0.1
9	15–45	150	0.2	0.1
10	15–45	150	0.3	0.1
11	15–45	150	0.2	0.1
12	15–45	350	0.2	0.1
13	15–45	500	0.2	0.1
14	15–45	600	0.2	0.1

Table 5. Parametric design plan for the sharp tool geometry numerical simulations.

Test ID	Sharp Tool Angle θ_t (°)	Cutting Speed (m/min)	Feed Rate (mm/rev)
1	35–65	250	0.2
2	35–65	500	0.2
3	35–65	750	0.2
4	35–65	1000	0.2
5	35–65	150	0.1
6	35–65	150	0.2
7	35–65	150	0.3

The first numerical simulation tests were conducted to study only the effects of chamfer and sharp tool geometry on the different machining characteristics. For the chamfer tool design, the chamfer width (w_n) was varied and the other machining and tool parameters were kept constant. Then, chamfer angle (γ_n) was varied, with the other machining and tool parameters kept constant. For the sharp tool design, θ_t was varied and the other parameters were kept constant.

Throughout the study, the clearance angle (α) was kept constant at α = 15°. For the chamfer tool, the rake angle (γ) was kept constant (γ = 10°) for different tested chamfer widths. For the sharp tool, the rake angle (γ_c) changes when the sharp angle (θ_t) varies.

In order to illustrate the influence of chamfer width, chamfer angle and sharp angle on cutting temperature, effective stress and tool wear, four tool designs were proposed. Figures 3–5 display the schematic representations of the cutting tool geometries with different chamfer widths chamfer angles and sharp angles, respectively.

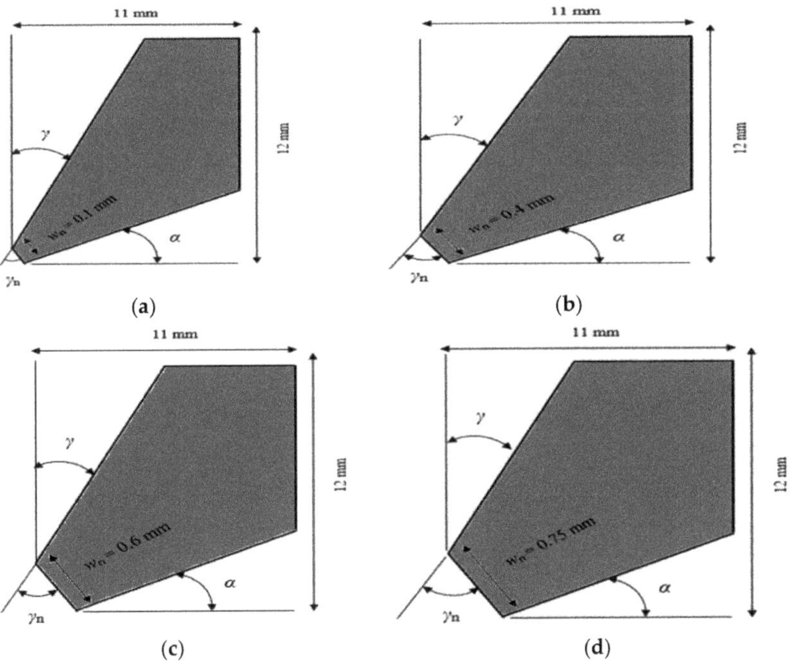

Figure 3. Chamfer cutting tool geometries with different chamfer widths w_n: (**a**) 0.1 mm, (**b**) 0.4 mm, (**c**) 0.6 mm, (**d**) 0.75 mm.

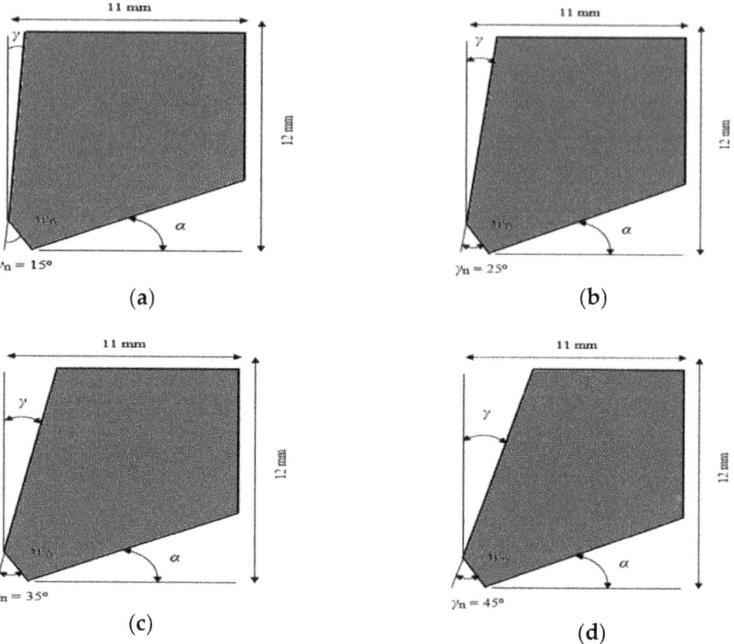

Figure 4. Chamfer cutting tool geometries with different chamfer angles γ_n: (**a**) 15°, (**b**) 25°, (**c**) 35°, (**d**) 45°.

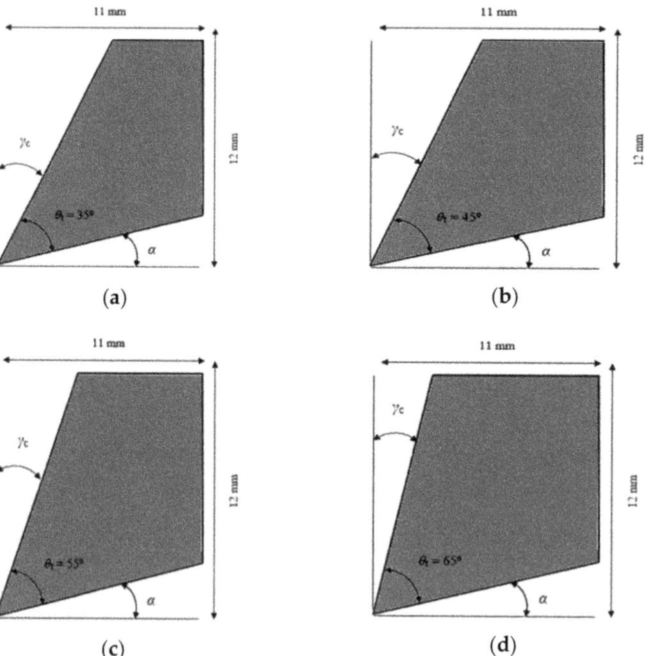

Figure 5. Sharp cutting tool geometries with different sharp angles θ_t: (**a**) 35°, (**b**) 45°, (**c**) 55°, (**d**) 65°.

3. Results and Discussion

3.1. Influence of Tool Geometry on Machining Process Performance Indicators

The simulated machining characteristics were cutting force, temperature, stress and tool wear. The transient mode was selected in order to evaluate the evolution of forces, temperature and stress in the model with increments of time. The machining force results comprise cutting forces and thrust forces. Figure 6 shows the variation of these forces as a function of time for chamfer and sharp geometries, respectively, at V = 150 m/min, w_n = 0.10 mm, f = 0.2 mm/rev, γ_n =10° and at θ_t = 45°.

The numerical calculation of the machining force was carried out in the plane defined by the cutting direction and the one perpendicular to it. Thus, the cutting and thrust forces were obtained in the cutting direction and the thrust force in the direction perpendicular to the cutting one. For every numerical test, the simulated forces reached the steady state rapidly as shown in Figure 6. The total workpiece length set took about 1 ms to be completed; that is why the drop of forces is noted at the end of the cutting process.

It may be seen from Figure 6 that for both the chamfer and sharp geometries, the cutting forces are higher than the thrust forces because the cutting forces are applied in the machining direction and represent 70 to 80% of the total force, which contributes to determining the global power necessary for performing the machining process [21]. Additionally, the cutting and thrust forces for the chamfer tool (apparent negative rake angle) are higher than those for the sharp tool.

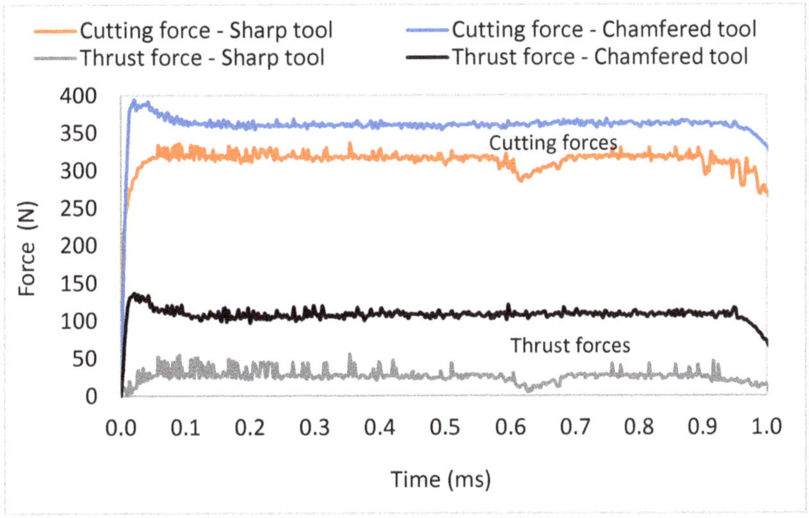

Figure 6. Variation in force versus time during cutting process for chamfer tool geometry at V = 150 m/min, w_n = 0.1 mm, f = 0.2 mm/rev, γ_n = 10 ° and sharp tool geometry θ_t = 45°.

Figure 7 shows the variation of the workpiece and tool temperatures in the cutting zone as a function of the cutting time. The steady state for the temperature is reached at about 0.4 ms, depending on the cutting parameters used. That is to say, the maximum temperature is attained at about half of the sample length (30 mm) when using a sharp or chamfered edge tool. The tool temperature was higher when using a chamfered tool as compared to that obtained with a sharp tool; this can be explained by the higher forces and stresses acting on the tool when using a chamfered tool and also by the high deformation accompanying the cutting process with negative effective rake angle near the tool tip.

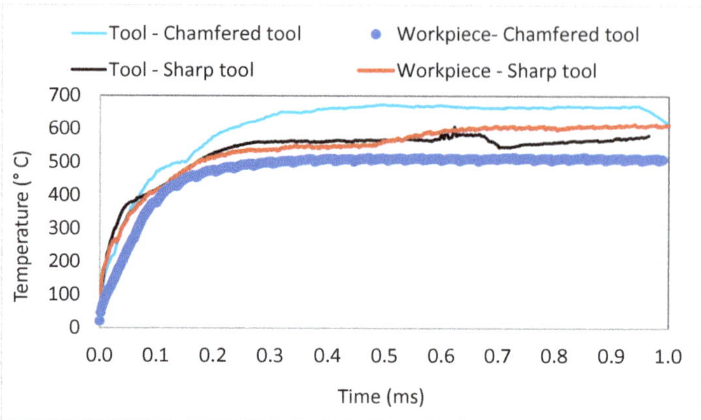

Figure 7. Variation in maximum temperature versus time during cutting process for chamfer tool geometry at V = 150 m/min, w_n = 0.1 mm, f = 0.2 mm/rev, γ_n = 10 ° and sharp tool geometry with θ_t = 45°.

Figure 8 shows the distribution of cutting temperature in the model at V = 150 m/min, f = 0.2 mm/rev, γ_n =10° for different chamfer widths. From Figure 8 it may be noted that the cutting temperature increases when the chamfer width increases. For chamfer widths from 0.1 mm to 0.75 mm, the increase in cutting temperature is around 10%. The higher temperatures were located in the chip-tool contact zone due to the higher heat generated in this area. Additionally, the more the chamfer width increased, the more the maximum temperatures were pronounced in this zone, extending to the rake face.

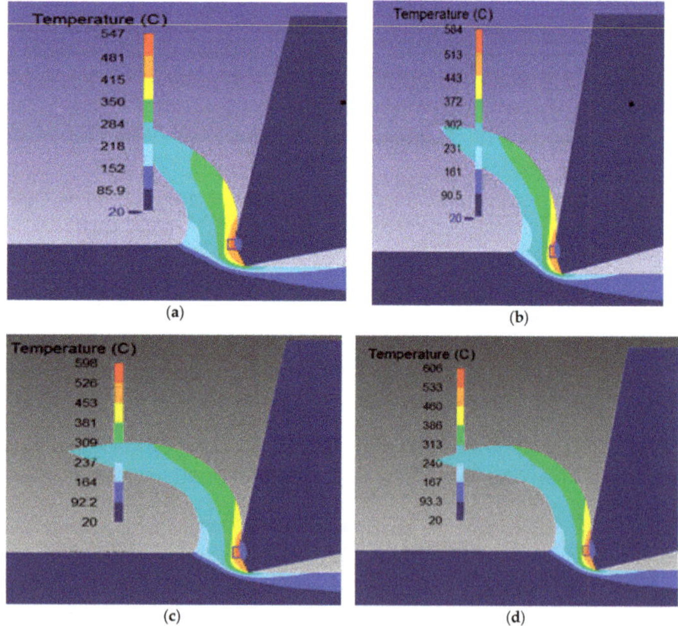

Figure 8. Distribution of cutting temperature in the model at V = 150 m/min, f = 0.2 mm/rev, γ_n = 10° for the chamfer widths w_n: (**a**) 0.1 mm, (**b**) 0.4 mm, (**c**) 0.6 mm, (**d**) 0.75 mm.

Figure 9 shows the distribution of effective stress in the model at V = 150 m/min, f = 0.2 mm/rev and γ_n =10° for different chamfer widths. It may be seen from the figure that the effective stress increases when the chamfer width increases. This result shows that a higher chamfer width produces a larger contact stress with the workpiece [22]. Changing the chamfer width from 0.1 mm to 0.75 mm produced a small increase in cutting stress of about 3%. The cutting stresses were maximum in the shear plane. Tagiuri et al. [15] also observed this result previously.

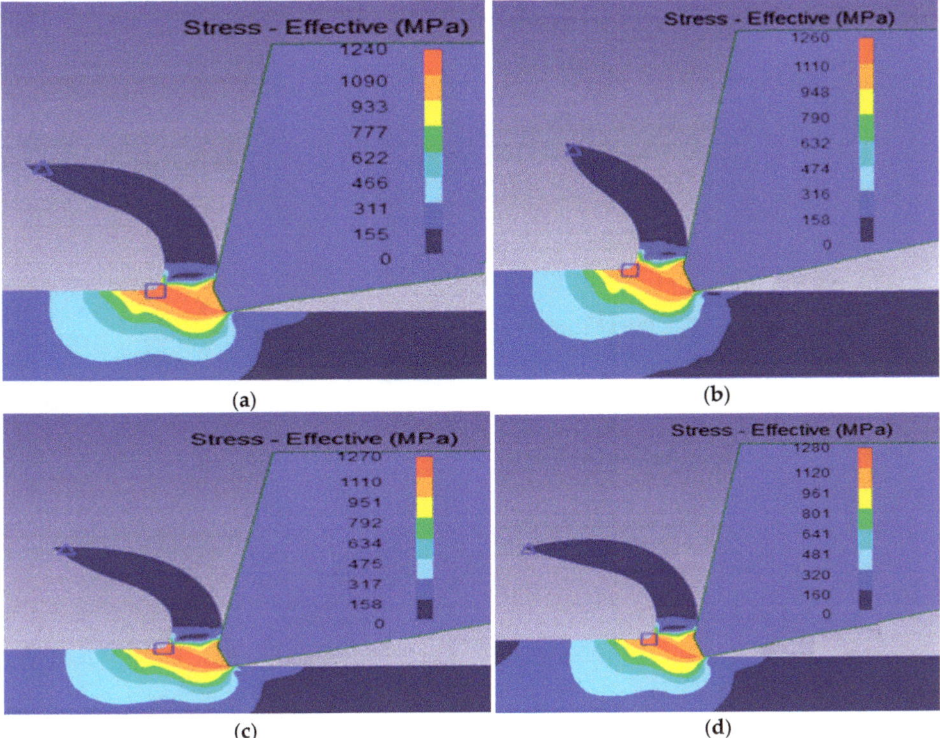

Figure 9. Distribution of effective stress in the model at V = 150 m/min, f = 0.2 mm/rev and γ_n = 10° for the chamfer widths w_n: (**a**) 0.1 mm, (**b**) 0.4 mm, (**c**) 0.6 mm, (**d**) 0.75 mm.

Figure 10 shows the distribution of wear depth for different chamfer widths during the cutting simulation at V = 150 m/min, f = 0.2 mm/rev and γ_n =10 °. Due to the microscale nature of the obtained total wear, the images were zoomed in on the chamfer width area in order to show clearly the wear depth zone. From Figure 10, it is observed that the tool wear depth increases with increasing chamfer width. Increasing from a chamfer width of 0.1 mm to 0.4, 0.6 and 0.75 mm, the increases in tool wear depth were 74%, 98% and 11%, respectively. While there was a large increase from 0.1 mm to 0.6 mm, the increase noted was small from 0.6 mm to 0.75 mm. The tool wear is higher on the chamfer width of the tool at the tool–chip contact area. These variations are due to the pressure balance applied in this area, caused by the opposing effects [4]. Otherwise, while chamfer width increases from 0.1 to 0.6 mm, the increase in pressure increases the wear depth. From 0.6 to 0.75 mm, the pressure decreases and consequently the wear depth is reduced.

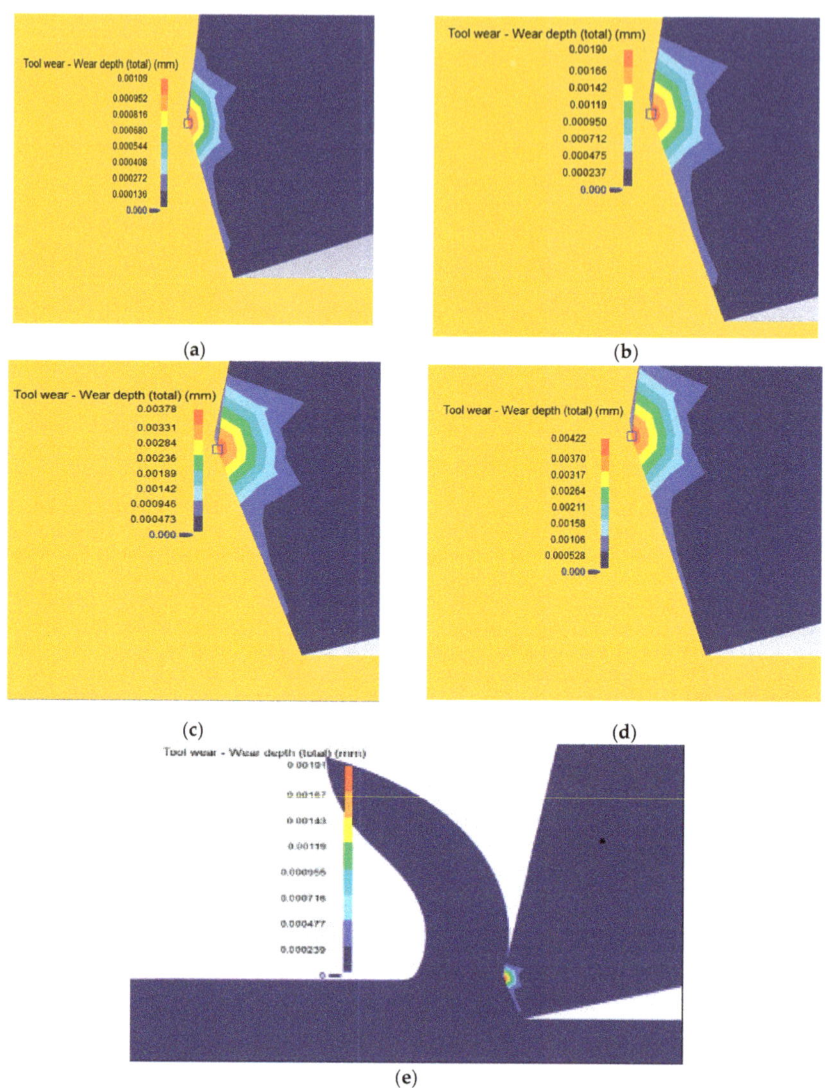

Figure 10. Distribution of tool wear in the model at V = 150 m/min, f = 0.2 mm/rev and γ_n =10° for the chamfer widths w_n: (**a**) 0.1 mm, (**b**) 0.4 mm, (**c**) 0.6 mm, (**d**) 0.75 mm, (**e**) zoomed of (**b**).

Figure 11 shows the distribution of cutting temperature for different chamfer angles during the cutting simulation at V = 150 m/min, f = 0.3 mm/rev and w_n = 0.1 mm. From Figure 11, it can be noted that the cutting temperature increases with increasing chamfer angle. The increase is approximately 3% in the range of 517 ± 13 °C for the maximum temperature. From chamfer angle 25 to 35°, the cutting temperature remains nearly constant. Additionally, the more the chamfer angle increases, more the rake angle decreases. This creates longer contact between the chip and cutting tool, which produces much heat in these locations [4,17].

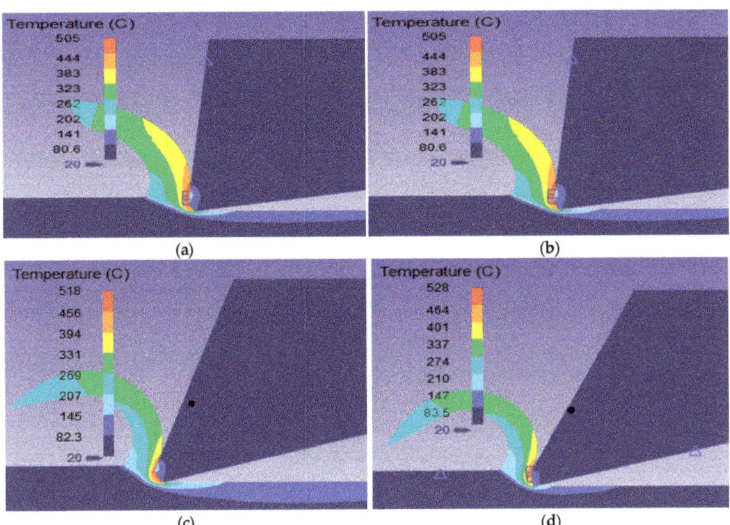

Figure 11. Distribution of cutting temperature in the model at V = 150 m/min, f = 0.3 mm/rev and w_n = 0.1 mm for the chamfer angles γ_n: **(a)** 15°, **(b)** 25°, **(c)** 35°, **(d)** 45°.

Figure 12 shows the distribution of effective stress for different chamfer angles during the cutting simulation at V = 150 m/min, f = 0.3 mm/rev and w_n = 0.1 mm. An examination of the figure shows that the effective stress increases when the chamfer angle increases from 15 to 25° and from 35 to 45° but it decreases from 25 to 35° as more thrust forces are applied than cutting force in these cases [17]. The variation in cutting stress is around 2%. Thus, it is reasonable to conclude that the chamfer angle does not have a significant effect on the cutting stress.

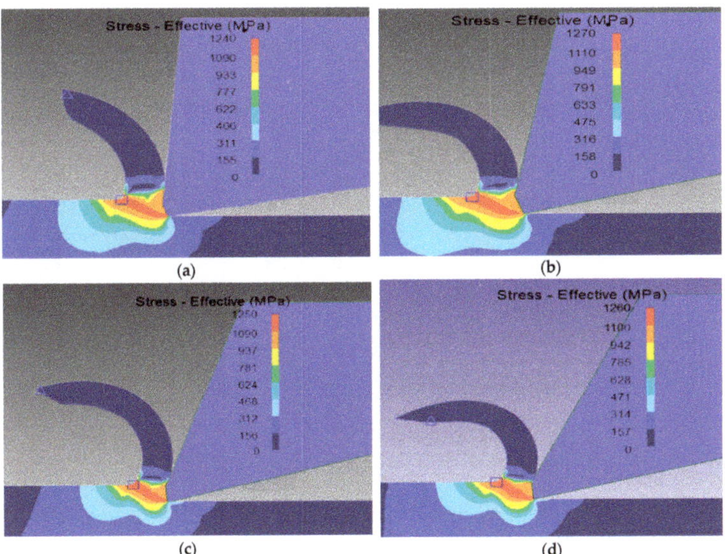

Figure 12. Distribution of effective stress in the model at V = 150 m/min, f = 0.3 mm/rev and w_n = 0.1 mm for the chamfer angles γ_n: **(a)** 15°, **(b)** 25°, **(c)** 35°, **(d)** 45°.

Figure 13 illustrates the distribution of cutting temperature in the model at V = 150 m/min, f = 0.3 mm/rev for different sharp angles. As may be seen, the cutting temperature increases with increasing sharp angle. The rate of increase is approximately 2 to 5% in the range of 426 ± 15 °C for the maximum temperature. It is also observed that the more the sharp angle increases, the more the rake angle decreases, which creates greater contact between the chip and the cutting tool; this results in much heat in this area. The maximum temperatures are located on the contact area between the chip and rake face. The more the sharp angle increases, the more the maximum temperature is pronounced in this zone, extending to the chip as well.

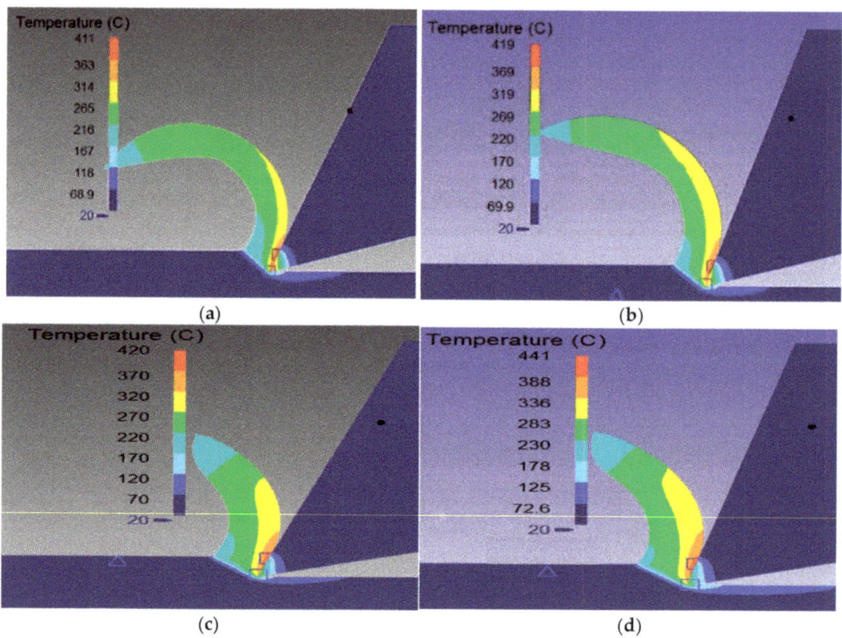

(a) (b) (c) (d)

Figure 13. Distribution of cutting temperature at V = 150 m/min and f = 0.3 mm/rev for the sharp angles θ_t: (**a**) 35°, (**b**) 45°, (**c**) 55°, (**d**) 65°.

Figure 14 shows the distribution of effective stress for different sharp tool angles during the cutting simulation at V = 150 m/min and f = 0.3 mm/rev. From Figure 13, it is noted that there is a slight increase in cutting stress when moving from a sharp angle of 45° to 65°. The increase is around 1% for maximum stress. For the values from 35° to 45°, however, the cutting stress variation is about 4%.

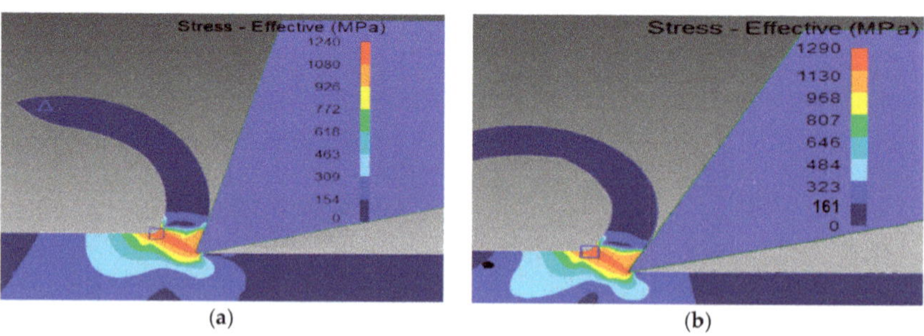

(a) (b)

Figure 14. *Cont.*

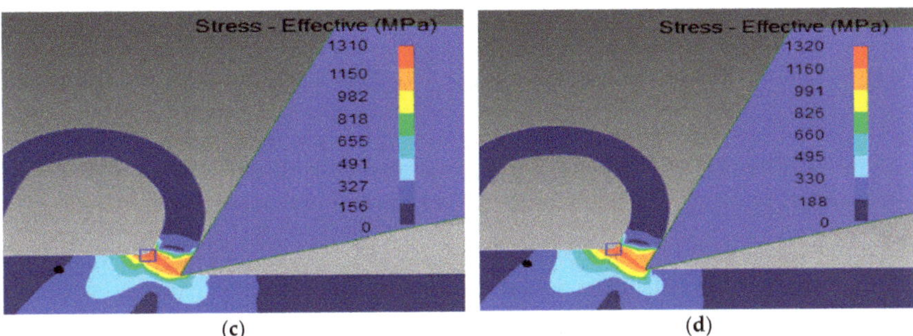

(c) (d)

Figure 14. Distribution of effective stress in the model at V = 150 m/min and f = 0.3 mm/rev for the sharp angles θ_t: (**a**) 35°, (**b**) 45°, (**c**) 55°, (**d**) 65°.

Figure 15 illustrates the distribution of tool wear depth for different tool sharp angles during the cutting simulation at V = 150 m/min and f = 0.3 mm/rev. From Figure 15 it is observed that wear depth increases with increasing sharp angle and chamfer angle. For the sharp tool, the tool wear occurred on the rake face, whereas it occurs on the chamfer width for the chamfer tool. As discussed earlier, the more the tool angle (sharp or chamfer angle) increases, the more the rake angle decreases, which creates longer contact between chip and cutting tool which presses the chips against the cutting tool in the contact area [12]. Hence, both the rake face of the sharp tool and the corner at the chamfer width become more stressed, which causes significant wear at these respective locations. Furthermore, tool wear depths for the sharp tool were observed to be higher than those for the chamfer tool, due to the lower rake angles created.

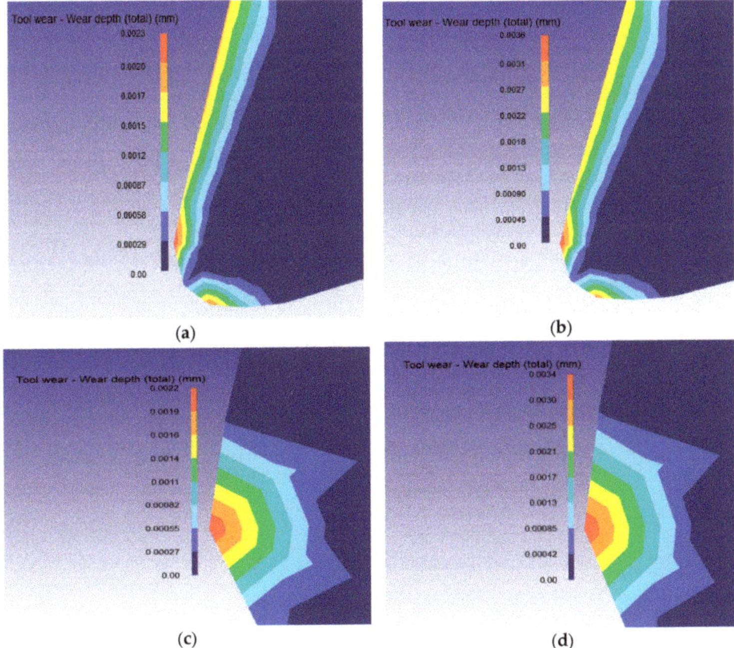

(a) (b)

(c) (d)

Figure 15. Distribution of tool wear depth in the model at V = 150 m/min and f = 0.3 mm/rev, for the cutting tool angles: (**a**) sharp edge; θ_t = 45°, (**b**) sharp edge; θ_t = 55°, (**c**) chamfered edge; γ_n = 25°, (**d**) chamfered edge; γ_n = 35°.

3.2. Analysis of Variance (ANOVA)

In order to identify the main parameters that influence the different machining performance characteristics, such as temperature, effective stress, chip thickness and tool wear, it is necessary to study statistically the above-investigated interactive effects. This statistical study consists of performing an analysis of variance (ANOVA) test based on the Taguchi design method, using the Statgraphics Centurion software in our case. The design of experiments (DOE) developed consisted in selecting four factors with different levels. Two types of DOEs were designed for the chamfer tool geometry and the sharp tool geometry, respectively. Tables 6 and 7 present the two designs of experiments for the two tool-type geometries, respectively.

Table 6. Design of experiments for the chamfer tool.

Factors	Level 1	Level 2	Level 3	Level 4
V: Cutting speed (m/min)	250	350	500	600
f: Feed rate (mm/rev)	0.1	0.2	0.3	-
w_n: Chamfer width (mm)	0.15	0.25	0.35	0.45
γ_n: Chamfer angle (°)	15	25	35	45

Table 7. Design of experiments for the sharp tool.

Factors	Level 1	Level 2	Level 3	Level 4
V: Cutting speed (m/min)	250	350	500	600
f: Feed rate (mm/rev)	0.1	0.2	0.3	-
θ_t: Sharp angle (°)	35	45	55	65

Accordingly, two ANOVA analyses were carried out, one for the chamfer tool and the other for the sharp tool to obtain the response parameters of temperature and stress. The significant parameters were identified using the condition that the p-values must be less than 0.05 at the 95.0% confidence level. Several models were analyzed and only those statistically valid at the 95% confidence level were retained.

3.2.1. ANOVA for Chamfer Tool

The obtained results showed that the feed rate, the cutting speed and their interactions were the most significant parameters affecting temperature, stress and wear depth. In addition, it can be noted that the interactions between the cutting speed and chamfer width and that between the cutting speed and chamfer angle are also most significant for temperature.

A first general linear model was analyzed and the non-statistically significant terms or interactions were removed for the analysis. Table 8 summarizes the final ANOVA parameters retained for the temperature when using a chamfered tool. The cutting speed was the most influential factor (F-value = 80.74), followed by the chamfer width (F-value = 42.03). The cutting speed × feed rate interaction exhibits the most important term with an F-value of 18.81.

Table 8. Analysis of variance for temperature as a function of cutting speed (V), feed rate (f) and tool geometry (γ_n and w_n) when using a chamfered tool.

Source	Df	Sum of Square	Mean Square	F-Value	p-Value
Model	7	2.974248×10^7	4.24891×10^6	833.85	0.0000
V: Cutting speed	1	411,429	411,429	80.74	0.0000
f: Feed rate	1	57,900.1	57,900.1	11.36	0.0016
w_n: Chamfer width	1	50,961.4	50,961.4	10.00	0.0029
γ_n: Chamfer angle	1	214,141	214,141	42.03	0.0000
$V \times f$	1	95,862.4	95,862.4	18.81	0.0001
$V \times \gamma_n$	1	74,028.0	74,028.0	14.53	0.0005
$w_n \times \gamma_n$	1	56,705.8	56,705.8	11.13	0.0018
Residual Error	41	208,918.0	5095.56		
Total	48	2.99513×10^7			

Table 9 summarizes the ANOVA results of the stress on the tool when machining with a chamfered tool. The tool chamfer angle did not show a statistically significant influence on the result (for the 95% confidence interval considered). This was also the case for the interactions between the chamfer angle and the other factors: cutting speed, feed rate and tool chamfer width. Therefore, this factor and its interactions were removed from the analysis. It can thus be seen from Table 9 that the stress on the cutting tool is influenced mainly by the cutting speed, the chamfer width and feed rate, plus their interactions.

Table 9. Analysis of variance for stress as a function of cutting speed (V), feed rate (f) and tool edge chamfer with (w_n).

Source	Df	Sum of Square	Mean Square	F-Value	p-Value
Model	6	8.34658×10^7	1.3911×10^7	1401.19	0.0000
V: Cutting speed	1	1.0384×10^6	1.0384×10^6	104.59	0.0000
f: Feed rate	1	495,304	495,304	49.89	0.0000
w_n: Chamfer width	1	674,986	674,986	67.99	0.0000
$V \times f$	1	170,323	170,323	17.16	0.0002
$V \times w_n$	1	252,041	252,041	25.39	0.0000
$f \times w_n$	1	107,954	107,954	10.87	0.0020
Residual Error	42	416,974	9927.94		
Total	48	8.38828×10^7			

3.2.2. ANOVA for Sharp Tool

The performance of the sharp tool was analyzed as a function of the cutting parameters (cutting speed and feed rate) and tool angle (θ_t). Table 10 summarizes the results of analysis of variance obtained for the temperature, considering a general linear model and before removing non-statistically significant factors and terms. The tool angle and its interactions with cutting and feed rate were not statistically significant at the 95% confidence interval, and hence these terms were removed from the final model.

Table 10. Analysis of variance for temperature as a function of cutting speed (V), feed rate (f) and sharp tool angle (θ_t)—Original model before suppression of factors.

Source	Df	Sum of Square	Mean Square	F-Value	p-Value
Model	6	722,164	120,361.0	52.94	0.0000
V: Cutting speed	3	22,731.0	22731	10	0.0029
f: Feed rate	2	36,024.6	36,024.6	15.84	0.0003
θ_t: Sharp tool angle	3	1331.57	1331.57	0.59	0.4485
$V \times f$	6	9055.13	9055.13	3.98	0.0526
$V \times \theta_t$	9	1564.35	1564.35	0.69	0.4116
$f \times \theta_t$	6	0.00110	0.00110	0	0.9994
Residual Error	41	93,221.5	2273.7		
Total	47	815,385.0	-		

The cutting speed, the feed rate and the cutting speed x feed rate interaction were the most significant terms. This was confirmed by Qasim et al. [23] during the turning of AISI 1045 steel using carbide cutting sharp tools. They reported that the cutting speed and the feed rate are the most significant factors for cutting temperature and cutting force, respectively.

For the stress on the cutting tool, the results of the analysis of variance are presented in Table 11. Once again, the tool angle and its interactions with cutting speed and feed rates were found to be not statistically significant when using a general linear model with a constant. Some effects were found, however, when considering a general linear model without a constant, as will be seen in the equations and response surface analysis presented in the next section.

Table 11. Analysis of variance for stress as a function of cutting speed (*V*), feed rate (*f*) and sharp tool angle (θ_t).

Source	Df	Sum of Square	Mean Square	F-Value	*p*-Value
Model	6	7.55473×10^7	1.2591×10^7	3392.62	0.0000
V: Cutting speed	1	329,927	329,927	88.9	0.0001
f: Feed rate	1	167,344	167,344	45.09	0.0001
θ_t: Sharp tool angle	1	604,873	604,873	162.98	0.00001
$V \times f$	1	42,829.7	42,829.7	11.54	0.0015
$V \times \theta_t$	1	129,005	129,005	34.76	0.00001
$f \times \theta_t$	1	50,784.1	50,784.1	13.68	0.0006
Residual Error	42	155,877	2273.7		
Total	48	7.57×10^7	-		

In order to show the most influencing parameters that affect the cutting temperature and the effective stress during orthogonal cutting, the above results can be explicitly given in terms of the effect order. Table 12 summarizes the influential parameters in terms of importance—with 1 being most influential and 4 being the least. It may be seen that the cutting speed and feed rate are the most influential parameters affecting the different machining characteristics studied.

Table 12. Summary of most influential parameters in terms of importance (1 = most, 4 = least).

Parameters	Chamfer Tool		Sharp Tool	
	Temperature	Stress	Temperature	Stress
V: Cutting speed	1	1	2	2
f: Feed rate	-	3	1	3
*T**: Tool geometry	2	2	-	1
$V \times f$: Interaction Speed × Feed rate	3	3	3	4
$V \times T$: Interaction speed × Tool geometry	4	4	-	
$f \times T$: Interaction feed × Tool geometry		-		

*T**: Tool geometry: Chamfer width (w_n) or chamfer angle (γ_n); sharp tool angle for sharp tool (θ_t).

3.3. Response Surface Analysis

The results obtained from the ANOVA tests may be expressed through mathematical and statistical equations using regression models to establish response functions fitting the available data. The following empirical models were obtained from statistical analysis:

For the chamfer cutting tool, the regression equations for temperature and stress are presented in Equations (7) and (8):

$$\text{Temperature} = 1.467\ V + 1172 f + 719.78\ w_n + 16.463\ \gamma_n - 3.490\ Vf - 0.0208\ V\gamma_n - 23.635\ w_n\gamma_n \quad R^2 = 99\% \tag{7}$$

$$\text{Stress} = 2.21154\ V + 3565.15 f + 2713.88\ w_n - 4.53194\ Vf - 3.84915\ V w_n - 4596.42 f w_n \quad R^2 = 99.5\% \tag{8}$$

For the sharp cutting tool, the regression equations of temperature and stress predicted from statistical analysis are given in Equations (9) and (10). The analysis of the wear did not show statistical significance at the 95% confidence interval.

$$\text{Temperature} = 229 + 0.650\ V + 1867 f - 1.25\ Vf \quad R^2 = 88\% \tag{9}$$

$$\text{Stress} = 1.902\ V + 2993.72 f + 17.763\ \theta_t - 2.5146\ Vf - 0.02486\ V\theta_t - 30.06 f\theta_t \quad R^2 = 99\% \tag{10}$$

The general linear model for the estimation of the stress for the sharp tool gave a coefficient of correlation that was not acceptable (68%). For this reason, a model with no constant was preferred. The predicted results shown in Equations (7)–(10) fitted well with the results obtained by simulation as displayed in Figures 16 and 17.

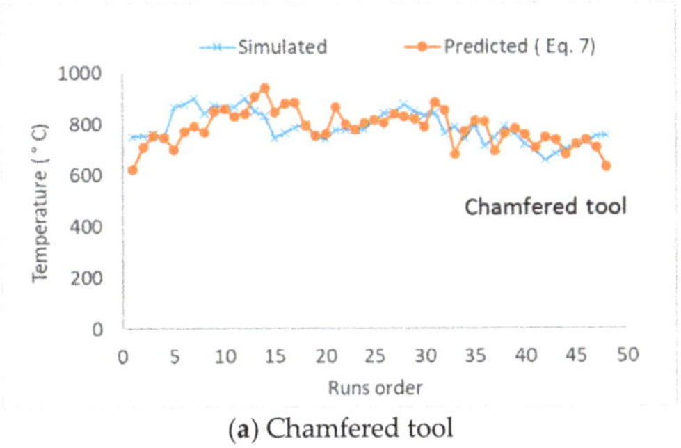

(**a**) Chamfered tool

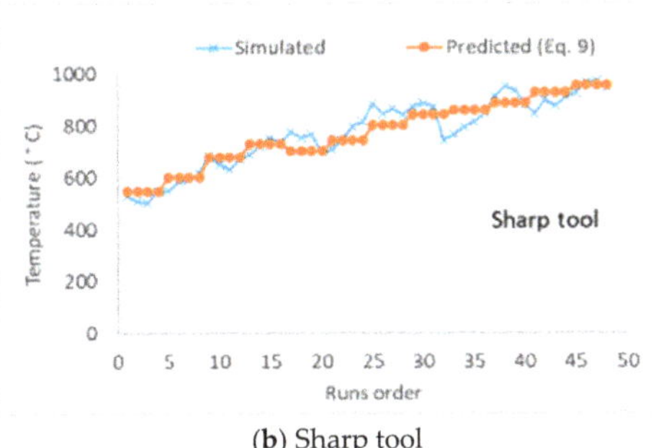

(**b**) Sharp tool

Figure 16. Comparison of predicted and simulated temperatures: (**a**) chambered tool and (**b**) sharp tool.

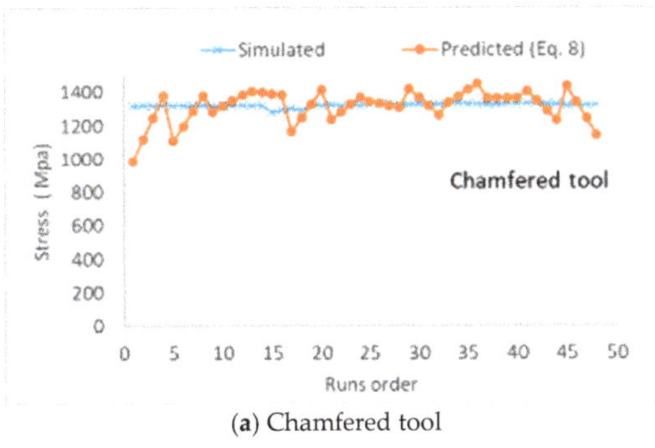

(**a**) Chamfered tool

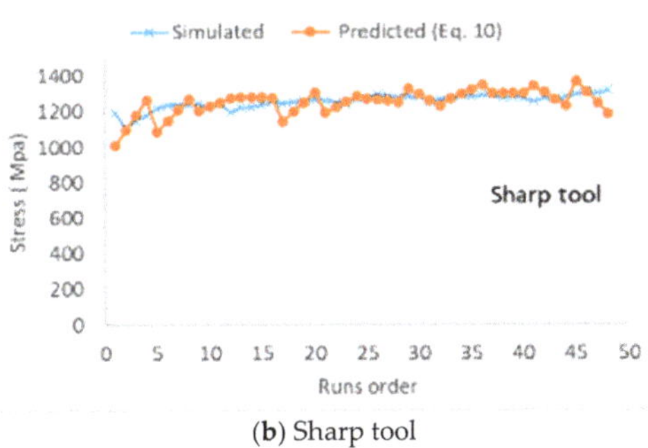

(**b**) Sharp tool

Figure 17. Comparison of predicted and simulated stress: (**a**) chambered tool and (**b**) sharp tool.

In order to predict the unknown response values and the corresponding cutting conditions, 3D statistical graphs or 3D surface plots were produced to interpret the effects of the continuous parameters on the response values based on the fitted model. Figure 18 shows the surface plots of the parametric effects on temperature for the chamfered cutting tool. This figure shows the interaction between the chamfered tool geometry and the machining parameters (cutting speed and feed rate) on the cutting temperature. For low chamfer angles (15° and 20°; Figure 18a,b), the maximum temperature occurs at a higher cutting speed and low feed rate setting (at speed about 650 m/min and a feed rate of about 0.1 mm/rev). At a higher value of chamfer angle (40°; Figure 18d), the maximum values of cutting temperatures occurred in irregular settings of cutting speed and feed rates as depicted in Figure 18d. This is due to the interactive effects between the tool edge preparation and the cutting parameters, as also noted in Equation (8). As both the chamfered tool width (w_n) and the chamfered angle (γ_n) influenced the temperature (Equation (8)), a 3D effect of cutting speed, feed rate and chamfer width (w_n) on the cutting temperature was also analyzed and is presented in Figure 19. It is seen from Figure 19 that the effect of chamfer width (w_n) on temperature is moderate, as compared to those of the cutting speed and feed rate.

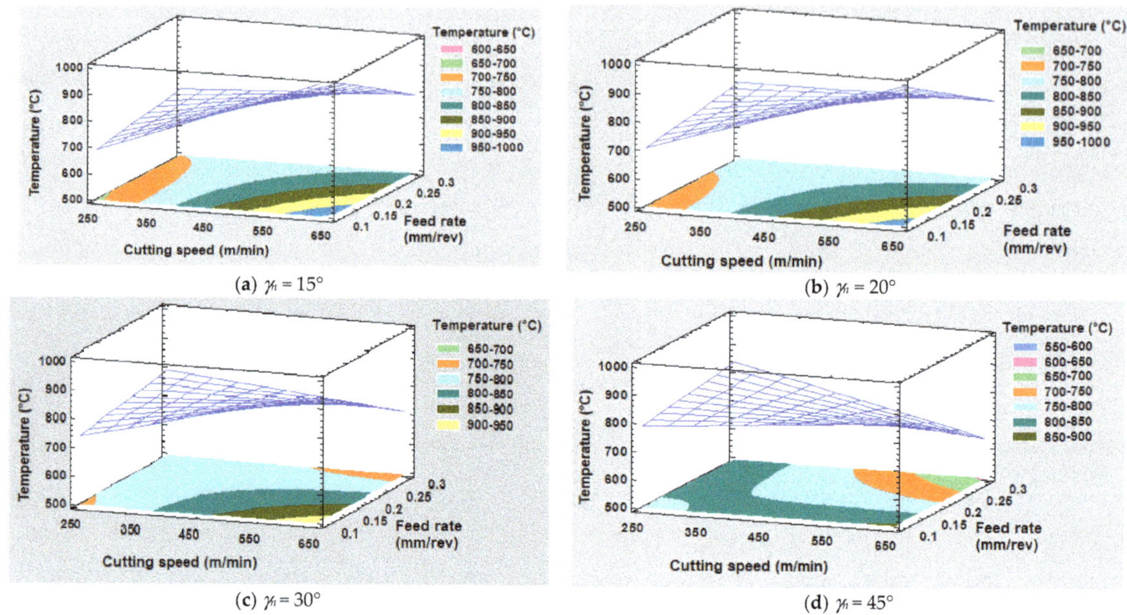

Figure 18. 3D Surface plots of parametric effects of cutting speed, feed rate and chamfered tool angle (γ_n) on temperature (°C) when using a chamfered tool (w_n = 0.34 mm).

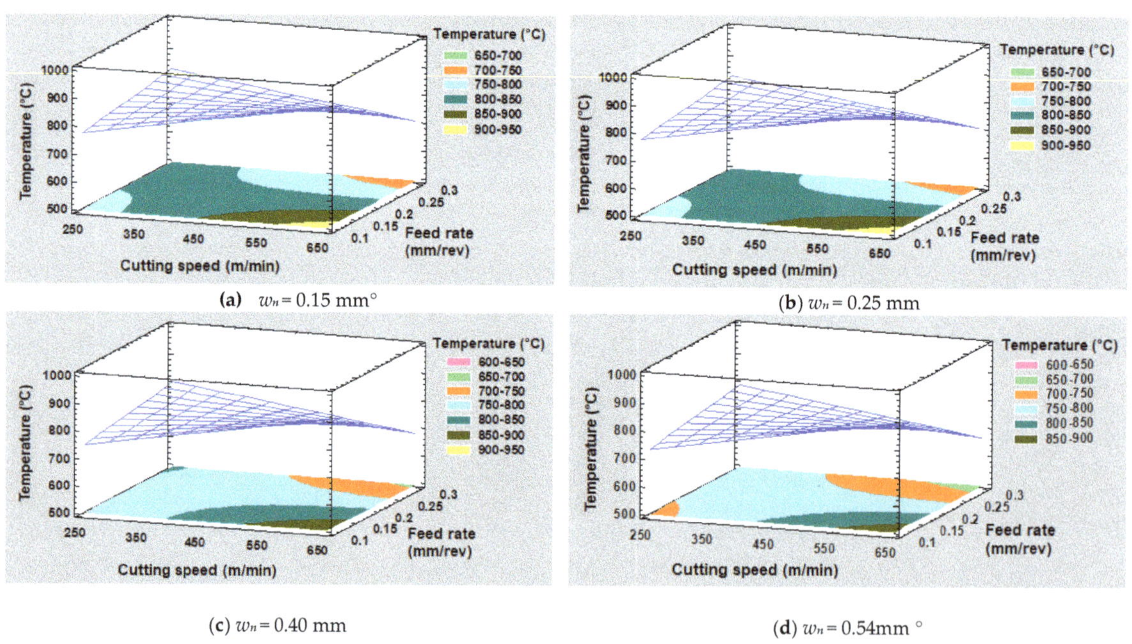

Figure 19. 3D Surface plots of parametric effects of cutting speed, feed rate and chamfered tool length (w_n) on temperature (°C) when using a chamfered tool (γ_n = 35°).

A 3D surface plot of cutting speed and feed rate on the cutting stress on chamfered tool is presented in Figure 20. This stress is mostly influenced by the cutting speed and the feed rate and, in second order, by the chamfer width and its interactions with cutting speed and feed rate (see Equation (9)). For chamfer width below 0.25 mm, the maximum stress takes place at high cutting and low feed rate (Figure 20a,b), while for higher values of chamfer width (0.35 mm and 0.50 mm), the maximum stress occurs at high feed rate and low cutting speed but also at high cutting speed and low feed rate settings. The stresses are higher for the large feed rate, as the chip section is thicker and the forces high.

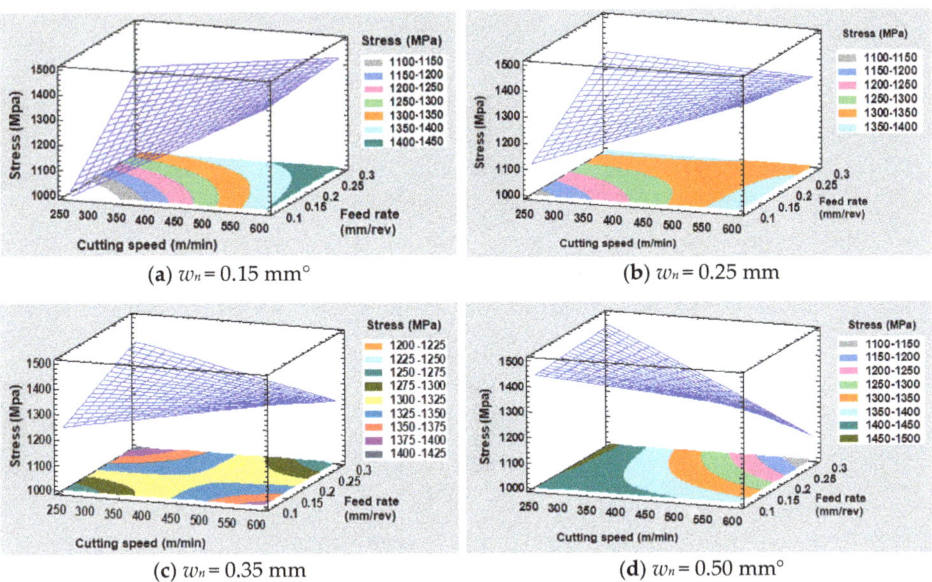

(**a**) $w_n = 0.15$ mm° (**b**) $w_n = 0.25$ mm

(**c**) $w_n = 0.35$ mm (**d**) $w_n = 0.50$ mm°

Figure 20. 3D Surface plots of parametric effects of cutting speed, feed rate and chamfered tool length (w_n) on the stress when using a chamfered tool ($\gamma_n = 35°$).

For a sharp tool, the maximum temperature occurred when using a high feed rate and a high cutting speed (Figure 21a,b) independently of the sharp tool angle used. At this setting, the stress on the cutting tool is high (Figure 22a,b).

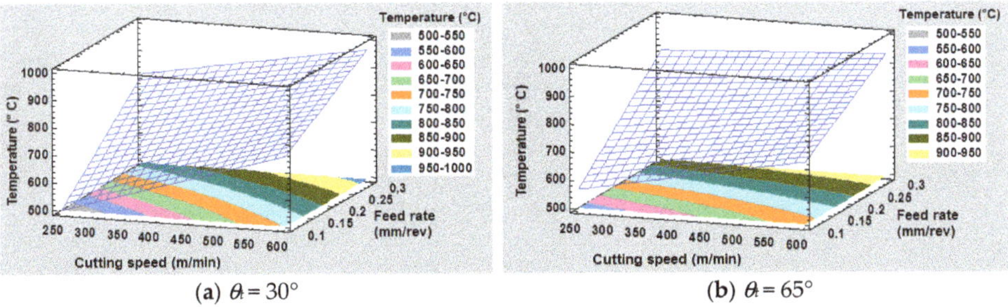

(**a**) $\theta_t = 30°$ (**b**) $\theta_t = 65°$

Figure 21. 3D Surface plots of parametric effects of cutting speed, feed rate and sharp tool angle (θ_t) on the temperature (°C) when using a sharp tool ($\gamma_n = 35°$).

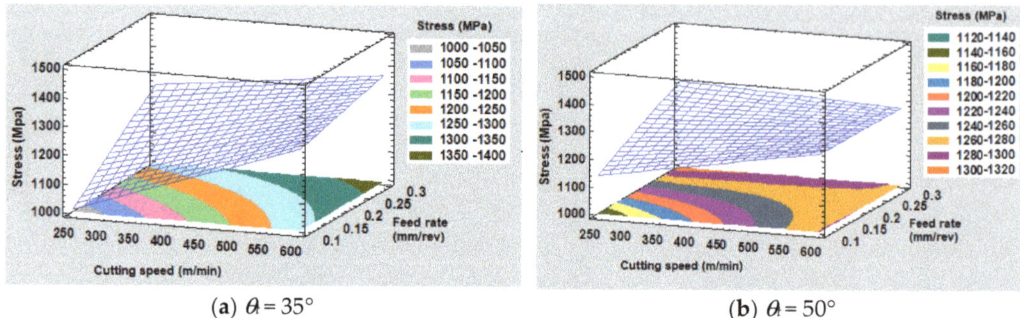

(**a**) $\theta_t = 35°$ (**b**) $\theta_t = 50°$

Figure 22. 3D Surface plots of parametric effects of cutting speed, feed rate and sharp tool angle (θ_t) on the stress (MPa) when using a sharp tool (γ_n = 35°).

4. Validation of Results

The obtained numerical results for chamfered tools were validated using the experimental data obtained (see Table 13) by Khalili and Safaei [17] who conducted a study on the effect of tool edge preparation on 1045 steel for a feed rate of 0.20 mm/rev and using chamfered carbide tools with different chamfer widths. The results of effective stress are very comparable (errors of about 4.7 to 6.9 %), while those of the temperature are slightly more different (errors varying from 10–15%).

Table 13. Comparison of numerical and experimental results for chamfer tool.

	Numerical Results (V = 250 m/min)			**Experimental Data [17] (V = 100–200 m/min)**			**Errors**		
Parameter			Chamfer width w_n (mm)						
Characteristics	0.1	0.2	0.3	0.1	0.2	0.3	0.1	0.2	0.3
Temperature (°C)	741.23	750.32	799.35	637	675	680	14.1%	10.0%	14.9%
Cutting stress (MPa)	1320.87	1325.48	1290.75	1230	1225	1230	6.9%	7.6%	4.7%

The obtained numerical and statistical analysis results obtained with chamfered tools are also in good agreement with the study by Ucun et al. [23], who conducted a similar numerical study on AISI 1045 steel with similar properties using ceramic tools and cutting speeds ranging from 100 m/min to 400 m/min and feed rates from 0.1–0.5 mm/rev and with a depth cut of 1 mm. The numerical results of [23] were validated by experimental data from the literature [4]. The experimental conditions have been described in this study [4]. Ucun et al. [23] obtained a cutting temperature of about 800 °C and they found that the tool chamfer lengths (0.1 mm–0.3 mm) and chamfer angle (5 °C–20 °C) do not have a significant effect on the cutting temperature but may increase the machining forces. Although a straight comparison cannot be made since ceramic tool has different heat transfer properties, it remains that the effect of the tool geometry (chamfer parameters) in similar machining conditions can be compared, as it follows the same trend and leads to the same conclusion as in the present study.

For orthogonal cutting of AISI 1045 with sharp carbide tool, the closest comparison we found was that of the work by Qasim et al. [24]. They compared the process performance for a large range of cutting speeds (200–630 m/min) and cutting feed rates (0.1 to 0.2 mm/rev). The results obtained in the present study are close those of Qasim et al. [24], as shown in Table 14, in which the tool angle is ignored as it did not affect our results. The differences observed could be explained by the small variations in the materials data used in the two cases.

Table 14. Comparison of numerical and experimental results for sharp tool.

	Temperature	Temperature [24]	Error
Variable feed rates (cutting speeds ranging from 200 m/min to 630 m/min)			
0.1 mm /rev	629	700	10%
0.2 mm /rev	793	765	−4%
Variable cutting speeds (feed rates ranging from 0.1to 0.2 m/rev)			
200 m/min	677.5	600	−13%
600 m/min	830	800	−4%

5. Conclusions

In the present research study, the interactive effects between chamfer width, chamfer angle, sharp angle, and cutting speed and feed rate on cutting temperature and effective stress were investigated during the milling of AISI 1045 steel using the finite element method and employing DEFORM-2D software. The statistical analysis of numerical simulation results using ANOVA was carried out in order to determine significant parameters. From the results obtained, the following points may be summarized:

- Statistical analysis showed that the effective stress and cutting temperature are mainly influenced by the cutting speed, the feed rate and their interaction for the chamfer tool and by the feed rate for the sharp tool. For the chamfer tool, the interactions between cutting speed and chamfer width and between cutting speed and chamfer angle have a significant influence on the cutting temperature. Whereas for the sharp tool, the cutting speed and the interaction between the sharp tool angle and feed rate are important. Therefore, for a given tool edge geometry, the cutting speed and the feed rate should be chosen wisely.
- The main effects of the tool edge preparation parameters tested did not show significant influence on the tested machining process performance (temperature, stress), especially for chamfered tools. The performance of these tools is mainly influenced by the cutting speed, the feed rate and their interactions. In contrast, the tool with chamfer edge led to high temperature and high effective stress as compared to the sharp edge tool.
- For the chamfer tool, the cutting speed and the tool chamfer width were found to be the most influential factors, followed by the interaction between the cutting speed and the feed rate and then by the interaction between the cutting speed and the tool chamfer width. The maximum temperature obtained on chamfered tools ranged from 700 °C to 900 °C, while the minimum was about 550 °C, depending on the feed rates and speeds used. The effective stress varied less and was about 1320 MPa.
- For the sharp tool, the tool angle has a slight influence on temperature and stress. The values of tool angle affected the 3D surface responses of maximum temperature and effective stress when both the cutting speed and the feed rates were varied. The maximum temperature reached was 970 °C while the minimum was about 550 °C, depending on the feed rate and speeds used. The effective stress varied less and was about 1250 MPa.

Author Contributions: Conceptualization and methodology, Z.A.M.T. and V.S.; simulation, formal analysis, writing—original draft, Z.A.M.T.; resources, V.S.; supervision, writing—review and editing, V.S. and T.-M.D.; and final revision and editing, A.M.S. All authors have read and agreed to the published version of the manuscript.

Funding: This research received no external funding. The APC was funded by Victor Songmene's research funds. The authors wish to thank the Ministry of Higher Education and Scientific Research in Libya in collaboration with the Canadian Bureau for International Education (CBIE) for providing the Ph.D. scholarship for the first author (Z.A.M.T.).

Data Availability Statement: Data available on request due to privacy or ethical restrictions.

Conflicts of Interest: The authors declare no conflict of interest.

Nomenclature

Symbol	Unit	Meaning
w_n	mm	Chamfer width
γ_n	deg	Chamfer angle
θ_t	deg	Sharp angle
V	m/min	Cutting speed
f	mm/rev	Feed rate
α	deg	Clearance angle
γ, γ_c	deg	Rake angle
R^2	%	Correlation coefficient

References

1. Rodriguez, C.J.C. Cutting Edge Preparation of Precision Cutting Tools by Applying Micro-Abrasive Jet Machining and Brushing. Ph.D. Thesis, Kassel University, Kassel, Germany, 2009.
2. Shnfir, M.; Olufayo, O.A.; Jomaa, W.; Songmene, V. Machinability Study of Hardened 1045 Steel When Milling with Ceramic Cutting Inserts. *Materials* **2019**, *12*, 3974. [CrossRef] [PubMed]
3. Denkena, B.; Biermann, D. Cutting edge geometries. *CIRP Ann.* **2014**, *63*, 631–653. [CrossRef]
4. Yen, Y.C.; Jain, A.; Altan, T. A finite element analysis of orthogonal machining using different tool edge geometries. *J. Mater. Process. Technol.* **2004**, *146*, 72–81. [CrossRef]
5. Cheng, X.; Zha, X.; Jiang, F. Optimizing the geometric parameters of cutting edge for rough machining Fe-Cr-Ni stainless steel. *Int. J. Adv. Manuf. Technol.* **2016**, *85*, 683–693. [CrossRef]
6. Emamian, A. The Effect of Tool Edge Radius on Cutting Conditions Based on Updated Lagrangian Formulation in Finite Element Method. Master's Thesis, McMaster University, Hamilton, ON, Canada, 2018.
7. Daoud, M.; Chatelain, J.F.; Bouzid, A. Effect of rake angle on Johnson-Cook material constants and their impact on cutting process parameters of Al2024-T3 alloy machining simulation. *Adv. Manuf. Technol.* **2015**, *81*, 1987–1997. [CrossRef]
8. Davoudinejad, A.; Noordin, M.Y. Effect of cutting-edge preparation on tool performance in hard-turning of DF-3. tool steel with ceramic tools. *J. Mech. Sci. Technol.* **2014**, *28*, 4727–4736. [CrossRef]
9. Gao, P.; Liang, Z.; Wang, X.; Li, S.; Zhou, T. Effects of different chamfered cutting edges of micro end mill on cutting performance. *Int. J. Adv. Manuf. Technol.* **2018**, *96*, 1215–1224. [CrossRef]
10. Wan, L.; Wang, D.; Gao, Y. Investigations on the Effects of Different Tool Edge Geometries in the Finite Element Simulation of Machining. *J. Mech. Eng.* **2015**, *61*, 157–166. [CrossRef]
11. Gregório, A.V.L.; Silva, T.E.F.; Reis, A.P.; de Jesus, A.M.P.; Rosa, P.A.R. A Methodology for Tribo-Mechanical Characterization of Metallic Alloys under Extreme Loading and Temperature Conditions Typical of Metal Cutting Processes. *J. Manuf. Mater. Process.* **2022**, *6*, 46. [CrossRef]
12. Javidikia, M.; Sadeghifar, M.; Songmene, V.; Jahazi, M. On the impacts of tool geometry and cutting conditions in straight turning of aluminum alloys 6061-T6: An experimentally validated numerical study. *Int. J. Adv. Manuf. Technol.* **2020**, *106*, 4547–4565. [CrossRef]
13. Zhuang, K.; Weng, J.; Zhu, D.; Ding, H. Analytical modeling and experimental validation of cutting forces considering edge effects and size effects with round chamfered ceramic tools. *J. Manuf. Sci. Eng.* **2018**, *140*, 081012. [CrossRef]
14. Altintas, Y.; Ren, H. Mechanics of machining with chamfered tools. *J. Manuf. Sci. Eng.* **2000**, *122*, 650–659.
15. Tagiuri, Z.A.M.; Dao, T.-M.; Samuel, A.M.; Songmene, V. A Numerical Model for Predicting the Effect of Tool Nose Radius on Machining Process Performance during Orthogonal Cutting of AISI 1045 Steel. *Materials* **2022**, *15*, 3369. [CrossRef] [PubMed]
16. Chowdhury, M.A.; Khalil, M.K.; Nuruzzaman, D.M.; Rahaman, M.L. The effect of sliding speed and normal load on friction and wear property of aluminum. *IJMME Int. J. Eng. Sci.* **2011**, *11*, 45–49.
17. Khalili, K.; Safaei, M. FEM analysis of edge preparation for chamfered tools. *Int. J. Mater. Form.* **2009**, *2*, 217–224. [CrossRef]
18. Engineering ToolBox. Convective Heat Transfer. 2003. Available online: https://www.engineeringtoolbox.com/convective-heat-transfer-d_430.html (accessed on 14 August 2022).
19. Jaspers, S.P.F.C.; Dautzenberg, J.H.; Taminiau, D.A. Temperature Measurement in Orthogonal Metal Cutting, International Journal of Adv. *Manuf. Technol.* **1998**, *14*, 7–12. [CrossRef]
20. Afrasiabi, M.; Saelzer, J.; Berger, S.; Iovkov, I.; Klippel, H.; Röthlin, M.; Zabel, A.; Biermann, D.; Wegener, K. A Numerical-Experimental Study on Orthogonal Cutting of AISI 1045 Steel and Ti6Al4V Alloy: SPH and FEM Modeling with Newly Identified Friction Coefficients. *Metals* **2021**, *11*, 1683. [CrossRef]
21. Marinov, V. ME 364 Manufacturing Technology Lecture Notes. indd. Available online: https://www.yumpu.com/en/document/view/17440102/me-364-manufacturing-technology-lecture-notes-department-of- (accessed on 14 August 2022).
22. Liu, C.; Wang, Z.; Zhang, G.; Liu, L. The effect of cutting speed on residual stresses when orthogonal cutting TC4. *EPJ Web Conf.* **2015**, *94*, 1035. [CrossRef]

23. Ucun, I.; Aslantas, K.; Ucun, I. Finite Element Modeling of Machining of AISI 1045 with Ceramic Cutting Tool. Available online: https://www.academia.edu/27298543/Finite_Element_Modeling_of_Machining_of_Aisi_1045_with_Ceramic_Cutting_Tool (accessed on 14 August 2022).
24. Qasim, A.; Nisar, S.; Shah, A.; Khalid, M.S.; Sheikh, M.A. Optimization of process parameters for machining of AISI-1045 steel using Taguchi design and ANOVA. *Simul. Model. Pract. Theory* **2015**, *59*, 36–51. [CrossRef]

 processes

Article

Experimental Investigation on the Machinability of PCBN Chamfered Tool in Dry Turning of Gray Cast Iron

Ganggang Yin [1,2,*], Jianyun Shen [3], Ze Wu [2], Xian Wu [3,*] and Feng Jiang [4]

1 Kistler Innovative Technology China Ltd., Shanghai 201107, China
2 School of Mechanical Engineering, Southeast University, Nanjing 211189, China
3 College of Mechanical Engineering and Automation, Huaqiao University, Xiamen 361021, China
4 Institute of Manufacturing Engineering, Huaqiao University, Xiamen 361021, China
* Correspondence: ganggang.yin@kistler.com (G.Y.); xianwu@hqu.edu.cn (X.W.)

Abstract: Polycrystalline cubic boron nitride (PCBN) tools are widely used for hard machining of various ferrous materials. The edge structure of the PCBN cutting tool greatly affects the machining performance. In this paper, dry turning experiments were conducted on gray cast iron with a PCBN chamfered tool. Both the cutting temperature and the cutting force were measured, and then the surface quality and tool wear mechanisms were analyzed in detail. It was found that the cutting temperature and cutting force increased with the increase in feed rate, depth of cut, and cutting speed. The surface roughness firstly decreased, and then increased with an increase in feed rate. The minimum surface roughness was obtained with a feed rate of 0.15 mm/r which exceeded the tool chamfer width. The PCBN tool wear mode was mainly micro notches on the rake face and micro chipping on the tool chamfer, while the adhesion wear mechanism was the main tool wear mechanism.

Keywords: PCBN tool; gray cast iron; surface quality; tool wear

Citation: Yin, G.; Shen, J.; Wu, Z.; Wu, X.; Jiang, F. Experimental Investigation on the Machinability of PCBN Chamfered Tool in Dry Turning of Gray Cast Iron. *Processes* **2022**, *10*, 1547. https://doi.org/10.3390/pr10081547

Academic Editor: Raul D.S.G. Campilho

Received: 24 June 2022
Accepted: 31 July 2022
Published: 7 August 2022

1. Introduction

Cast iron materials are widely used in the automotive industry, e.g., in engine boxes and brake discs [1,2]. In terms of hard machining cast iron, Martinho et al. investigated the continuous dry turning of gray cast iron using ceramic tools with and without diamond coating; the coated tools presented higher main cutting force and lower flank wear [3,4]. Heck et al. found the different wear behaviors of PCBN cutting tools during machining of compacted graphite iron compared to gray cast iron [5]. Grzesik et al. complexly assessed cutting force, cutting temperature, tool wear progress, and surface roughness in orthogonal turning of nodular cast iron with CBN tools with a chamfer width of 100 µm [6]. Chen et al. studied the influence of material properties on the machining performance in machining of high-chromium white cast iron with CBN tools, and it was found that the micro hardness of workpiece material showed significant impacts on the tool wear, tool life, cutting forces and surface quality [7].

The PCBN material has many excellent properties, such as high hardness, high temperature resistance, good chemical stability, and low friction coefficient [8,9]. These excellent properties make it become the most perfect cutting tool material for hard machining of various steel materials [10,11], such as hardened steel, cast iron, bearing steel, and superalloys. As one of the super-hard cutting tools, PCBN cutting tools have been widely applied in aerospace, automobile, and construction machinery manufacturing. There is extensive literature focused on the fabrication and applications of PCBN cutting tools [12,13]. Regarding the preparation of CBN tool materials, Yun et al. designed and fabricated CBN tool materials with a graded structure to improve material strength and toughness [14]. Mo et al. found that the increase in CBN percentage increased the hardness and wear resistance of PCBN composite, and then increased tool life in the cutting of hard-bearing steel GCr15 [15].

Tamang et al. investigated the brazing of CBN material to tungsten carbide by microwave hybrid heating by applying a paste of Ag–Cu–In–Ti alloy [16]. Fei et al. attempted the pulsed magnetic treatment on CBN tools to improve the cutting performance and found a relatively lower cutting force and less wear area [17]. In terms of the applications of CBN cutting tools, Manoj et al. experimentally studied the machinability of AISI D6 tool steel using low-percentage-CBN tools. They analyzed the effect of machining parameters on cutting temperature, cutting force, and surface roughness, and they found that crater wear, micro chipping, and cutting edge breakage were the main tool wear mechanisms during continuous turning [18]. Gutnichenko et al. investigated the hard turning of helical gear hubs made of low-carbon alloyed steel after heat treatment with hardened up to HRC 60–63 using PCBN cutting tools; it was found that the main tool wear characterizations were flank wear and crater formation, and different strategies were also proposed to monitor tool wear and predict tool life [19]. Schultheiss et al. evaluated the tool wear mechanisms in the machining of gray cast iron with PCBN tools, the results showed that flank and notch wear were the main wear mechanisms, and the non-chamfered tools exhibited the higher notch wear [20].

The tool cutting edge geometry greatly affects the cutting performance, such as the tool nose radius and cutting edge passivation [21,22]. Kumar et al. investigated the effect of cutting parameters and tool geometry in hard machining of hardened AISI H13 material; it was found that the surface roughness decreased while the cutting forces increased with increasing tool nose radius [23]. The chamfered tool structure has a narrow edge ground on the tool rake face with a negative rake angle to enhance tool cutting edge strength. A chamfered tool can improve tool life and is widely used in the cutting tool treatment for rough machining [24,25]. Grzesik et al. developed an empirical model of friction for oblique cutting with CBN chamfered tools; it was found that the chamfered CBN tool decreased the friction coefficient [26]. Tang et al. developed a special chamfer on cutting tools and used it to suppress tool wear in the machining of hard material [27]. Souza et al. investigated the effect of edge preparation of PCBN tools on the wear performance during orthogonal turning of high-speed steels; the PCBN tools with ground land and honed edge radius, as well as a high CBN content, showed chipping in the rake face [28]. Ventura et al. studied the effect of cutting edge geometry on the tool wear behavior of CBN tools during interrupted turning of hardened steel; the results indicated that the chamfered cutting edge reinforces the cutting edge without excessively increasing the mechanical and thermal loads [29]. In summary, a reasonable tool geometry design is helpful to obtain better machining quality and tool life.

In this paper, to assess the machinability of the PCBN chamfered tool in hard machining of gray cast iron, dry turning experiments were performed on gray cast iron with the PCBN cutting tools. The effect of machining parameters on the cutting temperature and cutting force was measured. Then, the machined surface roughness and tool wear were studied in detail. The related machining mechanisms were analyzed to understand the cutting process according to the experimental results. The results are helpful for machining parameter selection in the machining of brittle materials with PCBN cutting tools.

2. Experimental Procedure

Due to their excellent properties, PCBN tools are commonly used in the machining of cast iron materials. The used cutting tool for rough machining in this work was the PCBN cutting tool (grade BN500, Sumitomo, Tokyo, Japan) prepared by Xiamen Xiazhi Technology Tool Co., Ltd., Xiamen, China, as shown in Figure 1. The grain size of CBN materials was about 1–2 μm, and the measured volume proportion of CBN material in the PCBN tools was about 56%. The PCBN tool presented a nose radius of 1.5 mm and cutting edge radius of 25 μm. The tool rake angle and clearance angles were 0° and 5°, while the cutting edge angle is 45°. The tool chamfer with a width of 0.1 mm and a negative rake angle of −20° was prepared on the rake face of the PCBN tool, as shown in Figure 1a.

The tool chamfer is helpful to enhance the cutting edge, and then avoid the premature chipping of the tool cutting edge in the machining of the brittle materials.

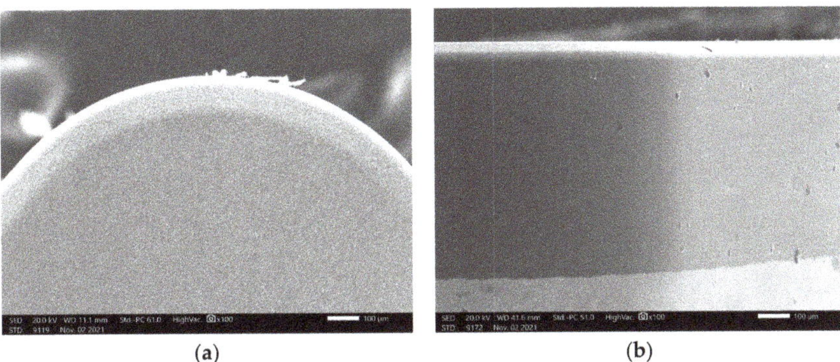

Figure 1. PCBN cutting tool: (a) rake face; (b) flank face.

Dry turning experiments were conducted on a CNC turning machine tool (CAK6150, Shenyang Machine Tool Co., Ltd., Shenyang, China) as shown in Figure 2a. This machine tool is widely used to manufacture rotary parts in various applications. The maximum spindle rotating speed was 2000 rpm. The workpiece material was gray cast iron HT200, which is widely used for the brake disc and engine cylinder in the automotive industry [20,30]. The material properties of the workpiece are listed in Table 1. The workpiece samples were cut into small pieces with dimensions of Ø 50 × 250 mm, and then clamped in a three-jaw chuck machine tool. In the turning experiments, the cutting temperature was measured using a thermal imager (TIM 200, Micro Epsilon, Germany), with a temperature measurement range of −20–900 °C. The cutting force was recorded using a dynamometer (9257B, Kistler, Winterthur, Switzerland), which is widely used in various applications for cutting force measurement [31]. The cutting force measurement system was composed of a dynamometer, connecting cable, charge amplifier (5080A, Kistler, Winterthur, Switzerland), DAQ system, and PC, as shown in Figure 2b. The sampling frequency used in cutting force measurement was 20 kHz to obtain the complete cutting force signal without distortion.

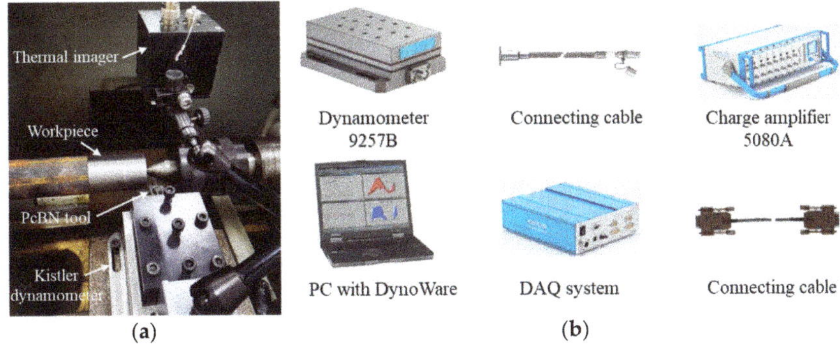

Figure 2. Experimental setup: (a) turning experiment, (b) cutting force measurement system.

In the experiment, the workpiece was pre-machined to ensure a smooth surface using a cemented carbide tool [32], and then the continuous turning experiments were performed. The detailed turning parameters are shown in Table 2. In single-factor experiments, four different levels were set for the machining parameters of spindle speed, feed rate, and depth of cut, according to the existing literatures [10,33]. The corresponding cutting speeds for different spindle rotating speeds were 110, 157, 204, and 251 m/min. The experiments were divided into two stages; during single-factor turning experiments in the first stage, the feed travel for every level of machining parameters was 20 mm, while the total feed travel for every variable of the machining parameters was 80 mm. Then, to test the tool wear mechanism, the PCBN cutting tool was used for the continuous tool wear experiment to reach the feed travel of 4000 mm with fixed machining parameters (v = 204 m/min, a_p = 0.2 mm, and f = 0.15 mm). After the experiment, the workpiece was cleaned using an ultrasonic cleaner, and then the surface roughness was inspected with a roughness meter (3220, Beijing Shidaizhifeng Instrument Co., Ltd., Beijing, China) at different positions along the feed direction. The sampling length and cutoff wavelength were 2.5 mm in the surface roughness measurement. Each experiment was repeated three times, and the averaging values were adapted as the results. The tool wear was inspected using a scanning electron microscope (SEM, JSM-IT500, JEOL, Tokyo, Japan).

Table 1. Material properties of gray cast iron HT200.

Properties	Value
Density (g/mm^3)	7.2
Hardness (HB)	193
Tensile strength (MPa)	320
Elasticity modulus (GPa)	135
Poisson ratio	0.3

Table 2. Turning parameters.

Parameters	Value
Spindle speed n (rpm)	700, 1000, 1300, 1600
Depth of cut a_p (mm)	0.1, 0.2, 0.3, 0.4
Feed rate f (mm/r)	0.05, 0.1, 0.15, 0.2

3. Results and Discussions

3.1. The Cutting Temperature

In addition to the thermal image, the used thermal imager provides a camera to record a video image of the cutting process, as shown in Figure 3. The emissivity of the workpiece material was calibrated to about 0.5 in this work using a digital display heating platform. Combining the thermal image and video image, it was easier to observe the temperature field distribution of the cutting process. It was observed that the chips formed at the cutting zone with fragment shapes and were then sprayed around. In the recorded thermal image, the maximum temperature is exhibited in the cutting zone, and the sprayed chips present a second temperature source with heat dissipation during the spraying process. The mean temperature in the contact zone between the tool and workpiece is difficult to acquire; hence, during the turning experiment, the maximum temperature measured in the cutting zone and displayed in the upper right corner was adopted as the cutting temperature.

The cutting temperature dramatically affects the cutting process and the machining quality. Figure 4 shows the effect of machining parameters on the measured cutting temperature. The results show that the cutting temperature increased from 226 °C to 429 °C with the feed rate increasing from 0.05 mm/r to 0.2 mm/r. Its increasing rate was rapid at first but decreased for a feed rate larger than 0.1 mm/r. The cutting temperature increased from 278 °C to 423 °C with the cutting speed increasing from 100 m/min to 251 m/min. The initial increasing rate slowed after the cutting speed exceeded 157 m/min. This was probably due to the the heat dissipation area being similar under the same depth of cut when the cutting temperature was relatively higher with more extensive machining parameters, while the temperature difference with the surrounding environment became larger. Hence, the larger temperature difference caused more heat dissipation and resulted in a relatively low temperature increase. The cutting temperature almost linearly increased from 257 °C to 369 °C with an increase in the depth of cut, while the heat dissipation area also increasing with the depth of cut.

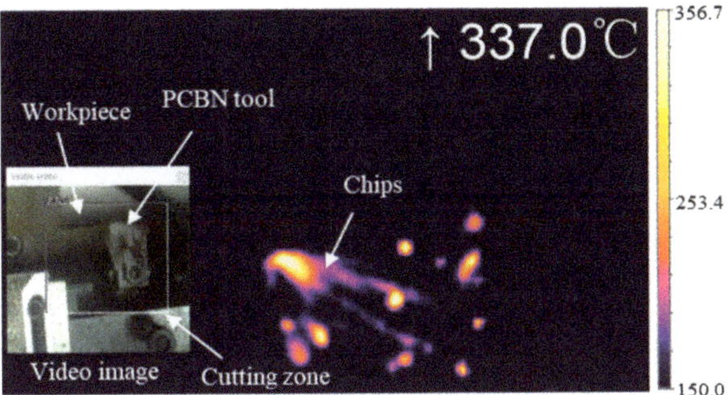

Figure 3. Cutting temperature measurement.

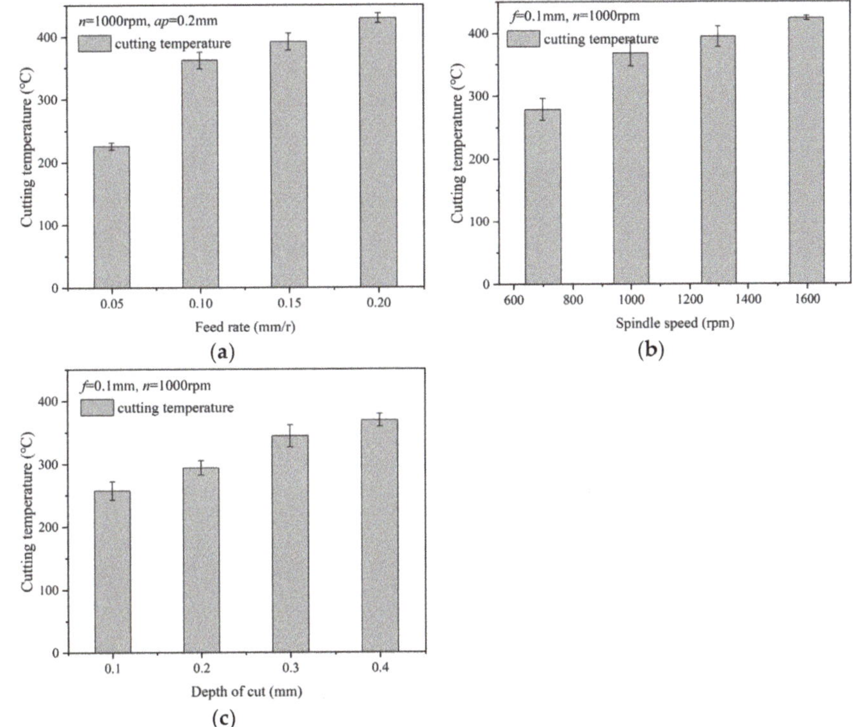

Figure 4. Effect of machining parameters on cutting temperature: (**a**) effect of feed rate on cutting temperature; (**b**) effect of cutting speed on cutting temperature; (**c**) effect of depth of cut on cutting temperature.

3.2. The Cutting Force

Figure 5 shows the recorded cutting force signal in continuous turning experiments with different feed rates. The three cutting force components in order of magnitude were the tangential force (main cutting force) Fz, the radial force (passive force) Fy, and the axial force (feed force) Fx. The radial force Fy was similar to the tangential force Fx. In the turning process, the depth of cut was much lower than the tool nose radius, and only a small segment of the tool nose arc took part in cutting process. The contact condition resulted in the equivalent tool cutting edge angle becoming small. Hence, the component of the resultant cutting force in the radial direction Fy increased. From the results, the cut in and cut out process can be clearly observed, and the cutting force signal waveform was more stable under a large feed rate than a low feed rate of 0.05 mm/r. Under the low feed rate of 0.05 mm/r, the profound ploughing effect led to an unstable cutting process because of the small, undeformed chip thickness. The unstable cutting process resulted in noticeable fluctuations in the cutting force signal. The mean value of the cutting force in the stable cutting process was adopted to quantitatively analyze the cutting force results.

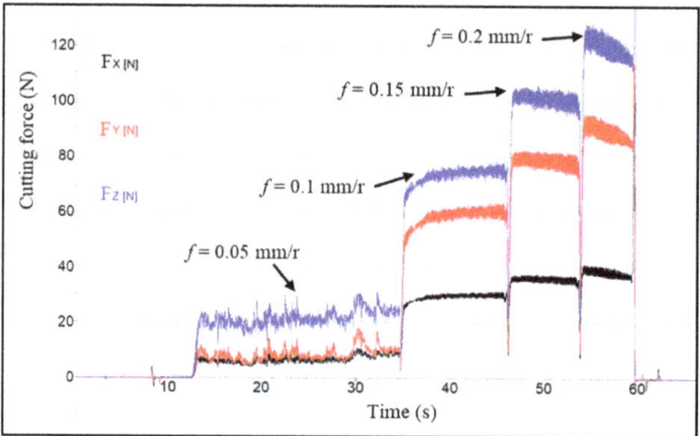

Figure 5. The cutting force signal.

Figure 6 shows the cutting force varying with the machining parameters. The cutting forces Fx, Fy, and Fz increased from 6 N, 8 N, and 20.5 N to 36.5 N, 86.5 N, and 115.5 N, respectively, with the increase in feed rate. The increment in axial force Fx was much smaller than that in tangential force Fz and radial force Fy. A similar phenomenon to the cutting temperature was observed, whereby the increasing rate was rapid at first but decreased for feed rates larger than 0.1 mm/r. The cutting process was continuous turning in this work, and the adhesion material on the tool surface was small at the first stage. A lower tool adhesion caused a relatively low cutting force.

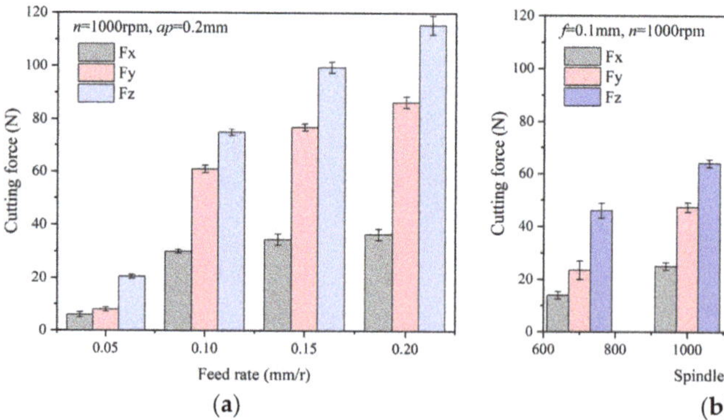

Figure 6. *Cont.*

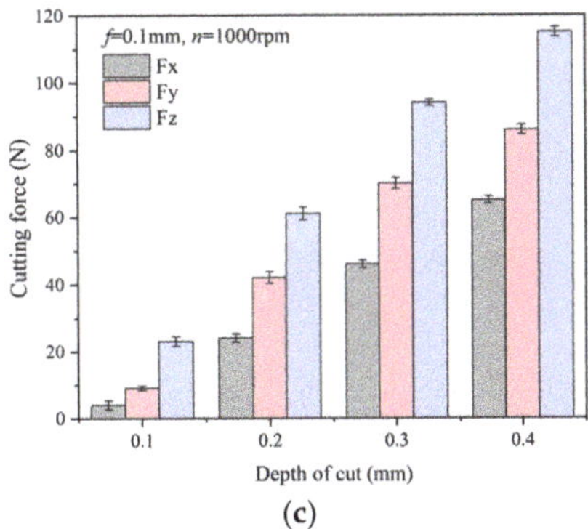

(c)

Figure 6. Effect of machining parameters on cutting force: (**a**) effect of feed rate on cutting force; (**b**) effect of cutting speed on cutting force; (**c**) effect of depth of cut on cutting force.

Although the cutting temperature increased with an increase in cutting speed, the cutting force increased gradually. A probable reason is that this cutting temperature was in the built-up edge (BUE) formation scope. A serious BUE in the cutting process increases the friction and material removal, resulting in a larger cutting force. The cutting speed should be further increased to avoid the BUE. The cutting forces Fx, Fy, and Fz increased from 4 N, 9 N, and 23 N to 65 N, 86 N, and 115 N, respectively, when the depth of cut increased from 0.1 mm to 0.4 mm. This is similar to the cutting temperature, whereby the cutting force linearly increased with an increase in the depth of cut.

3.3. The Surface Quality

Figure 7 shows the surface roughness Ra and surface morphology under different feed rates. It was found that the surface roughness firstly decreased and then increased with an increase in feed rate. The built-up edge was formed on the tool rake face due to the adhesion materials. When the built-up edge along the tool nose arc was uneven, the actual undeformed chip thickness along the tool nose arc was also uneven. Then, the uneven undeformed chip thickness led to transverse micro tears perpendicular to the cutting direction [34]. These micro tears on the surface morphology confirmed the formation of built-up edge in the cutting process.

Surface roughness is determined by the physical factors and theoretical residual area, which is proportional to the feed rate and tool geometry. As shown in Figure 8, the main physical factors significantly influencing the surface roughness included micro burrs and micro bulges induced by the ploughing effect and material plastic side flow, along with micro tears caused by the built-up edge [35]. When the feed rate was minimal, the ploughing effect and BUE formation were severe in the cutting process. This caused uneven scratch marks and many micro tears on the machined surface morphology, resulting in a large surface roughness. However, since the feed rate increased to larger than 0.2 mm, these physical factors were weakened, but the theoretical residual area became significant in terms of the surface morphology and was a dominant factor for the surface roughness. The large theoretical residual area induced a large surface roughness. The minimum surface roughness was obtained with a feed rate of 0.15 mm/r.

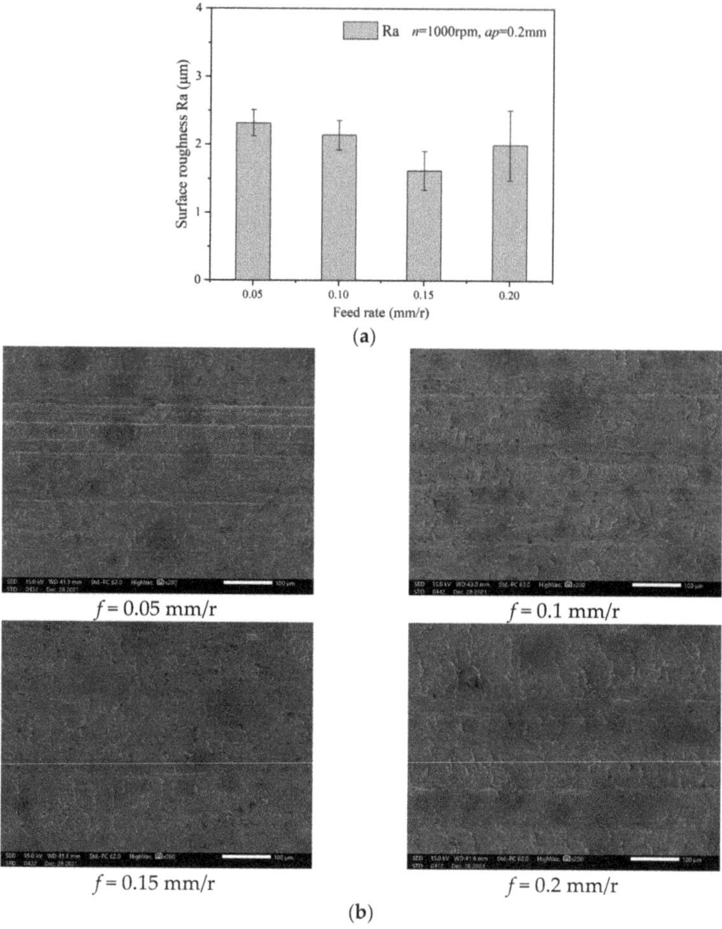

Figure 7. Effect of feed rate on surface quality: (**a**) effect of feed rate on surface roughness; (**b**) surface morphology varying with different feed rates.

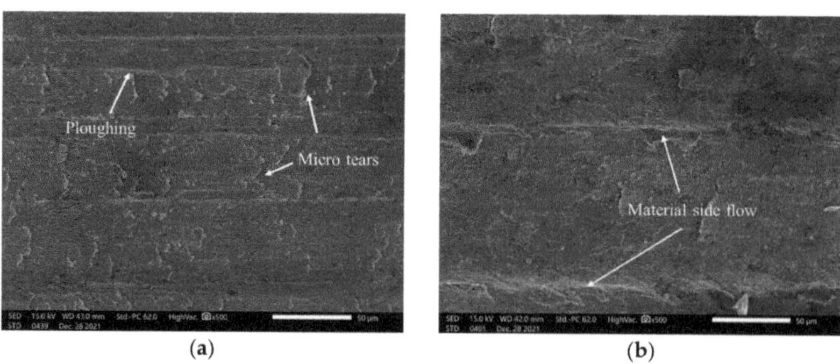

Figure 8. Enlarged view of surface morphology: (**a**) micro tears on surface morphology; (**b**) material side flow on surface morphology.

The variations of surface roughness and surface morphology with cutting speed are shown in Figure 9. It is shown that the surface roughness value gradually decreased with an increase in cutting speed. According to the surface morphology, the scratch marks on the machined surface decreased under the higher cutting speed, and a better surface roughness was obtained. This was probably due to the BUE formed in cutting process becoming the protrusion part in front of the tool tip. This led to the actual rake angle changing from a negative rake angle (equal to angle of tool chamfer) to a positive rake angle. This phenomenon improved the effective negative rake angle on the tool chamfer in the cutting process, which was equivalent to increasing the sharpness of the cutting edge. The sharper cutting edge could decrease the ploughing effect and material plastic side flow in the cutting process. Hence, more minor scratch marks were formed on the surface, as shown in Figure 8b.

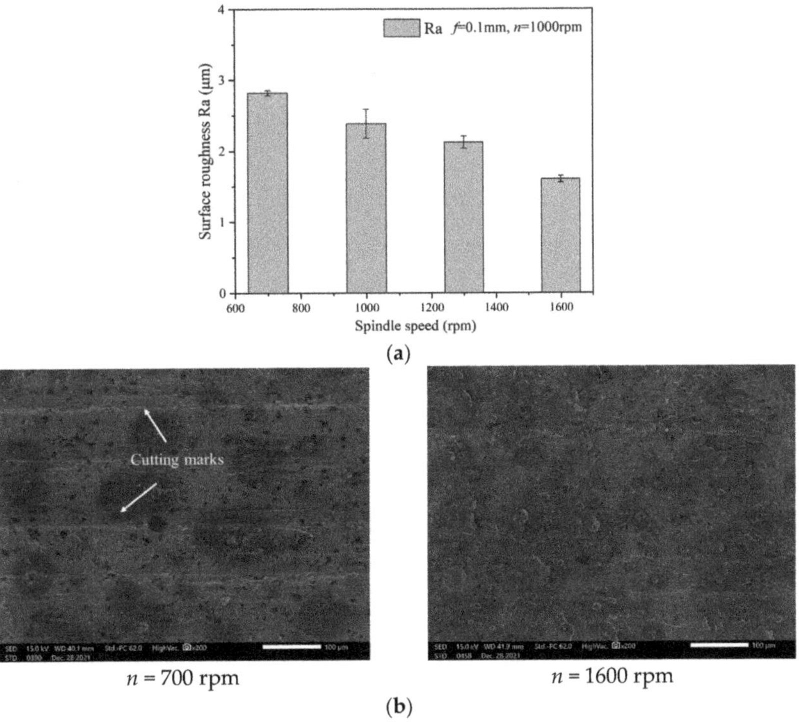

Figure 9. Effect of cutting speed on surface quality: (**a**) effect of cutting speed on surface roughness; (**b**) surface morphology varies with different cutting speed.

Figure 10 shows the surface roughness and surface morphology under different depths of cut. The surface roughness became larger with an increase in the depth of cut. The increasing rate was slow at first, before increasing when the depth of cut exceeded 0.3 mm. When the depth of cut was very small, the equivalent tool cutting edge angle was close to zero. The tool nose was helpful to squeeze the machined surface to be very flat, and a relatively low surface roughness was achieved. When the depth of cut reached 0.4 mm, on the basis of the surface morphology, it was found that the cutting marks became prominent, and scales appeared on the machined surface, which deteriorated the surface roughness.

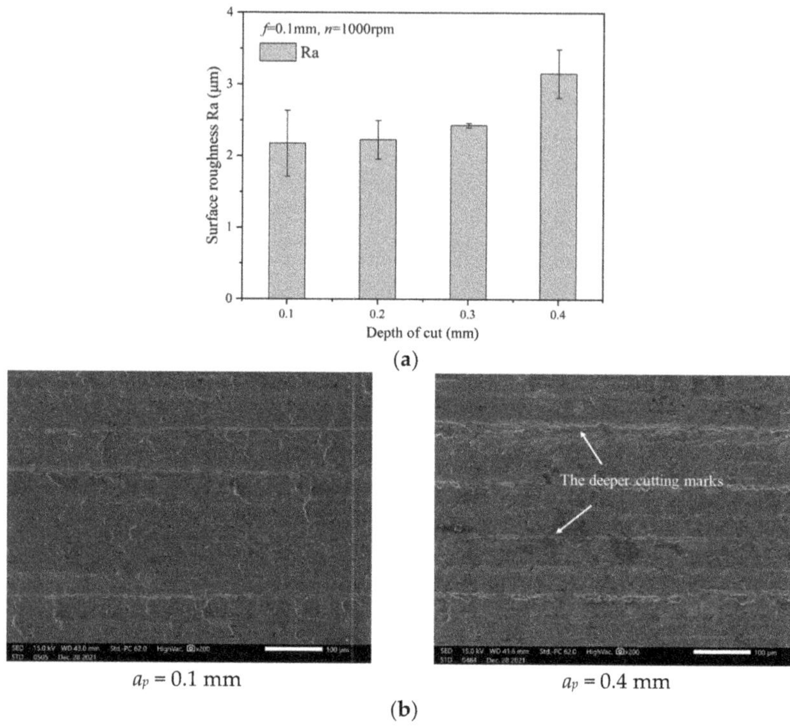

Figure 10. Effect of depth of cut on surface quality: (**a**) effect of depth of cut on surface roughness; (**b**) surface morphology varying with different depth of cuts.

3.4. The Tool Wear

Figure 11 shows the rake face and flank face morphology of the PCBN cutting tool with a cutting distance of 4 m. Compared with the original tool morphology in Figure 1, despite the high hardness difference between the tool material and gray cast iron, there was noticeable tool wear according to the tool morphology. The tool wear on the rake face was much more severe than on the flank face. It was found that the PCBN tool wear mode mainly involved the rake face wear, including some micro notches and many adhesion materials. The distribution of adhesion materials on the rake face was beyond the tool surface range that actually took part in the cutting process. Due to the severe adhesion material and micro chipping, the tool chamfer at the rake face was worn, rendering its observation challenging. However, there was minimal adhesion material on the flank face with hardly any tool wear. The observed tool wear characterizations are consistent with the literature [19].

To confirm the composition of the adhesion material, an EDS analysis was performed, and the results are shown in Figure 12. It was found that, in addition to the elements of the PCBN tool, the main elements of the adhesion material constituted Fe, Si, Mn, etc. These elements corresponded to the composition of gray cast iron, confirming that the adhesion material was the workpiece material. These adhesion materials on the tool surface could easily generate the built-up edge phenomenon in the cutting process.

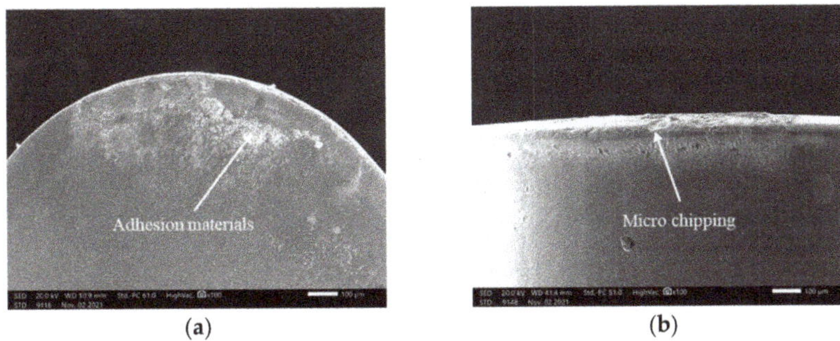

Figure 11. PCBN tool wear characteristics: (**a**) rake face; (**b**) flank face.

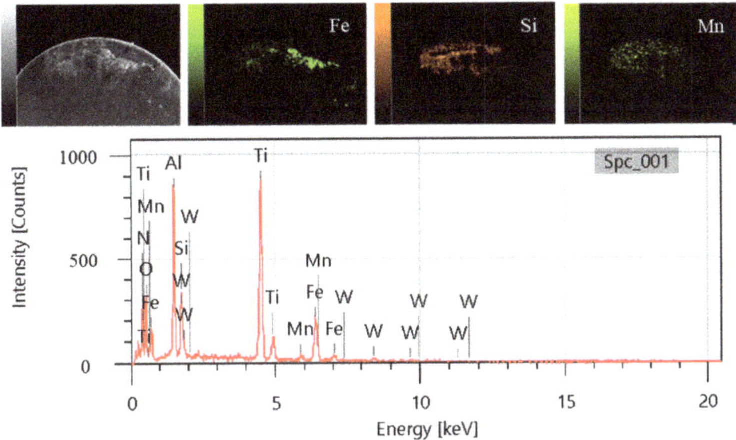

Figure 12. EDS analysis of adhesion material.

Figure 13 shows an enlarged view of the adhesion material on the rake face and flank face. It is revealed that the tool wear mechanism was mainly adhesion wear at the rake face and micro chipping at the tool chamfer. Under the action of the cutting force, the workpiece material and fragment chips were adhered and embedded on the tool surface, thus forming a bulge on the tool surface. These adhesion materials underwent serious friction and shear action with the continuous cutting process. Hence, some adhesion materials were fractured and fell off from tool surface. Simultaneously, this caused some tool material to spall and drop off from the surface, and then micro notches were formed on the rake face, as shown in Figure 13a. This adhesion wear mechanism would continue periodically throughout the cutting process. A negative rake angle caused very serious stress distribution in the local zone on the tool chamfer. At the same time, the adhesion wear could induce some micro cracks or notches on the too chamfer, which could further develop into micro chipping under a large stress distribution.

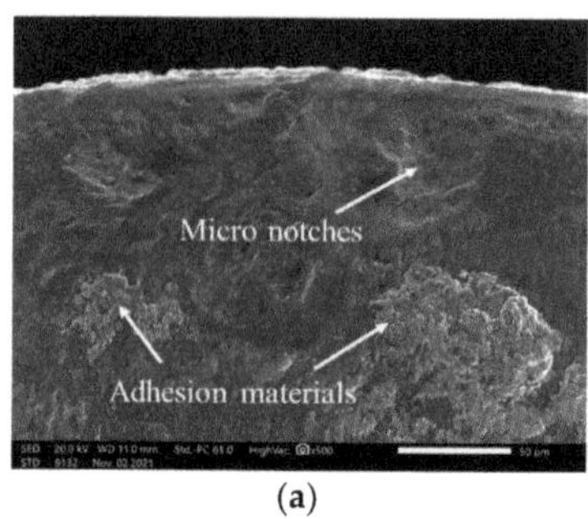

(a)

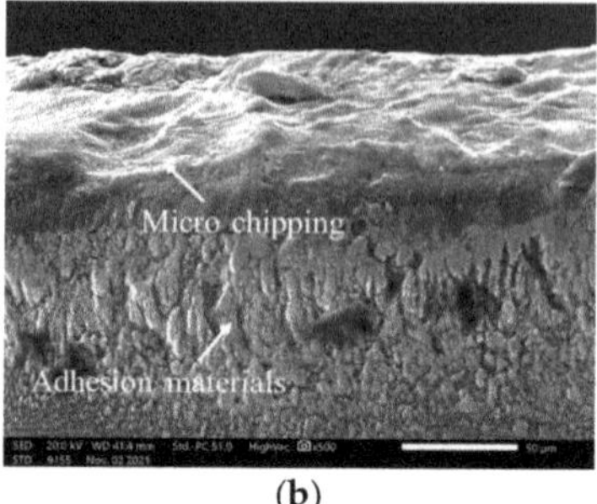

(b)

Figure 13. The adhesions on tool surface: (**a**) adhesion on rake face; (**b**) adhesion on clearance face.

4. Conclusions

This paper presented an experimental study on the drying turning of gray cast iron using a PCBN cutting tool. The following conclusions could be drawn:

1. Both the cutting temperature and the cutting force increased with the increase in feed rate and depth of cut. Although the cutting temperature increased with an increase in cutting speed, the built-up edge during the cutting process induced the cutting force to gradually increase with a larger cutting speed.
2. The surface roughness values were found to decrease at first, and then increase with an increase in feed rate; the minimum surface roughness value was obtained with a feed rate of 0.15 mm/r. The feed rate should be selected to exceed the tool chamfer width to obtain a good surface quality. The surface roughness gradually decreased with the increase in cutting speed, but increased with the increase in depth of cut. The equivalent tool cutting edge angle decreased at a low depth of cut, which was helpful to obtain a smooth surface and achieve a relatively low surface roughness.
3. The PCBN tool wear mode mainly involved micro notches on the rake face and micro chipping on the tool chamfer. During the cutting process, the workpiece material could easily adhere to and embed onto the tool surface, while the fall off of adhesion materials induced some tool material spalling on the tool surface. Micro chipping and adhesion wear were the main wear mechanism for the PCBN tool due to the negative rake angle on the tool chamfer.

Author Contributions: Experiments and paper writing, G.Y. and X.W.; Measurement, J.S. and Z.W.; Data analysis, F.J. All authors have read and agreed to the published version of the manuscript.

Funding: This research was funded by the National Natural Science Foundation of China, grant number 52075097 (W.Z.).

Institutional Review Board Statement: Not applicable.

Informed Consent Statement: Not applicable.

Data Availability Statement: Not applicable.

Conflicts of Interest: The authors declare no conflict of interest.

References

1. Ren, F.Z.; Li, F.J.; Liu, W.M.; Ma, Z.H.; Tian, B.H. Effect of inoculating addition on machinability of gray cast iron. *J. Rare Earths* **2009**, *27*, 294–299. [CrossRef]
2. Genga, R.M.; Cornish, L.A.; Woydt, M.; Vuuren, A.J.; Polese, C. Microstructure, mechanical and machining properties of LPS and SPS NbC cemented carbides for face-milling of grey cast iron. *Int. J. Refract. Met. Hard Mater.* **2018**, *73*, 111–120. [CrossRef]
3. Martinho, R.P.; Silva, F.J.G.; Baptista, A.P.M. Wear behaviour of uncoated and diamond coated Si3N4 tools under severe turning conditions. *Wear* **2007**, *263*, 1417–1422. [CrossRef]
4. Martinho, R.P.; Silva, F.J.G.; Baptista, A.P.M. Cutting forces and wear analysis of Si3N4 diamond coated tools in high speed machining. *Vacuum* **2008**, *82*, 1415–1420. [CrossRef]

5. Heck, M.; Ortner, H.M.; Flege, S.; Reuter, U.; Ensinger, W. Analytical investigations concerning the wear behavior of cutting tools used for the machining of compacted graphite iron and grey cast iron. *Int. J. Refract. Met. Hard Mater.* **2008**, *26*, 197–206. [CrossRef]
6. Grzesik, W.; Kiszka, P.; Kowalczyk, D.; Rech, J.; Claudin, C. Machining of nodular cast iron (PF-NCI) using CBN tools. *Procedia CIRP* **2012**, *1*, 483–487. [CrossRef]
7. Chen, L.; Zhou, J.; Bushlya, V.; Stahl, E.J. Influences of Micro Mechanical Property and Microstructure on Performance of Machining High Chromium White Cast Iron with cBN Tools. *Procedia CIRP* **2015**, *31*, 172–178. [CrossRef]
8. Volodymyr, B.; Filip, L.; Axel, B.; Hisham, A.; Mattias, T.; Stahla, J.E.; Rachid, M. Tool wear mechanisms of PcBN in machining Inconel 718, Analysis across multiple length scale. *CIRP Ann.* **2021**, *70*, 73–78.
9. Saketi, S.; Sveen, S.; Gunnarsson, S.; Saoubi, R.M.; Olsson, M. Wear of a high cBN content PCBN cutting tool during hard milling of powder metallurgy cold work tool steels. *Wear* **2015**, *332–333*, 752–761. [CrossRef]
10. Larissa, J.S.; Denis, B.; Rolf, B.S. Evaluation of layer adhered on PCBN tools during turning of AISI D2 steel. *Int. J. Refract. Met. Hard Mater.* **2019**, *84*, 104977.
11. Zhang, T.; Jiang, F.; Huang, H.; Lu, J.; Wu, Y.Q.; Jiang, Z.Y.; Xu, X.P. Towards understanding the brittle–ductile transition in the extreme manufacturing. *Int. J. Extrem. Manuf.* **2021**, *3*, 022001. [CrossRef]
12. Vallabh, D.P.; Anish, H.G. Analysis and modeling of surface roughness based on cutting parameters and tool nose radius in turning of AISI D2 steel using CBN tool. *Measurement* **2019**, *138*, 34–38.
13. Tatsuya, S.; Haruki, T.; Toshiyuki, E. Development of Novel CBN Cutting Tool for High Speed Machining of Inconel 718 Focusing on Coolant Behaviors. *Procedia Manuf.* **2017**, *10*, 436–442.
14. Yun, H.; Zou, B.; Wang, J.; Huang, C.; Xing, H.; Shi, Z.; Xue, K. Design and fabrication of graded cBN tool materials through high temperature high pressure method. *J. Alloys Compd.* **2020**, *832*, 154937. [CrossRef]
15. Mo, P.C.; Chen, J.; Zhang, Z.; Chen, C.; Pan, X.; Xiao, L.; Lin, F. The effect of cBN volume fraction on the performance of PCBN composite. *Int. J. Refract. Met. Hard Mater.* **2021**, *100*, 105643.
16. Tamang, S.; Aravindan, S. Brazing of cBN to WC-Co by Ag-Cu-In-Ti alloy through microwave hybrid heating for cutting tool application. *Mater. Lett.* **2019**, *254*, 145–148. [CrossRef]
17. Fei, H.; Wu, H.; Yang, X.; Xiong, J.; Zhang, L.; Chen, Z.; Jiang, K.; Liu, J. Pulsed magnetic field treatment of cBN tools for improved cutting performances. *J. Manuf. Processes* **2021**, *69*, 21–32. [CrossRef]
18. Manoj, N.; Rakesh, S.; Rajender, K. Investigating machinability of AISI D6 tool steel using CBN tools during hard turning. *Mater. Today Proc.* **2021**, *47*, 3960–3965.
19. Gutnichenko, O.; Nilsson, M.; Lindvall, R.; Bushlya, V.; Andersson, M. Improvement of tool utilization when hard turning with cBN tools at varying process parameters. *Wear* **2021**, *477*, 203900. [CrossRef]
20. Schultheiss, F.; Bushlya, V.; Lenrick, F.; Johansson, D.; Kristiansson, S.; Ståhl, J.-E. Tool Wear Mechanisms of pCBN tooling during High-Speed Machining of Gray Cast Iron. *Procedia CIRP* **2018**, *77*, 606–609. [CrossRef]
21. Henrik, P.; Filip, L.; Luiz, F.; Stahla, J.E.; Volodymyr, B. Wear mechanisms of PcBN tools when machining AISI 316L. *Ceram. Int.* **2021**, *47*, 31894–31906.
22. Zeng, K.; Wu, X.; Jiang, F.; Zhang, J.; Kong, J.; Shen, J.; Wu, H. Experimental research on micro hole drilling of polycrystalline Nd:YAG. *Ceram. Int.* **2022**, *48*, 9658–9666. [CrossRef]
23. Kumar, P.; Chauhan, S.R.; Aggarwal, A. Effects of cutting conditions, tool geometry and material hardness on machinability of AISI H13 using CBN tool. *Mater. Today Proc.* **2021**, *46*, 9217–9222. [CrossRef]
24. Gao, H.; Liu, X.; Chen, Z. Cutting Performance and Wear/Damage Characteristics of PCBN Tool in Hard Milling. *Appl. Sci.* **2019**, *9*, 772. [CrossRef]
25. Li, S.Y.; Chen, T.; Qiu, C.Z.; Wang, D.Y.; Liu, X.L. Experimental investigation of high-speed hard turning by PCBN tooling with strengthened edge. *Int. J. Adv. Manuf. Technol.* **2017**, *92*, 3785–3793. [CrossRef]
26. Grzesik, W.; Zak, K. Friction quantification in the oblique cutting with CBN chamfered tools. *Wear* **2013**, *304*, 36–42. [CrossRef]
27. Tang, X.; Nakamoto, K.; Obata, K.; Takeuchi, Y. Ultraprecision micromachining of hard material with tool wear suppression by using diamond tool with special chamfer. *CIRP Ann.* **2013**, *62*, 51–54. [CrossRef]
28. Souza, D.J.A.; Weingaertner, W.L.; Schroeter, R.B.; Teixeira, C.R. Influence of the cutting edge micro-geometry of PCBN tools on the flank wear in orthogonal quenched and tempered turning M2 steel. *J. Braz. Soc. Mech. Sci. Eng.* **2014**, *36*, 763–774. [CrossRef]
29. Ventura, C.E.H.; Koehler, J.; Denkena, B. Influence of cutting edge geometry on tool wear performance in interrupted hard turning. *J. Manuf. Process* **2015**, *19*, 129–134. [CrossRef]
30. Chen, L.; Stahl, J.E.; Zhao, W.; Zhou, J. Assessment on abrasiveness of high chromium cast iron material on the wear performance of PCBN cutting tools in dry machining. *J. Mater. Processing Technol.* **2018**, *255*, 110–120. [CrossRef]
31. Zhao, G.; Xin, L.; Li, L.; Zhang, Y.; He, N.; Jansen, H.N. Cutting force model and damage formation mechanism in milling of 70wt% Si/Al composite. *Chin. J. Aeronaut.* **2022**. [CrossRef]
32. Salim, C.; Mohamed, A.; Salim, B.; Ahmed, B.; Khaoula, S.; Abdelkrim, H. Coated CBN cutting tool performance in green turning of gray cast iron EN-GJL-250, modeling and optimization. *Int. J. Adv. Manuf. Technol.* **2021**, *113*, 3643–3665.
33. Wu, X.; Shen, J.; Jiang, F.; Wu, H.; Li, L. Study on the oxidation of WC-Co cemented carbide under different conditions. *Int. J. Refract. Met. Hard Mater.* **2021**, *94*, 105381. [CrossRef]

34. Samad, N.B.O.; Yigit, K. Investigating the influence of built-up edge on forces and surface roughness in micro scale orthogonal machining of titanium alloyTi6Al4V. *J. Mater. Processing Technol.* **2016**, *235*, 28–40.
35. Wu, X.; Li, L.; Zhao, M.; He, N. Experimental investigation of specific cutting energy and surface quality based on negative effective rake angle in micro turning. *Int. J. Adv. Manuf. Technol.* **2016**, *82*, 1941–1947. [CrossRef]

 *processes*

Article

Effects of Coating Parameters of Hot Filament Chemical Vapour Deposition on Tool Wear in Micro-Drilling of High-Frequency Printed Circuit Board

Fung Ming Kwok [1], Zhanwen Sun [1,2], Wai Sze Yip [1], Kwong Yu David Kwok [3,*] and Suet To [1,*]

1 State Key Laboratory in Ultra-Precision Machining Technology, Department of Industrial and Systems Engineering, The Hong Kong Polytechnic University, Kowloon, Hong Kong, China; oscar.kwok@connect.polyu.hk (F.M.K.); zw.sun@gdut.edu.cn (Z.S.); 13903620r@connect.polyu.hk (W.S.Y.)

2 State Key Laboratory of Precision Electronic Manufacturing Technology and Equipment, Guangdong University of Technology, Guangzhou 510006, China

3 Techmart (Shenzhen) Limited, 2F, Block 2, No. 2 Chongqing Road, Qiaotou Community, Fuyong Subdistrict, Bao'an District, Shenzhen 518104, China

* Correspondence: davidkwok@techmart.com.hk (K.Y.D.K.); sandy.to@polyu.edu.hk (S.T.); Tel.: +852-2766-6587 (S.T.); Fax: +852-2764-7657 (S.T.)

Abstract: High-frequency and high-speed printed circuit boards (PCBs) are made of ceramic particles and anisotropic fibres, which are difficult to machine. In most cases, severe tool wear occurs when drilling high-frequency PCBs. To protect the substrate of the drills, diamond films are typically fabricated on the drills using hot filament chemical vapour deposition (HFCVD). This study investigates the coating characteristics of drills with respect to different HFCVD processing parameters and the coating characteristics following wear from machining high-frequency PCBs. The results show that the methane concentration, processing time and temperature all have a significant effect on the grain size and coating thickness of the diamond film. The grain size of the film obviously decreases as does the methane concentration, while the coating thickness increases. By drilling high-frequency PCBs with drills with nanocrystalline and microcrystalline grain sizes, it is discovered that drills with nanocrystalline films have a longer tool life than drills with microcrystalline films. The maximum length of the flank wear of the nanocrystalline diamond-coated drill is nearly 90% less than microcrystalline diamond-coated tools. Moreover, drills with thinner films wear at a faster rate than drills with thicker films. The findings highlight the effects of HFCVD parameters for coated drills that process high-frequency PCBs, thereby contributing to the production of high quality PCBs for industry and academia.

Keywords: high-frequency PCB; drilling; coating technology; tool wear; hot filament chemical vapor deposition

Citation: Kwok, F.M.; Sun, Z.; Yip, W.S.; Kwok, K.Y.D.; To, S. Effects of Coating Parameters of Hot Filament Chemical Vapour Deposition on Tool Wear in Micro-Drilling of High-Frequency Printed Circuit Board. *Processes* **2022**, *10*, 1466. https://doi.org/10.3390/pr10081466

Academic Editor: Antonino Recca

Received: 28 June 2022
Accepted: 25 July 2022
Published: 27 July 2022

1. Introduction

As the world is accelerating into the 5G era, there are higher requirements for PCB transmission speeds and thermal design considerations. The new PCB design incorporates ceramic particles, which help increase the heat dissipation rate [1], but, in turn, lower the machinability of the workpiece. Researchers have extensively studied the relationship between drill coating thickness and grain when machining the PCBs of the third and fourth generations, such as the copper-clad resin-based FR-4 [2–5]. However, the relationships between the drill tool wear rate, the diamond coating thickness and grain size when machining PCBs of the fifth generation have still not been fully explored. Currently, high-frequency and high-speed PCBs have a wide range of needs and applications in 5G mobile communication technology due to their superior heat resistance and lower transmission loss when compared with conventional PCBs [6]. High-frequency and high-speed PCBs are made from a composite material that is composed of copper-clad laminate, ceramic particles, anisotropic fibres and epoxy resin [7,8]. In the manufacturing of high-frequency PCBs, micro drills are commonly used to fabricate micro holes on layers of printed circuit boards because of their high efficiency and the low cost. However, due to the high anisotropy and low thermal conductivity of general PCB drills, the rate of tool wear and hole quality are exceptionally high. PCB drills have to be capable of withstanding abrasive wear from ceramic fillers and adhesive wear from molten resin [8] adhering to the drills under high-temperature processing conditions. Friction and heat are generated during the drilling process. An increase in drilling temperature inside the hole causes the resin to melt, adhering to the wall of the hole and the drill tip and affecting the subsequent rotation of the drill bit. These two types of deterioration are primarily responsible for the tool becoming worn and broken before the end of its expected tool life.

To protect the substrate of the drills, the HFCVD technique is generally employed to fabricate diamond films on the drills. The novelty of the coating drill research is to determine the suitable coating thickness and grain size for high-speed board machining. It has been demonstrated that 4000 holes can be fabricated by HFCVD-coated drills without obvious drill wear, which is more than three times the lifespan of conventional non-coated drills [9]. Notably, different pretreatment methods and deposition parameters result in variations in the coating layer thickness and grain size of the diamond film, which can range from micrometres to nanometres. The main methods to fabricate microcrystalline diamond (MCD) and nanocrystalline diamond (NCD) film layers require gas mixtures and additional sources of energy such as thermal energy or plasma for the gas mixtures to dissociate and deposit on the substrate. While both coating types are applied to protect PCB drills, each has its own differential advantage. As MCD films and NCD films have different topographies and frictional coefficients, their cutting performance is different [10]. It is therefore important to identify the most critical processing condition in order to fabricate the optimum coating thickness and grain size of a drill for machining high-speed boards. Dawedeit et al. [11] found that the Young's modulus is 950 GPa for the MCD layer and 730 GPa for the NCD layer, and that their grain sizes are inversely proportional to the workpiece roughness measured by atomic force microscopy (AFM). Chen et al. [12] discovered that the friction coefficient is 0.126 for the MCD and 0.076 for the NCD. Williams [13] stated that the friction coefficient of NCD-coated surfaces can be as low as those of single crystal diamonds. The MCD film can, relatively, adhere better to the substrate surface with higher wear resistance, resulting in better tool life. The NCD film has a relatively lower surface roughness, resulting in work with more refined surface finishes. The processing of a high-speed board necessitates a drill with high wear resistance for machining PCB components with ceramic fillers and a low surface roughness to reduce the possibility of the tool adhering to the molten resin.

In addition to affecting the mechanical properties of the thin films, the thickness of the coating layer influences the sharpness of the cutting edge, thus negatively impacting the drilling performance. Rare scientific journals have investigated the effects of coating thickness and grain size on the processing of high-frequency and high-speed PCBs. The purpose of this study was to investigate the quality of diamond films with varying MCD and NCD grain sizes and coating thicknesses in order to determine the effects of grain size and coating thickness on tool wear rates when drilling high-frequency PCBs, thereby enhancing the drilling performance of high-frequency PCBs.

2. Theory—Fabrication of Drills with MCD and NCD by HFCVD

Prior to the diamond film deposition, the surface of the PCB drills should be treated with Murakami's solution ($KOH:K_3(Fe(CN)_6:H_2O$ = 1:1:10) and Caro's acid ($H_2O_2 + H_2SO_4$) [14]. The Murakami treatment enables the WC to dissolve into the chemical solution and increase the surface roughness of the substrate surface. Caro acid is used to remove cobalt from the surface, preventing graphitisation during the diamond film growth process.

In the HFCVD diamond coating process, chemical gases are utilized as a carbon source and filaments are utilized to generate heat. The heat induces chemical reactions that generate diamond films on the surfaces of substrates. The HFCVD diamond coating process requires a hot filament to react with the carbon source input gas methane (CH_4). The most common filament materials for HFCVD are tungsten (W) and tantalum (Ta). As given in Figure 1a, the carbon source CH_4 and hydrogen (H_2) pass through the hot filament at the start of the deposition process. When the hot pure tantalum filament begins to react with the mixed gas (methane and hydrogen), it causes filament carbonisation and embrittlement. After passing through the hot filament, the mixed gas receives enough thermal energy to break hydrogen bonds (H_2, CH_4) into atomic hydrogen (H^0), as illustrated in Equation (1). This dissociation into atomic forms enables diamond growth. Hydrogen atoms (H^0) can interact further with small hydrocarbon molecules (mostly CH_4) to form methyl radicals (CH_3) and diamond growth, according to Equation (2). As illustrated schematically in Figure 1a,b, the following chemical equations [15] apply:

$$H_2 \leftrightarrow H^0 + H^0 \quad (1)$$

$$CH_4 + H^0 \rightarrow CH_3 + H_2 \quad (2)$$

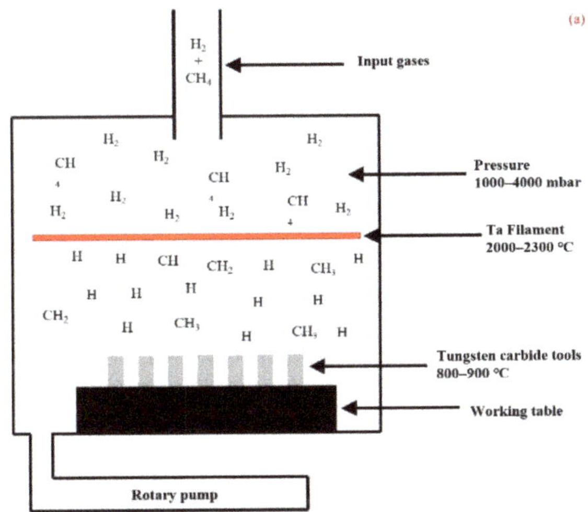

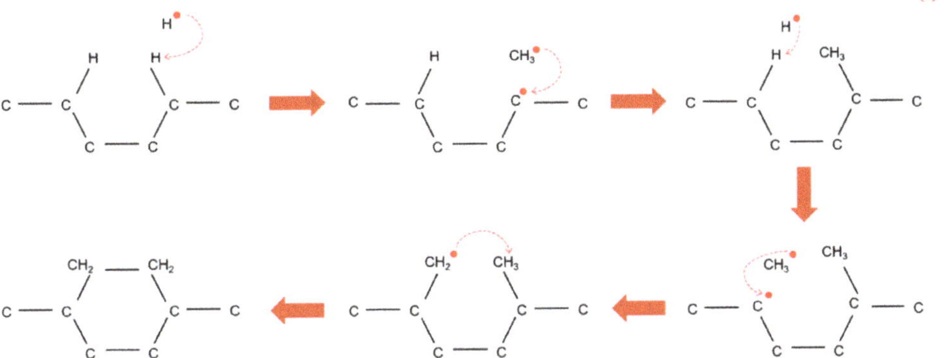

Figure 1. Schematics of (**a**) HFCVD machine layout and endothermic process of bond breaking, (**b**) reaction of the chemical gases to form diamond layer on the substrate surface.

In this study, the diamond film on PCB drills is grown using the hot filament chemical vapour deposition (HFCVD) method. Several processing parameters, including temperature, methane concentrations and deposition durations, can be adjusted to stabilize the growth of the film. Different grain sizes and different coating thicknesses can be formed on the drills.

3. Experimental Procedures and Setup

In this study, PCB drills (∅1.1, 6% Co) tungsten carbide (WC) with grain size of 0.5–0.8 μm were selected. Drills of microcrystalline diamond (MCD) film and nanocrystalline diamond (NCD) film with coating thickness of 3 μm, 6 μm, 9 μm were investigated in this study, as shown in Table 1. The following parameters were specified for deposition: tool temperature between 800–900 °C; methane 20–80 sccm and deposition duration between 10–24 h. The temperature, deposition durations and CH_4 concentration parameters were varied to investigate different grain growth rate and size of diamond films.

Table 1. The coating type and thickness of the drills.

Samples	Coating Type	Coating Thickness (μm)
M1	MCD	3
M2	MCD	6
M3	MCD	9
N1	NCD	3
N2	NCD	6
N3	NCD	9

After diamond films had been deposited on the drills, scanning electron microscope (SEM) was used to observe the coating thickness and grain size of the coating films. The drills were then tested on Rogers 4350B high-frequency PCB using VEGA-D2CMSL drilling machine with feed rate of 2.4 m/min and spindle speed of 45,000 r/min. Boards were stacked together that consisted of 5 boards, 3 of which were phenolic boards with a thickness of 0.5 mm, and 2 of which were high-frequency PCB with a thickness of 0.8 mm, as shown in Figure 2. The condition of the drill (wearing) was observed when every 1000 holes were drilled. The micro drills were ultrasonically washed by acetone to remove the dirt after drilling. Then, wear was observed by digital microscope of KEYENCE VHX-7000 (e.g., flank wear at the major cutting edges). Surface morphology of the films was measured by atomic force microscopy (AFM) with size 10 μm × 10 μm sample regions; three different regions were measured in order to obtain the average value. The adhesion characteristics of films with different coating layer thickness and grain size were analysed by Rockwell indentation with a test force of 588 N.

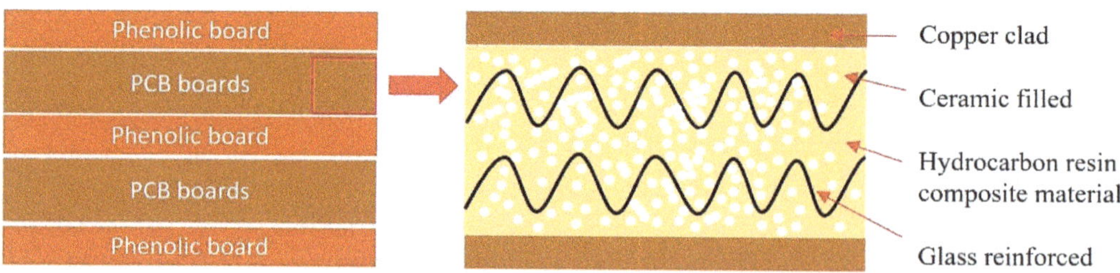

Figure 2. Layout of board stacking during machining and PCB's structure.

4. Results and Discussion

4.1. Effects of Coating Parameters on Grain Size and Coating Thickness

To investigate the control of the HFCVD-coated MCD and NCD films' layer thicknesses and grain sizes, the influences of temperature, methane concentration and deposition time on the coating thickness and grain size were studied. The micro-topographies and cross-sectional profiles of the MCD and NCD films with 3 μm, 6 μm and 9 μm coating thicknesses are shown in Figures 3 and 4. M1–M3 drills with MCD films exhibit sharp columnar crystal growth with clear grains of size between 2–4 μm while N1–N3 drills with NCD films exhibit cluster growth of nanosized diamond grains with a cauliflower-like morphology. The grain size and growth rates are different in accordance with the variety in methane, deposition duration and temperature, and the relationships are shown in Figure 4.

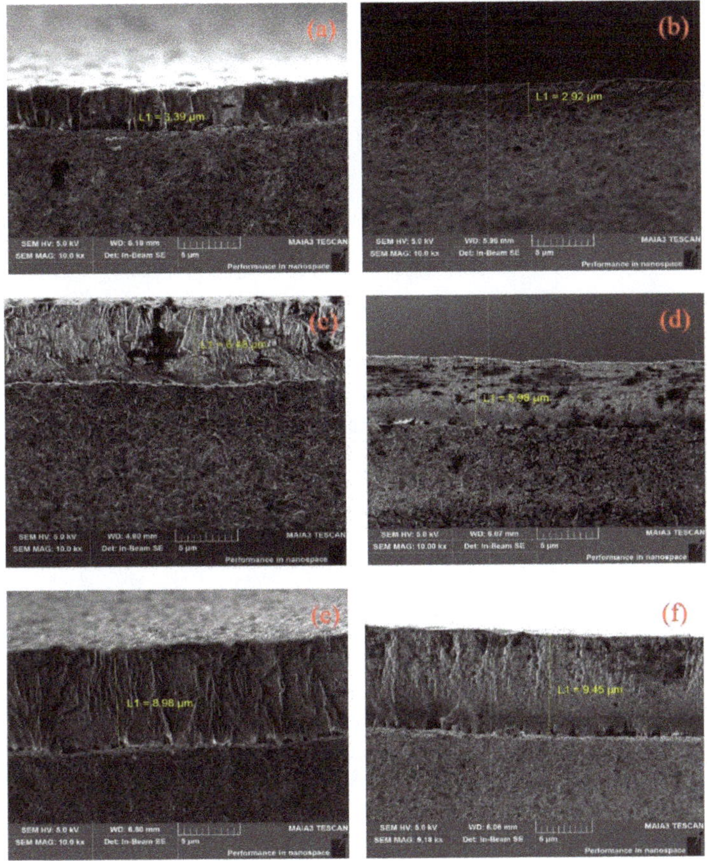

Figure 3. Cross-sectional profile of the diamond-coated drills: (**a**) M1; (**b**) N1; (**c**) M2; (**d**) N2; (**e**) M3; (**f**) N3.

As shown in Figure 5a, an increase in methane concentration leads to a decrease in the grain size and an increase in coating thickness. The increased methane concentration enhances secondary nucleation [16,17] as there is not enough hydrogen to etch the non-diamond phase of the drill and to increase the rate of ion bombardment. In this case, the newly formed diamond film layer has a smaller grain size than its previous layer. More methane leads to more methyl radicals incorporated in the coating region, resulting in increases in the growth rate of the diamond film [18]. As represented in Figure 5b, as the duration of diamond film deposition in the HFCVD machine chamber increases, the average grain size and number of grains decrease [19]. In accordance with the experimental findings, the coating thickness increases from 1 μm to 9 μm as the duration increases from 4 to 24 h. As shown in Figure 5c, increasing the substrate temperature could increase the dissociation speed of H_2, which results in a higher growth rate of the film/grain [20,21].

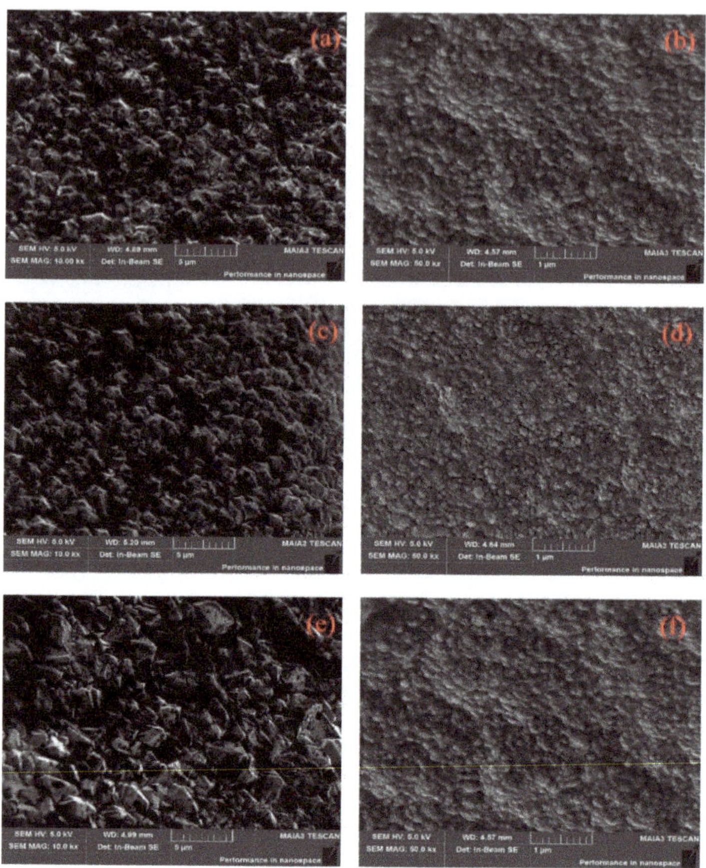

Figure 4. Surface morphology of diamond-coated drills: (**a**) M1; (**b**) N1; (**c**) M2; (**d**) N2; (**e**) M3; (**f**) N3.

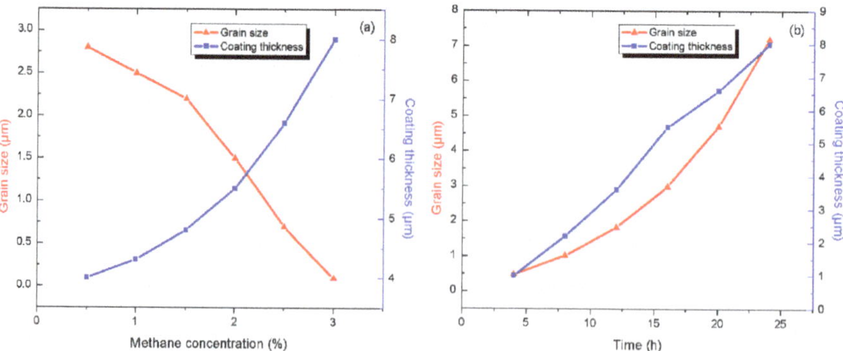

Figure 5. *Cont.*

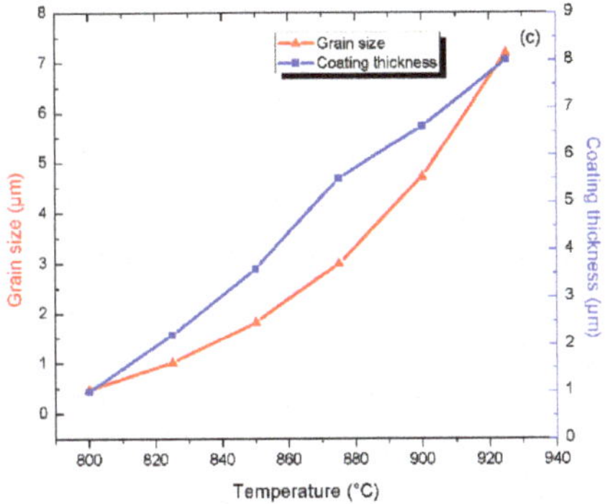

Figure 5. Variation in grain size and coating thickness with (**a**) methane concentration, (**b**) time and (**c**) temperature.

4.2. Effect of Coating Thickness and Grain Size on Adhesion Strength

Rockwell A with the load of 588 N was selected to evaluate the diamond coating adhesion on WC-Co substrates. Figure 6 shows SEM images of the Rockwell indentations on the tool surfaces with 3 μm, 6 μm and 9 μm MCD and NCD diamond coating layers. Two major defective results were observed: crack propagation and coating delamination. The diameters of the cracks upon delamination measured on six samples, M1–M3 and N1–N3, were 120 μm, 110 μm, 101 μm, 158 μm, 135 μm and 130 μm, respectively. The deformation area can be ranked as: M3, M2, M1, N3, N2 and N1. Cracks were observed on both the MCD and NCD films with 3 μm and 6 μm thicknesses. No crack propagation was initiated on the surfaces of samples with 9 μm MCD (M3) and NCD coating thickness (N3). Coating delamination occurred on MCD and NCD films with 3 μm and 6 μm thickness. The samples with 3 μm coating layers showed greater coating delamination than the 6 μm coating layers, implying that coating adhesion is stronger in thicker films. In summary, coating thickness was the critical factor influencing the adhesive strength of the diamond film, as neither M3 nor N3 observed crack propagation after indentation.

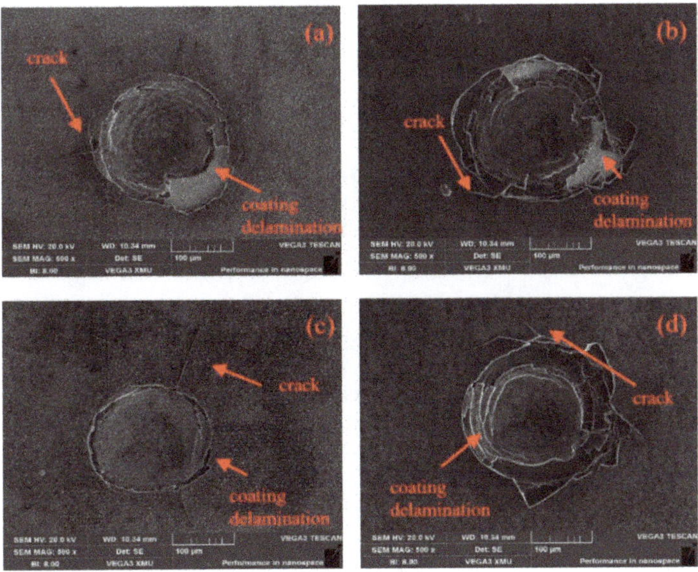

Figure 6. *Cont.*

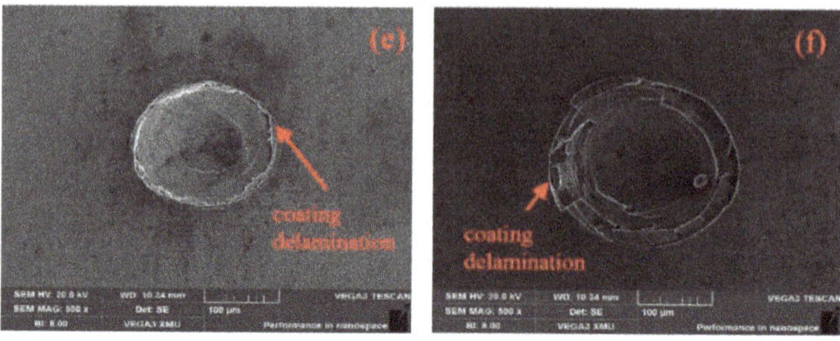

Figure 6. Rockwell A results of diamond-coated drills: (**a**) M1; (**b**) N1; (**c**) M2; (**d**) N2; (**e**) M3; (**f**) N3.

4.3. Effect of Coating Thickness and Grain Size on Surface Roughness

Just as depicted by the SEM images in Figure 4, the AFM images in Figure 7 show the same structures of the MCD and NCD films. AFM images of the diamond films with the size 10 μm × 10 μm are demonstrated in Figure 7. The means of the surface roughness of the M1–M3 and N1–N3 drills were 91.4 nm, 107.6 nm, 123.0 nm, 22.3 nm, 25.7 nm and 27.8 nm, respectively. MCD films appeared to have higher surface roughness due to the uneven grain sizes. NCD films appeared to have lower surface roughness due to smaller grains and lower coefficient of friction. According to the results, coating thickness is not the primary factor affecting surface roughness, with surface roughness increasing by 32 nm from M1 to M3 and 5 nm from N1 to N3. The results indicate that film grain size is the main influence on surface roughness, whereas coating thickness has a minimal impact among other factors shown in this study.

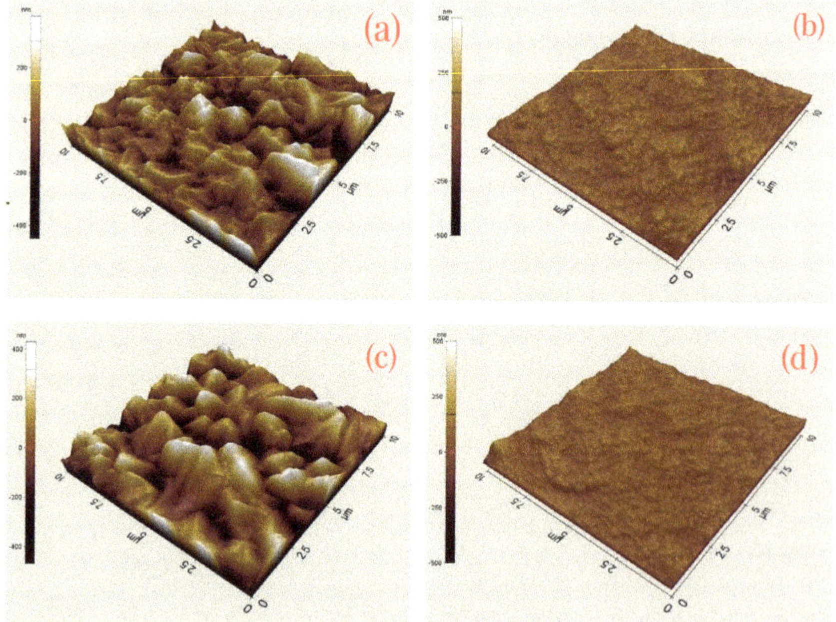

Figure 7. *Cont.*

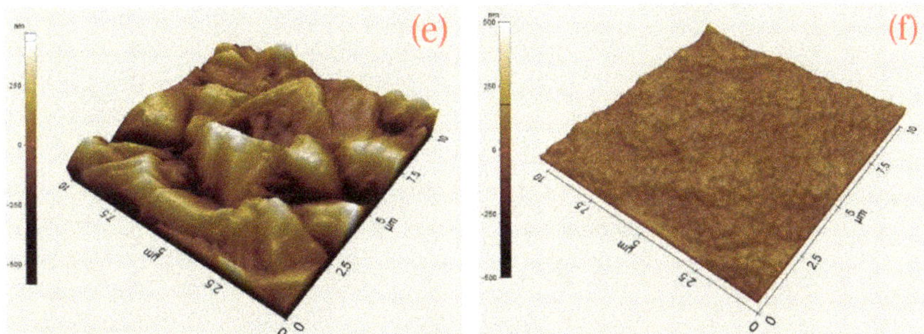

Figure 7. AFM images of diamond-coated drills: (**a**) M1; (**b**) N1; (**c**) M2; (**d**) N2; (**e**) M3; (**f**) N3.

4.4. Investigation on Tool Wear

To study the tool wear states, six drills with different grain sizes and film thicknesses were used to manufacture micro holes on high-frequency PCBs. As shown in Figure 8, significant abrasive wear was observed on the flank face of both the MCD and the NCD drill with a 3 μm coating thickness after processing 1000 holes, causing coating delamination. The 3 μm NCD-coated drill (N1) exhibited more damage than the 3 μm MCD-coated drill (M1) as the substrate material of the cutting edge was also removed. This indicates that compared with MCD drills, NCD drills have a longer tool life due to the lower adhesion of the diamond film on the WC-Co substrate.

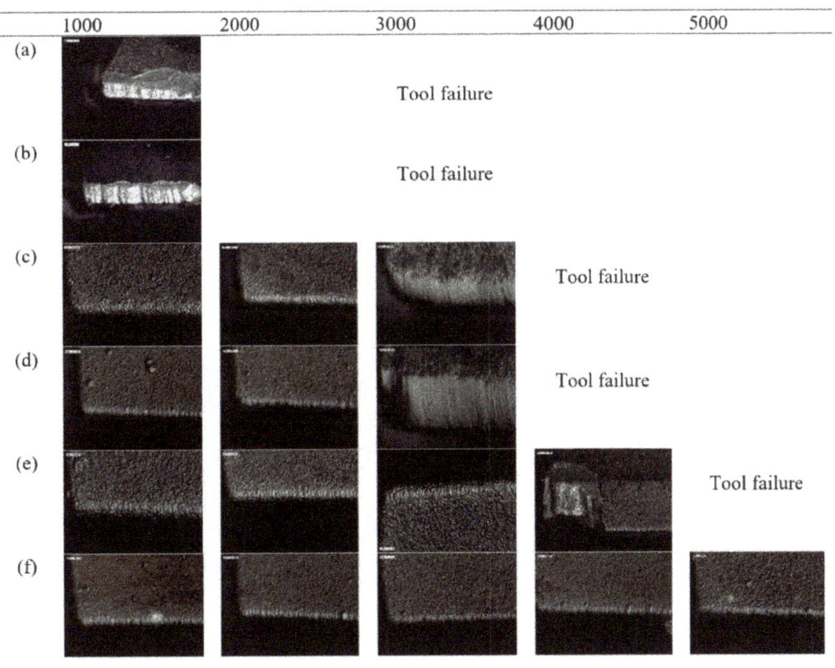

Figure 8. Flank wear states after drilling 1000—5000 holes for diamond-coated drills: (**a**) M1; (**b**) N1; (**c**) M2; (**d**) N2; (**e**) M3; (**f**) N3.

A comparison of the NCD drill wear between coating thicknesses of 6 μm and 9 μm (N2, N3) after drilling 3000 holes was conducted. Even though significant abrasive tool wear can be observed on the NCD drills with a 6 μm coating thickness (N2). The tool wear of N3 increased steadily during the experiment, with the maximum lengths of flank wear for N3 from 1000 to 5000 holes being 9.54 μm, 10.12 μm, 10.68 μm, 11.05 μm and 12.17 μm, respectively. No obvious tool wear or coating delamination occurred for 9 μm (N3) after drilling 5000 holes. It could be explained that the NCD film layer's thickness is sufficient to resist both abrasive wear and adhesive wear from processing 5000 holes due to its adhesion characteristics and smooth surface.

In drilling high-frequency PCBs, drills need to withstand abrasive wear from the ceramic layer and adhesive wear from the molten resin adhering to the drill following friction and heat generated from the process [2]. Figure 9 shows the maximum length of the flank wear of MCD and NCD drills with different coating thicknesses. It is learned that NCD-coated drills can outperform MCD-coated drills. In addition, the thicker the coating layer, the less chance the drilling force is distributed onto the drill substrate layer [22]. Thereby, localised delamination is reduced, and elasticity and the resistance to abrasive wear are increased. The tool wear of M1 and N1 is not illustrated on Figure 9 given the fact that the maximum length of flank wear of M1 and N1 cannot be measured due to the delamination of the tool cutting edge after drilling 1000 holes. The tools' lifespan can rank as: N3, M3, M2, N2, M1 and N1. The maximum length of the flank wear of the NCD drill is nearly 90% less than the MCD drill.

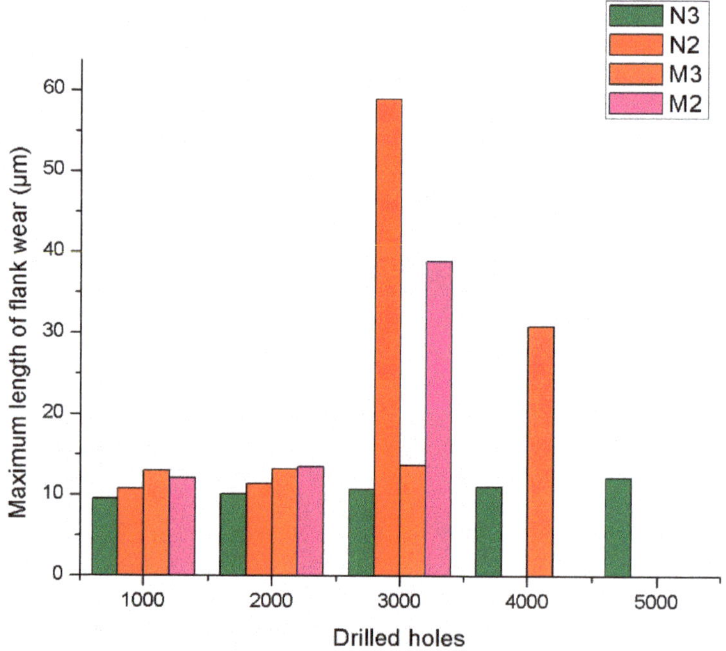

Figure 9. The maximum length of flank wear on NCD- and MCD-coated samples with respect to the number of holes drilled.

4.5. *Investigation on PCB Hole Wall Roughness*

Figure 10 describes the wall roughness of the PCB hole walls after being manufactured, which is one of the main criteria for judging the quality of finished PCBs [7]. Dimensional accuracy, the presence of entry and exit burrs, the smear of resin on the sides of holes and delamination are also characteristics of the workpiece to be examined. Drill film grain size and thickness determine the quality of finished holes. As shown in Figure 10, the results show that the surface roughness of the PCB holes decreases as the coating thickness increases. When 3 μm-MCD-coated drills (M1) are compared to 3-μm-NCD-coated drills

(N1), and 6-μm-MCD-coated drills (M2) are compared to 6-μm-NCD-coated drills (N2), the MCD-coated ones produce better hole quality than the NCD-coated ones. On the other hand, 9-μm-NCD-coated drills (N3) can drill more than 5000 holes with a smooth surface, whereas 9-μm-MCD-coated drills (M3) do not perform well after 3000 holes. The surface roughness of the PCB holes drilled by an NCD-coated drill is 30% better than MCD according to Figure 10. The quality of PCB holes is actually highly dependent on the drill condition. The roughness of the workpiece hole wall increases significantly as the tool deteriorates. Surface roughness of the drill coating layer has a direct impact on the surface roughness of the processed PCB hole wall. On the other hand, according to the AFM results shown in Figure 7, the surface roughness of the NCD film is much lower than the MCD film. Low surface roughness enhances the chip removal process: molten resin can be easily removed from the flute [4]. Therefore, the tool geometry is maintained and the possibilities of delamination and microcracks on the PCB workpiece are reduced.

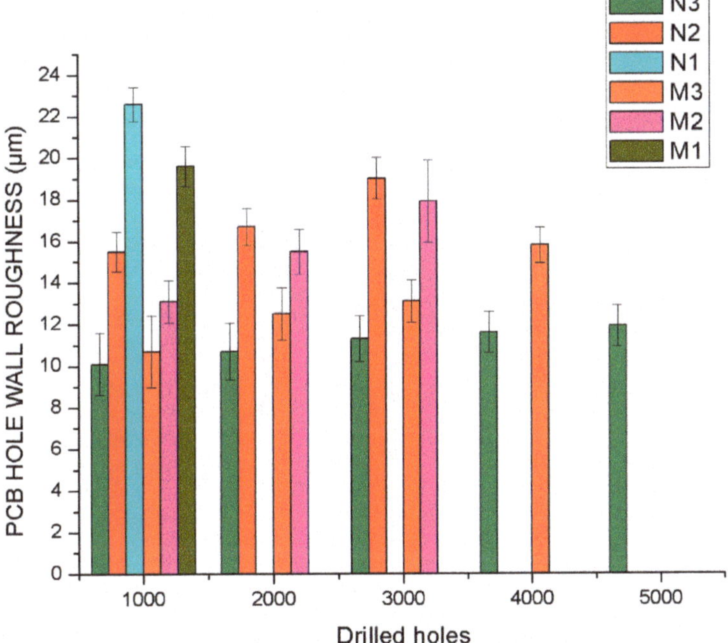

Figure 10. Surface roughness of PCB hole walls manufactured by NCD and MCD tools with respect to the number of holes drilled.

5. Conclusions

The purpose of this study is to determine the effects of the grain size and coating thickness of the diamond film on the drilling performance of high-frequency and high-speed printed circuit boards (PCB), thereby contributing to the relevant industries and academia. The major conclusions that can be drawn from this study are as follows:

1. As methane concentration increases, the grain size of the diamond film changes from microcrystalline to nanocrystalline, which is inversely proportional to temperature and time.
2. A thicker coating can protect the substrate of the drills well as the drilling force applied is spread across the coating layer instead of the substrate of the tool according to the Rockwell indentation, with no crack propagation or coating delamination between the diamond film and substrates when the thickness is 9 μm.

3. The advantage of a thicker diamond coating layer is that the drill is more durable and resistant to wear. For drills with a thicker coating layer, the drilling force applied is distributed throughout the coating layer. For drills with a thinner coating layer, the drilling force is supported by the substrate itself and the coating layer. Localised delamination is reduced as the adhesion is proportional to the film thickness.

Author Contributions: Conceptualization, F.M.K. and Z.S.; methodology, F.M.K.; software, F.M.K. and Z.S.; validation, W.S.Y. and S.T.; formal analysis, F.M.K.; investigation, F.M.K.; resources, S.T.; data curation, F.M.K. and Z.S.; writing—original draft preparation, F.M.K. and Z.S.; writing—review and editing, W.S.Y.; visualization, F.M.K., Z.S. and W.S.Y.; supervision, W.S.Y. and S.T.; funding acquisition, K.Y.D.K. and S.T. All authors have read and agreed to the published version of the manuscript.

Funding: This research was funded by the National Natural Science Foundation of China (No. 52005110, 51975128) and Innovation and Technology Commission (ITS/246/18FX).

Institutional Review Board Statement: Not applicable.

Informed Consent Statement: Not applicable.

Data Availability Statement: Not applicable.

Acknowledgments: This work was supported partially by the National Natural Science Foundation of China and the Innovation and Technology Commission.

Conflicts of Interest: The authors declare no conflict of interest.

References

1. Zhang, W.; Lu, C.; Ge, M.; Bu, F.; Zhang, J. Surface Modified and Gradation-Mixed Al2O3 as an Effective Filler for the Polyphenylene Oxide (PPO) Insulative Layer in Copper Clad Laminates. *J. Mater. Sci. Mater. Electron.* **2020**, *31*, 21602–21616. [CrossRef]
2. Wang, Y.; Zou, B.; Yin, G. Wear Mechanisms of Ti(C7N3)-Based Cermet Micro-Drill and Machining Quality during Ultra-High Speed Micro-Drilling Multi-Layered PCB Consisting of Copper Foil and Glass Fiber Reinforced Plastics. *Ceram. Int.* **2019**, *45*, 24578–24593. [CrossRef]
3. Zheng, L.; Wang, C.; Zhang, X.; Song, Y.; Zhang, L.; Wang, K. The Entry Drilling Process of Flexible Printed Circuit Board and Its Influence on Hole Quality. *Circuit World* **2015**, *41*, 147–153. [CrossRef]
4. Lei, X.L.; He, Y.; Sun, F.H. Optimization of CVD diamond coating type on micro drills in PCB machining. *Surf. Rev. Lett.* **2016**, *23*, 1–8. [CrossRef]
5. Zheng, L.; Wang, C.; Zhang, X.; Huang, X.; Song, Y.; Wang, K.; Zhang, L. The Tool-Wear Characteristics of Flexible Printed Circuit Board Micro-Drilling and Its Influence on Micro-Hole Quality. *Circuit World* **2016**, *42*, 162–169. [CrossRef]
6. Huang, X.; Wang, C.; Yang, T.; He, Y.; Li, Y.; Zheng, L. Wear Characteristics of Micro-Drill during Ultra-High Speed Drilling Multilayer PCB Consisting of Copper Foil and Ceramic Particle Filled GFRPs. *Procedia CIRP* **2020**, *101*, 326–329. [CrossRef]
7. Watanabe, H.; Tsuzaka, H.; Masuda, M. Microdrilling for Printed Circuit Boards (PCBs)-Influence of Radial Run-out of Microdrills on Hole Quality. *Precis. Eng.* **2008**, *32*, 329–335. [CrossRef]
8. Zheng, L.J.; Wang, C.Y.; Fu, L.Y.; Yang, L.P.; Qu, Y.P.; Song, Y.X. Wear Mechanisms of Micro-Drills during Dry High Speed Drilling of PCB. *J. Mater. Process. Technol.* **2012**, *212*, 1989–1997. [CrossRef]
9. Gäbler, J.; Schäfer, L.; Westermann, H. Chemical Vapour Deposition Diamond Coated Microtools for Grinding, Milling and Drilling. *Diam. Relat. Mater.* **2000**, *9*, 921–924. [CrossRef]
10. Wang, H.; Wang, C.; Wang, X.; Sun, F. Effects of Carbon Concentration and Gas Pressure with Hydrogen-Rich Gas Chemistry on Synthesis and Characterizations of HFCVD Diamond Films on WC-Co Substrates. *Surf. Coat. Technol.* **2021**, *409*, 126839. [CrossRef]
11. Dawedeit, C.; Kucheyev, S.O.; Shin, S.J.; Willey, T.M.; Bagge-Hansen, M.; Braun, T.; Wang, Y.M.; El-Dasher, B.S.; Teslich, N.E.; Biener, M.M.; et al. Grain Size Dependent Physical and Chemical Properties of Thick CVD Diamond Films for High Energy Density Physics Experiments. *Diam. Relat. Mater.* **2013**, *40*, 75–81. [CrossRef]
12. Chen, N.; Shen, B.; Yang, G.; Sun, F. Tribological and Cutting Behavior of Silicon Nitride Tools Coated with Monolayer- and Multilayer-Microcrystalline HFCVD Diamond Films. *Appl. Surf. Sci.* **2013**, *265*, 850–859. [CrossRef]
13. Williams, O.A. Nanocrystalline Diamond. *Diam. Relat. Mater.* **2011**, *20*, 621–640. [CrossRef]
14. Peters, M.G.; Cummings, R.H.; National Center for Manu-facturing Sciences. Methods for Coating Adherent Diamond Films on Cemented Tungsten Carbide Substrates. U.S. Patent 5,236,740, 17 August 1993.
15. Asmussen, J.; Reinhard, D.K. (Eds.) *Diamond Films Handbook*; CRC Press: Boca Raton, FL, USA, 2002; ISBN 0824795776.
16. Wei, Q.; Yu, Z.M.; Ashfold, M.N.R.; Chen, Z.; Wang, L.; Ma, L. Effects of Thickness and Cycle Parameters on Fretting Wear Behavior of CVD Diamond Coatings on Steel Substrates. *Surf. Coat. Technol.* **2010**, *205*, 158–167. [CrossRef]

17. Wiora, M.; Brühne, K.; Flöter, A.; Gluche, P.; Willey, T.M.; Kucheyev, S.O.; Van Buuren, A.W.; Hamza, A.V.; Biener, J.; Fecht, H.J. Grain Size Dependent Mechanical Properties of Nanocrystalline Diamond Films Grown by Hot-Filament CVD. *Diam. Relat. Mater.* **2009**, *18*, 927–930. [CrossRef]
18. Amaral, M.; Fernandes, A.J.S.; Vila, M.; Oliveira, F.J.; Silva, R.F. Growth Rate Improvements in the Hot-Filament CVD Deposition of Nanocrystalline Diamond. *Diam. Relat. Mater.* **2006**, *15*, 1822–1827. [CrossRef]
19. May, P.W.; Mankelevich, Y.A. From Ultrananocrystalline Diamond to Single Crystal Diamond Growth in Hot Filament and Microwave Plasma-Enhanced CVD Reactors: A Unified Model for Growth Rates and Grain Sizes. *J. Phys. Chem. C* **2008**, *112*, 12432–12441. [CrossRef]
20. Wang, X.; Zhang, T.; Shen, B.; Zhang, J.; Sun, F. Simulation and Experimental Research on the Substrate Temperature Distribution in HFCVD Diamond Film Growth on the Inner Hole Surface. *Surf. Coat. Technol.* **2013**, *219*, 109–118. [CrossRef]
21. Wei, Q.; Ashfold, M.N.R.; Mankelevich, Y.A.; Yu, Z.M.; Liu, P.Z.; Ma, L. Diamond Growth on WC-Co Substrates by Hot Filament Chemical Vapor Deposition: Effect of Filament-Substrate Separation. *Diam. Relat. Mater.* **2011**, *20*, 641–650. [CrossRef]
22. Wang, H.; Song, X.; Wang, X.; Sun, F. Fabrication, Tribological Properties and Cutting Performances of High-Quality Multilayer Graded MCD/NCD/UNCD Coated PCB End Mills. *Diam. Relat. Mater.* **2021**, *118*, 108505. [CrossRef]

Article

Influence of Material Microstructure on Machining Characteristics of OFHC Copper C102 in Orthogonal Micro-Turning

Chuan-Zhi Jing [1,2], Ji-Lai Wang [1,2,*], Xue Li [1,2], Yi-Fei Li [1,2] and Lu Han [1,2,*]

[1] Key Laboratory of High Efficiency and Clean Mechanical Manufacture of Ministry of Education, School of Mechanical Engineering, Shandong University, Jinan 250061, China; 201933971@mail.sdu.edu.cn (C.-Z.J.); 202034350@mail.sdu.edu.cn (X.L.); 201944031@mail.sdu.edu.cn (Y.-F.L.)

[2] National Demonstration Center for Experimental Mechanical Engineering Education, Shandong University, Jinan 250061, China

* Correspondence: jlwang@sdu.edu.cn (J.-L.W.); jl.wang@connect.polyu.hk (L.H.)

Abstract: Micro-cutting is different from conventional cutting in its mechanics. The workpiece material is not considered to be homogeneous in the micro-cutting process. As a result, it is critical to comprehend how microstructure affects surface integrity, cutting forces, and chip formation. In this paper, we experimented with micro-turning on oxygen-free high-conductivity (OFHC) copper with different microstructures after annealing. Feed rate parameters were smaller than, larger than, and equal to the grain size, respectively. Experimental results show that when the feed rates are equivalent to the grain size, the surface roughness of the machined surface is low and the width of the flake structure on the free surface of chips is minimal, and the explanations for these occurrences are connected to dislocation slip.

Keywords: micro-cutting; grain size; surface integrity; cutting forces; chip formation; OFHC copper C102

Citation: Jing, C.-Z.; Wang, J.-L.; Li, X.; Li, Y.-F.; Han, L. Influence of Material Microstructure on Machining Characteristics of OFHC Copper C102 in Orthogonal Micro-Turning. *Processes* **2022**, *10*, 741. https://doi.org/10.3390/pr10040741

Academic Editor: Jun Zhang

Received: 8 February 2022
Accepted: 6 April 2022
Published: 11 April 2022

1. Introduction

The fast development of micro-electro-mechanical systems (MEMS) has raised the bar for micro-component machining precision and molding quality [1–3]. As the workpiece size reduces, the mechanical properties of materials are highly affected by grain size and geometrical size; the mechanical characteristics of the material are different from macroscopic manufacturing. In conventional cutting, the workpiece material is perceived as isotropic, and the cutting tool nose radius is considered sharp, because the feed rate is not of the same order of magnitude as the tool tip [4–7]. The size of the workpiece is small in the micro-cutting process, the feed rates must be minimal because high feed rates will impact the stiffness and strength of the material. When feed rates are lower than grain size, chip formation may occur within the grain. The material will no longer be thought to be homogenous [8–11]. Traditional cutting theory will no longer be applicable. Therefore, in order to improve the surface quality after machining, it is necessary to study the influence of microstructure on the cutting mechanism during the micro-cutting process.

Kumar et al. [12] investigated the influence of cutting factors on micro-cutting in their study. The investigation revealed that the depth of cut was the most important factor influencing the surface roughness of the machined surface. During micro-cutting experiments on Ti-6AL-4V, Aslantas et al. [4] discovered that the surface roughness Sa and Sz increased with higher feeds, and this discovery was confirmed by SEM examination of chip morphology. Zhang et al. [13] discovered that microstructure could change the instability of the micro-topography of the machined surface. Vipindas et al. [14] conducted experiments on Ti-6Al-4V micro-milling and found that when the un-deformed cutting thickness was less than the minimum cutting thickness, the machined surface roughness caused by ploughing decreased with the increase of the un-deformed cutting thickness, and when the un-deformed cutting thickness was greater than the minimum cutting thickness, the machined surface roughness increased with the increase of the un-deformed cutting thickness. During micro-milling, Wu et al. [15] discovered that materials with fine grain sizes had higher cutting forces than those with large grain sizes. The reason for this is that grain boundaries can obstruct dislocation movement, and the smaller the grain size of the material, the higher the proportion of grain boundaries and, the greater the cutting forces. Cutting force analysis is critical in the investigation of the micro-cutting process. In the Komatsu's investigation, the specific cutting force of ultrafine grain steel was higher in small depths of cutting when compared to standard grain steel [16]. Based on the physics-based investigation, Siva et al. [17] quantified the influence of grain size, grain boundaries, and grain orientation on cutting forces. Jiang et al. [18] found that the degree of chip serration increased with grain size in cutting experiments on stainless steel 304 L. Wu et al. [19] found that the magnitude of cutting forces is affected by grain orientation and grain size. Popov et al. [20] discovered that the surface roughness of the workpiece dropped as the grain size decreased, from 0.5 μm when the grain size of the aluminum alloy was 100 μm to 0.15 μm when the grain size of the aluminum alloy was 0.6 μm.

Although much work has previously been carried out on the effect of microstructure on the machining characteristics of materials, few researchers have determined the machining parameters based on grain size. In this study, feed rates smaller than grain size, equal to grain size, and larger than grain size were used to further investigate the effect of grain size on material processing characteristics.

In this study, oxygen-free high-conductivity (OFHC) copper with different grain sizes was obtained by annealing at three different temperatures and holding times. Then, micro-turning experiments at various rates were conducted depending on grain size. When the feed rates were less than, equal to, and greater than grain size, the influence of grain size on machining characteristics was investigated. The cutting force in the micro-turning process was measured and studied using a dynamometer. The free surface of chips produced by machining was observed using a scanning electron microscope (SEM) (JXA-8530F PLUS, Tokyo, Japan). The surface roughness of the machined surface was observed by laser confocal microscope (VK-X200K, Osaka, Japan) to analyze the difference in machined surface roughness.

2. Experiment

2.1. Workpiece Material

Because commercial OFHC copper is commonly used in electronic components due to its superior electrical conductivity and machinability, it was chosen as the experimental material in this investigation. Different grain sizes were generated by annealing at 450 °C for 2 h, 600 °C for 2 h, and 750 °C for 3 h, respectively, to examine the influence of microstructure on surface integrity, chip formation, and cutting force in the micro-turning process. The workpiece material was annealed in tube vacuum furnaces. The annealing process was continuously supplied with high purity nitrogen to prevent the workpiece from being oxidized and affecting the experimental results.

After cold inlay, a polishing machine was utilized to grind and polish the samples. The samples were polished with 400, 600, 1200, and 2000# water sandpaper in turn. The surface of the sample was ground and polished until the scratches were consistent and uniform each time, and then a further polishing process was performed. Step by step, the samples were polished using alumina polishing solution with particle sizes of 1, 0.5, and 0.3 μm until the surface was smooth and free of visible scratches. For the corrosion solution, we prepared 10 g $FeCl_3$ + 30 mL HCL + 120 mL H_2O and used the immersion corrosion technique, using tweezers to keep the sample submerged in the corrosion solution, which underwent corrosion for about 5 s. An optical microscope was used to examine the grain size of the corroded sample. The metallographic microstructure is shown in Figure 1. Nano

Measurer software was used to measure and analyze the grain size. The average grain size after heat treatment was 30, 40, and 60 μm, respectively.

Figure 1. Microstructure of annealed OFHC copper. (**a**) 30 μm, (**b**) 40 μm, and (**c**) 60 μm.

2.2. Experimental Setup

As illustrated in Figure 2a, the orthogonal micro-turning experimental was conducted on a CKD6150H CNC lathe-machining center. The schematic diagram of orthogonal cutting is shown in Figure 2b. Because the feed rate in the micro-cutting process is very small, when the feed rate is less than the minimum chip thickness, the workpiece material does not produce chips, resulting in a severe ploughing and rubbing phenomenon on the surface of the workpiece [15]; when the feed rate is greater than the minimum cutting thickness, the workpiece material will only produce chips. Based on the previous studies' results [21–24], a PVD-coated carbide insert (TiAlN) with an edge radius of 25 μm was chosen as a cutting tool in this study, which met the experimental criteria. The rake angle of the cutting tool was 10°, and the relief angle was 11°. The nose radius of the tool was 0.19 mm. The workpiece was disc-shaped with a diameter of 40 mm and a thickness of 1 mm, with a hole in the center. The workpiece was clamped by the fixture and then cut with spindle rotation. The cutting force was measured by Kistler 9257B Dynamometer three-component piezoelectric dynamometers during the experiment, and the lubrication conditions for cutting were dry conditions.

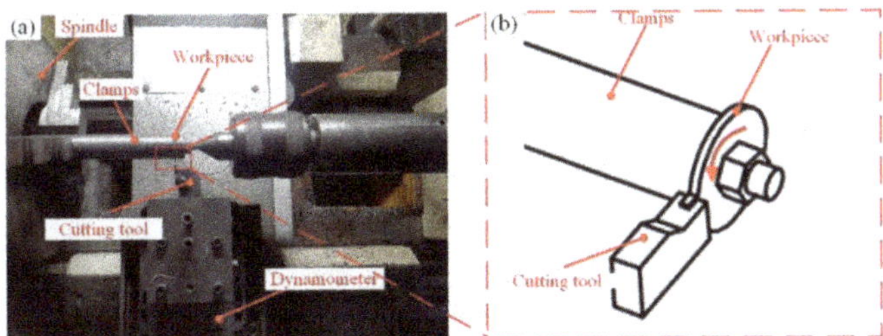

Figure 2. (**a**) Experimental setup, (**b**) schematic diagram of cutting.

Given the various microstructures of the workpiece after heat treatment (i.e., different grain sizes), feed rates were smaller than the grain size, equal to the grain size, and larger than the grain size, in that order. The cutting speed was 50 m/min. The cutting parameters are shown in Table 1. In particular, each set of intermediate feed rates is equal to the grain size. Cutting experiments with different feed rates were carried out on workpieces with the same microstructure to verify the effect of microstructure on the machining characteristics of OFHC copper. Subheadings may be used to split this section. It should provide a short and summary of the experimental data, their interpretation, and possible experimental inferences.

Table 1. The parameters of micro-cutting.

Machine Tool	CKD6150H Lathe-Machining Center
Cutting speed (m/min)	50
Cutting tool	PVD-coated carbide insert (TiAlN)
Workpiece material	OFHC copper
Feed rate (μm/rev)	20, 30, 40 30, 40, 50 50, 60, 70

After the experiment, scanning electron microscopy was used to observe the free surface of chips. The surface roughness was measured by using laser confocal microscopy to investigate the relationship between microstructure and surface integrity.

3. Analysis and Discussion Results

3.1. Machined Surface Integrity

After the experiment, the roughness of the machined surface was measured at five locations on the workpieces in order to avoid accidental errors in the measurement. The topography of the machined surface is shown in Figure 3. Figure 3 clearly shows the path of the turning tool on the surface of the workpiece. The toolpath on the machined surfaces of workpieces with a grain size of 60 μm was more prominent than the other two grain sizes. According to previous studies [5,24], the material flows to the two sides closest to the cutting edge during the cutting process; softer materials are prone to this flow phenomenon. Wang et al. [25] found that the larger the grain size of the material, the more likely it is that plastic flow will occur, and plastic flow increases surface roughness. Therefore, the most pronounced tool paths are found on machined surfaces of workpieces with a grain size of 60 μm.

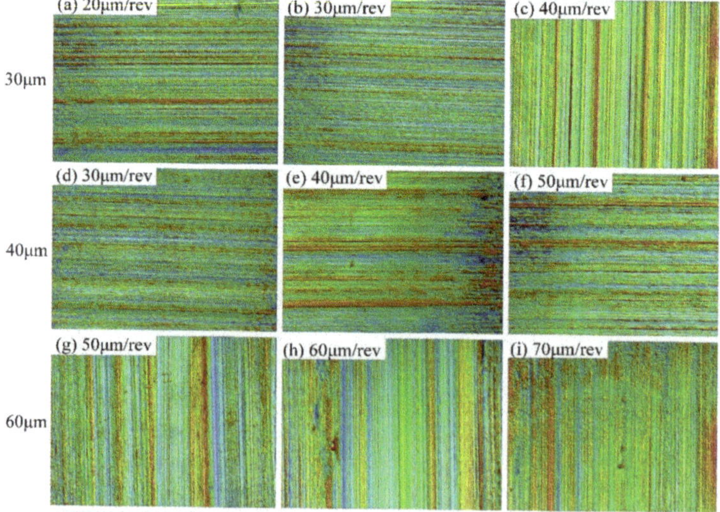

Figure 3. The topography of the machined surface. The grain size of the workpiece is 30 μm: (**a**) feed rate = 20 μm/rev, (**b**) feed rate = 30 μm/rev, (**c**) feed rate = 40 μm/rev; The grain size of the workpiece is 40 μm: (**d**) feed rate = 30 μm/rev, (**e**) feed rate = 40 μm/rev, (**f**) feed rate = 50 μm/rev; The grain size of the workpiece is 60 μm: (**g**) feed rate = 50 μm/rev, (**h**) feed rate = 60 μm/rev, (**i**) feed rate = 70 μm/rev.

Ra is based on the line profile method for roughness assessment. Ra is defined as the arithmetic mean deviation of the contour of the distance between points on the surface contour of the object, and the datum line and is mainly used to characterize one-dimensional profiles. Sa is based on the area profile method for roughness assessment. Sa is defined as the distance from the point on the surface profile of an object to the reference plane, which represents the arithmetic mean deviation of the area shape. Sa is mainly used to characterize two-dimensional profiles. It is more suitable when assessing surface roughness in the microscopic field. Therefore, Sa was used to characterize the roughness of machined surfaces in this study; the surface roughness values are shown in Figure 4. The Sa of the finished surface was shown to be the smallest when the feed rate was equal to the grain size for workpieces with the same grain sizes. When the feed rate was smaller than the grain size during the process of micro-turning, the cutting largely occurred inside the grains, the grains were over-twisted and ripped, and the surface roughness was severe. When the feed rate was equal to the grain size, the cutting mostly occurred at the grain boundaries, the grain was stripped from its position, and the surface roughness was slight; when the feed rate was larger than the grain size, the cutting situation was the same as when the feed rate was smaller than the grain size: the cutting occurred inside the grain. The results of laser confocal microscopy also showed that the machined surface was smoother, and there were fewer tool paths on the surface when the feed rate was equal to the grain size.

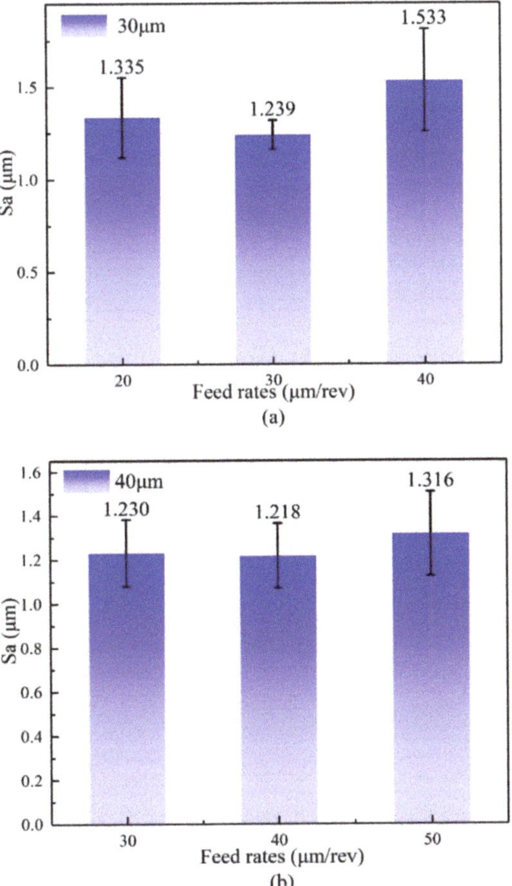

Figure 4. *Cont.*

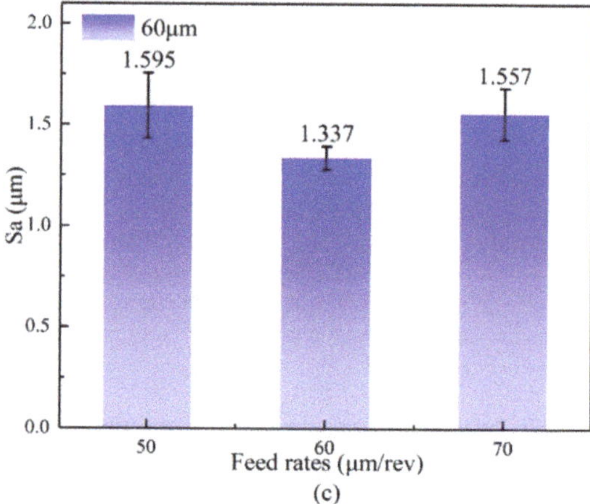

Figure 4. The roughness of machined surface at different feed rates. (**a**) Grain size = 30 μm, (**b**) Grain size = 40 μm, (**c**) Grain size = 60 μm.

3.2. Cutting Forces

To avoid chance errors, each group of cutting tests was repeated three times and the cutting forces were collected for the cutting process. The cutting forces Fp (radial force, perpendicular to the machined surface) and Fc (tangential force, parallel to the machined surface) for each group of cutting tests are measured by Kistler 9257B Dynamometer, which collects the force signals during the cutting process in real time and converts them into cutting force curves. The cutting force curves reflect the cutting forces created by the tool from the moment it touches the workpiece to the time it leaves the surface. The experimental cutting force curves were smoothed, and the study utilized the average of each group of cutting forces. The results are shown in Figure 5.

As clearly shown in Figure 5, the feed rate has an important influence on the cutting force of the same microstructure workpiece. Cutting forces increased in a linear relationship with feed rates. This implies that changes in macro-feed rates have a higher impact on cutting forces than changes in grain sizes. From Figure 5d,e, it can be seen that when the feed rate was 30 μm, the cutting force Fc of the workpiece with a grain size of 30 μm was less than that of the workpiece with a grain size of 40 μm. When the feed rate was 30 μm/rev, the turning tool cut the workpiece with a grain size of 30 μm, and the cutting path mostly passed through the grain boundary. However, the workpiece with a grain size of 40 μm was cut by the tool, and the cutting path mostly passed through the inside grain. Cutting inside the grain caused significant dislocation plugging, which in turn produced a large cutting force. The same phenomenon could be seen when the feed rate was 40 μm/rev.

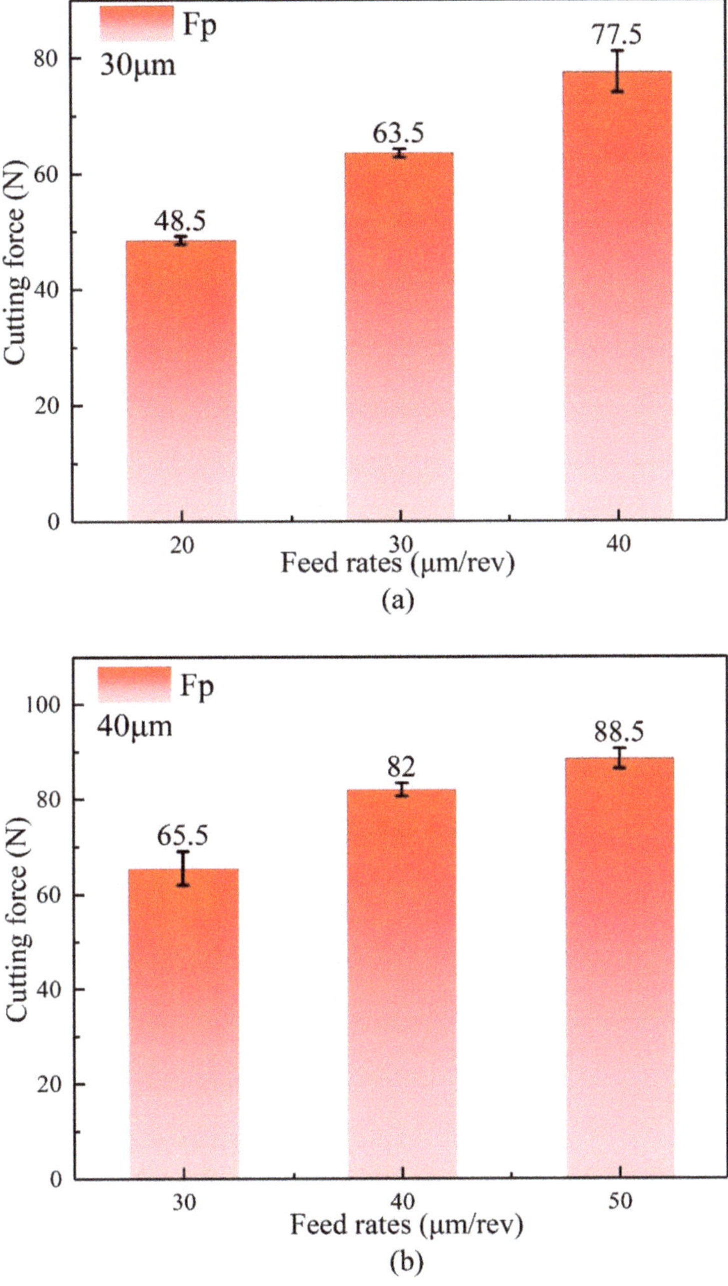

Figure 5. *Cont.*

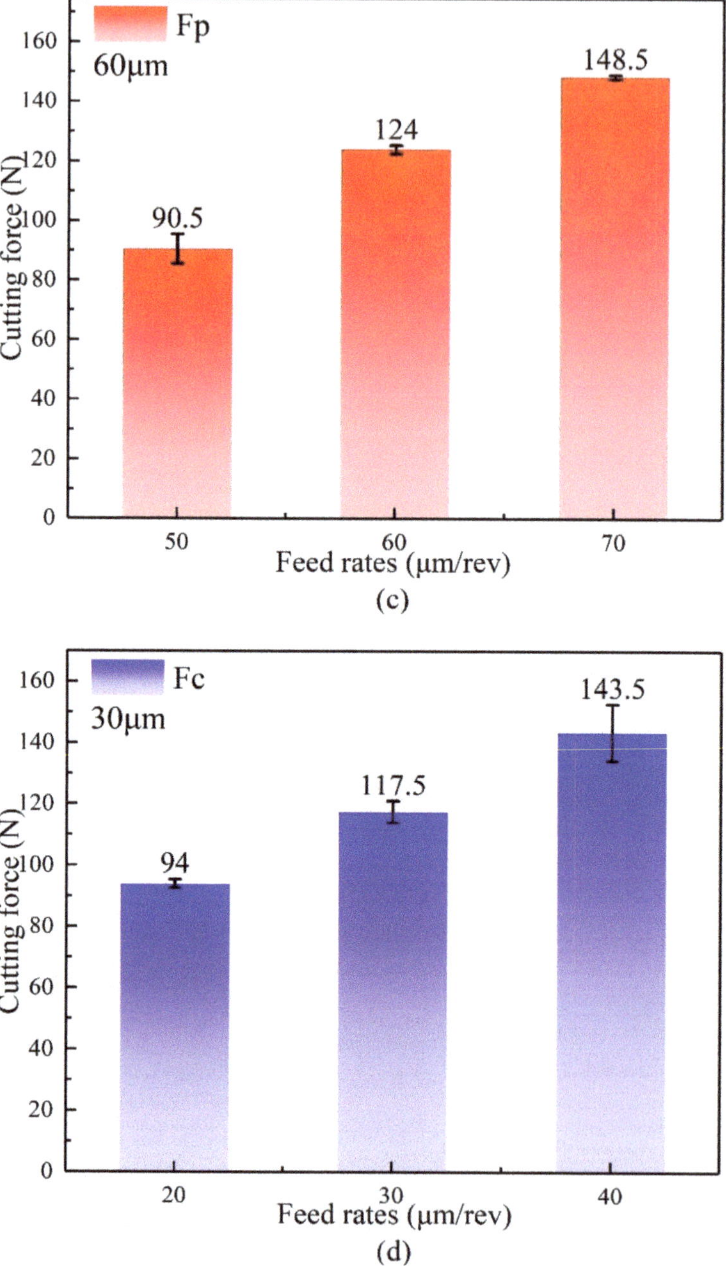

Figure 5. *Cont.*

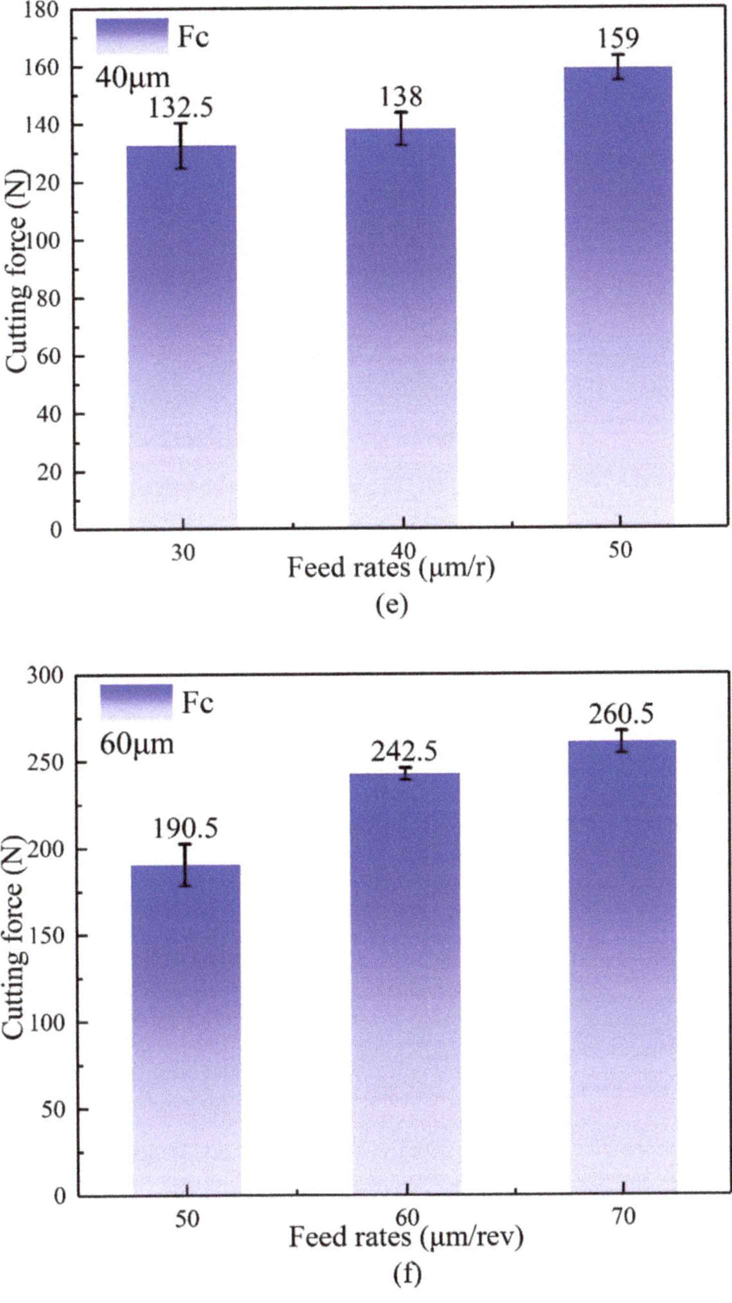

Figure 5. The cutting forces at different feeding rates. (**a–c**) Cutting forces Fp of Grain size = 30 μm, 40 μm, 60 μm, respectively; (**d–f**) Cutting forces Fc of Grain size = 30 μm, 40 μm, 60 μm, respectively.

3.3. Chip Formation

The chip morphology in micro-turning processes varies depending on the workpiece material and cutting settings. Due to its good ductility, OFHC copper is commonly processed to produce continuous ribbons of chips. Free surfaces of continuous ribbon chips commonly generate a lamellar pattern, which can be seen with a scanning electron microscope. The schematic diagram of the workpiece chip pattern is shown in Figure 6.

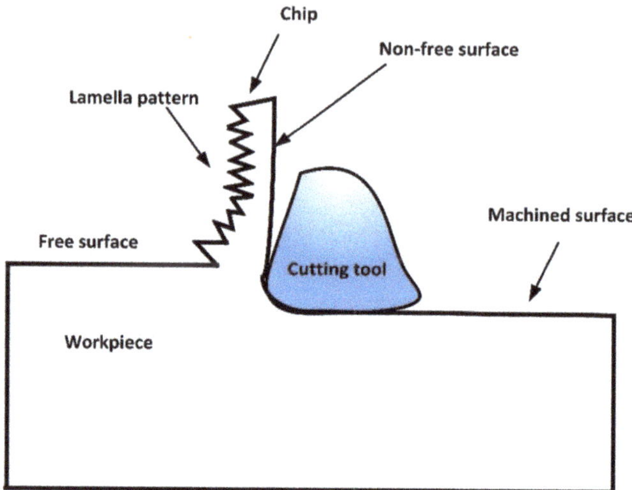

Figure 6. Schematic diagram of workpiece chip pattern.

The microscopic morphology of the free surface of chips is shown in Figure 7. The morphology of the lamellar pattern on the free surface is influenced by grain size and cutting settings, as can be observed. On the same chip, such lamellar folds were beautifully placed and about the same width, yet the chips were significantly different. The average width of the free surface lamellar structure was calculated by taking a line of the same length in the direction perpendicular to the lamellar structure at the same scale and determining the number of lamellar patterns on the line; the average width of the lamellar structure results are shown in Figure 8. As can be seen from Figure 8, when the grain sizes are 30 and 40 μm, the width of the lamellar patterns of chips whose feed rates were smaller than the grain size was about 1.5 times wider than when the feed rates were equal to the grain size, and the width of the lamellar patterns of chips whose feed rates were greater than the grain size was 2 times wider than when the feed rates were equal to the grain size.

When the feed rates are equivalent to the grain size, the breadth of the lamellar structures on the free surface is minimized, as seen in Figure 8. Figure 9a shows the microstructure of the chip and Figure 9b shows the microstructure of the machined workpiece. As shown in Figure 9a, the extrusion of the tool and the workpiece causes the original microstructure of the material to change and even form nanocrystals that could not be observed by corrosion during the cutting process. In Figure 9b, the microstructure of the machined workpiece was divided into two parts: region I and region II. Region II is far from the machined surface and is less affected by the machining, and the microstructure does not change much. Region I is close to the machined surface and is more affected by the machining, and the microstructure changes obviously. The microstructure of region I is the same as the microstructure of the chip in Figure 9a. Therefore, it is difficult to draw conclusions through metallographic observations. However, the experimental phenomenon can be indirectly explained by the findings of previous studies [5,26]: The continuous shear slip of the chip tends to alter the grain borders because the ability of grain boundaries to resist deformation is larger than the ability of the grain interior to resist deformation. When the feed rate is equal to the grain size, most of the micro-turning process occurs at the grain boundary and the distance of shear slip is small, so the width of the lamellar structure is small. When the feed is greater or less than the grain size, most micro-turning occurs inside the grain, the distance of the tool passing inside the grain is large, and the distance of cutting shear slip increases; therefore, the width of the lamellar structure is small.

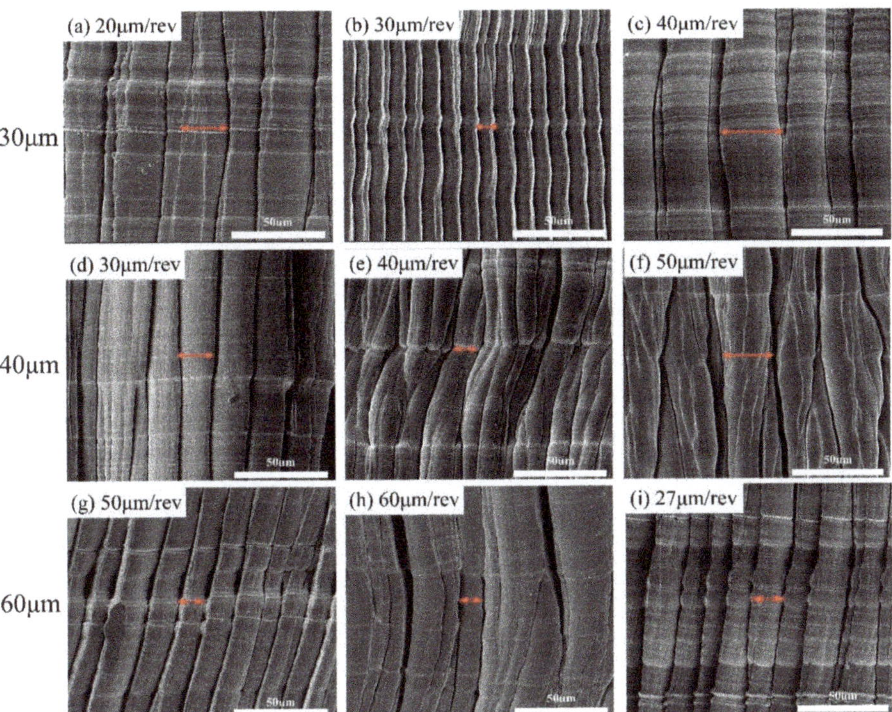

Figure 7. Micromorphology of free surface of workpiece chips at different feed rates. (**a–c**) Grain size = 30 μm, (**d–f**) Grain size = 40 μm, (**g–i**) Grain size = 60 μm.

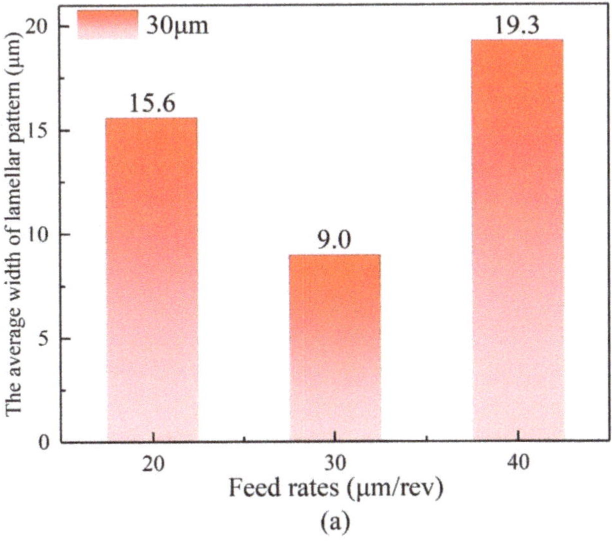

Figure 8. *Cont.*

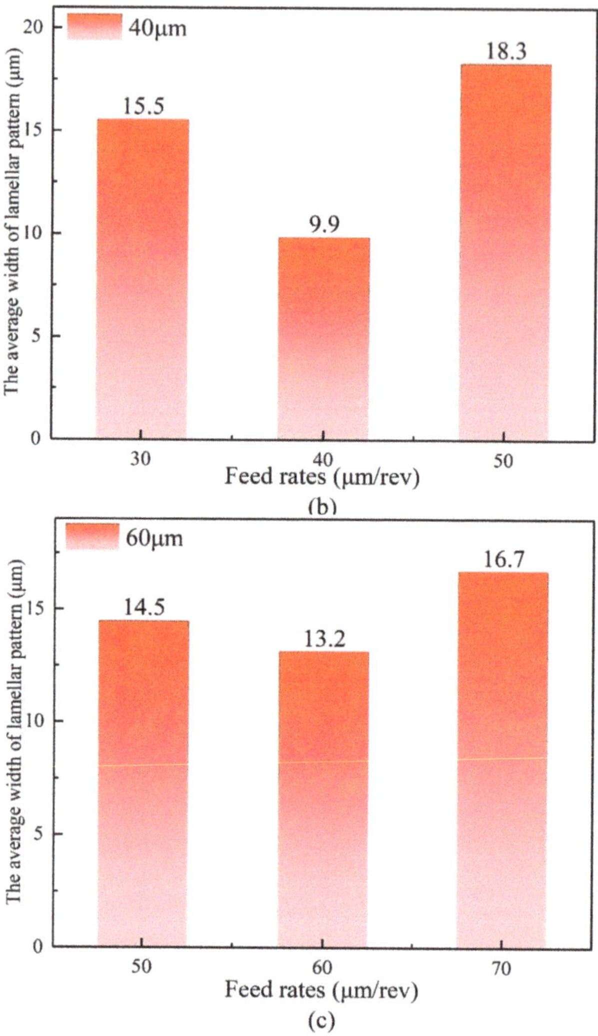

Figure 8. The average width of the lamellar structure: (**a**) 30 μm, (**b**) 40 μm, (**c**) 60 μm.

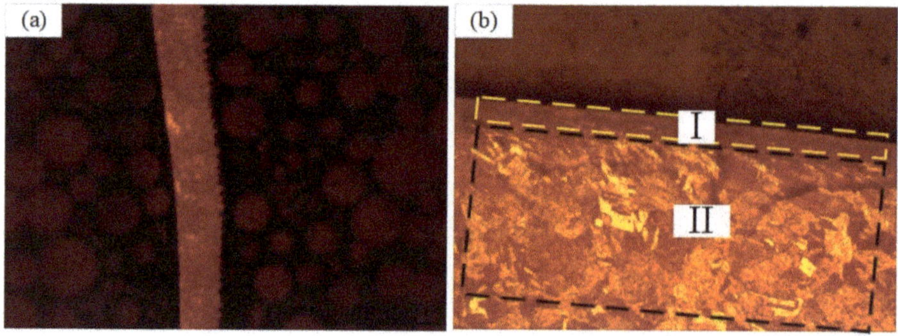

Figure 9. The microstructure of (**a**) the chip and (**b**) the machined workpiece.

4. Conclusions

In this paper, micro-turning experiments were performed on three different microstructures of OFHC copper to investigate the effects of micro-turning on cutting forces, surface integrity, and chip morphology for the same microstructure at feed rates smaller than, greater than, and equal to the grain size. The following are the key conclusions of this paper:

1. The toolpath on the machined surfaces of workpieces with a grain size of 60 μm were more prominent than the other two grain sizes. An explanation for this phenomenon is that material flows to the two sides closest to the cutting edge in the cutting process. The softer the material, the more prone it is to this flow phenomenon, and the workpiece with a grain size of 60 m is the softest. When the feed rate is equal to the grain size for the same grain size workpiece, the Sa of the surface completed is the smallest.
2. The cutting forces increased with increasing feed rates. This indicates that the change in feed rates during the cutting process has a higher influence on the cutting force than the location of the microstructure when the cutting process takes place. When the feed rate was 30 μm, the cutting force Fc of the workpiece with a grain size of 30 μm (cutting happened largely at the grain boundaries) was less than that of the workpiece with a grain size of 40 μm (cutting mostly occurred inside the grain). The reason for this is that cutting inside the grain induced a lot of dislocation plugs, which resulted in a lot of cutting force. Similar behavior may be observed when the feed rate is set to 40 μm/rev.
3. On the free surface, there were lamellar patterns that were nicely ordered and almost the same width on the same chip. The breadth of the lamellar patterns on the free surface was minimum when the feed rates were equal to the grain sizes. When the grain sizes were 30 μm and 40 μm, the width of the lamellar patterns of chips whose feed rates were smaller than the grain size was about 1.5 times wider than when the feed rates were equal to the grain size, and the width of the lamellar patterns of chips whose feed rates were greater than the grain size was about 2 times wider than when the feed rates were equal to the grain size.

Author Contributions: Methodology, J.-L.W.; investigation, C.-Z.J. and X.L.; data curation, Y.-F.L. and L.H. All authors have read and agreed to the published version of the manuscript.

Funding: The authors gratefully acknowledge funding supported by the Young Scholars Program of Shandong University, Undergraduate Education Teaching Reform and Research Programs of Shandong University (2021Y220), and Shandong Provincial Natural Science Foundation (ZR2019BEE062).

Institutional Review Board Statement: Not applicable.

Informed Consent Statement: Not applicable.

Data Availability Statement: Data is contained within the article.

Conflicts of Interest: The authors declare no conflict of interest.

References

1. Jing, C.; Wang, J.; Zhang, C.; Sun, Y.; Shi, Z. Influence of size effect on the dynamic mechanical properties of OFHC copper at micro-/meso-scales. *Int. J. Adv. Manuf. Technol.* **2022**, 1–15. [CrossRef]
2. Sun, X.; Liu, S.; Zhang, X.; Tao, Y.; Boczkaj, G.; Yoon, J.Y.; Xuan, X. Recent advances in hydrodynamic cavitation-based pretreatments of lignocellulosic biomass for valorization. *Bioresour. Technol.* **2021**, *345*, 126251. [CrossRef] [PubMed]
3. Xuan, X.; Wang, M.; Zhang, M.; Kaneti, Y.V.; Xu, X.; Sun, X.; Yamauchi, Y. Nanoarchitectonics of low-dimensional metal-organic frameworks toward photo/electrochemical CO_2 reduction reactions. *J. CO2 Util.* **2022**, *57*, 101883. [CrossRef]
4. Aslantas, K.; Danish, M.; Hasçelik, A.; Mia, M.; Gupta, M.; Ginta, T.; Ijaz, H. Investigations on surface roughness and tool wear characteristics in micro-turning of Ti-6Al-4V alloy. *Materials* **2020**, *13*, 2998. [CrossRef]
5. Yu, J.; Wang, G.; Rong, Y. Experimental study on the surface integrity and chip formation in the micro-cutting process. *Procedia Manuf.* **2015**, *1*, 655–662. [CrossRef]
6. Zhang, C.; Chen, S.; Wang, J.; Shi, Z.; Du, L. Reproducible Flexible SERS Substrates Inspired by Bionic Micro-Nano Hierarchical Structures of Rose Petals. *Adv. Mater. Interfaces* **2022**, 2102468. [CrossRef]

7. Li, Q.; Wu, Q.; Liu, J.; He, J.; Liu, S. Topology optimization of vibrating structures with frequency band constraints. *Struct. Multidiscip. Optim.* **2021**, *63*, 1203–1218. [CrossRef]
8. Bissacco, G.; Hansen, H.N.; Chiffre, L.D. Size Effects on Surface Generation in Micro Milling of Hardened Tool Steel. *CIRP Ann.* **2006**, *55*, 593–596. [CrossRef]
9. Zhang, C.; Chen, S.; Jiang, Z.; Shi, Z.; Wang, J.; Du, L. Highly sensitive and reproducible SERS substrates based on ordered micropyramid array and silver nanoparticles. *ACS Appl. Mater. Interfaces* **2021**, *13*, 29222–29229. [CrossRef]
10. Mian, A.J.; Driver, N.; Mativenga, P.T. A comparative study of material phase effects on micro-machinability of multiphase materials. *Int. J. Adv. Manuf. Technol.* **2010**, *50*, 163–174. [CrossRef]
11. Li, Q.; Sigmund, O.; Jensen, J.S.; Aage, N. Reduced-order methods for dynamic problems in topology optimization: A comparative study. *Comput. Methods Appl. Mech. Eng.* **2021**, *387*, 114149. [CrossRef]
12. Kumar, S.P.L. Measurement and uncertainty analysis of surface roughness and material removal rate in micro turning operation and process parameters optimization. *Measurement* **2019**, *140*, 538–547. [CrossRef]
13. Zhang, S.J.; To, S.; Cheung, C.F.; Zhu, Y.H. Microstructural changes of Zn–Al alloy influencing micro-topographical surface in micro-cutting. *Int. J. Adv. Manuf. Technol.* **2014**, *72*, 9–15. [CrossRef]
14. Vipindas, K.; Anand, K.N.; Mathew, J. Effect of cutting edge radius on micro end milling: Force analysis, surface roughness, and chip formation. *Int. J. Adv. Manuf. Technol.* **2018**, *97*, 711–722. [CrossRef]
15. Wu, X.; Li, L.; He, N.; Yao, C.; Zhao, M. Influence of the cutting edge radius and the material grain size on the cutting force in micro cutting. *Precis. Eng.* **2016**, *45*, 359–364. [CrossRef]
16. Komatsu, T.; Yoshino, T.; Matsumura, T.; Torizuka, S. Effect of crystal grain size in stainless steel on cutting process in micromilling. *Procedia Cirp* **2012**, *1*, 150–155. [CrossRef]
17. Venkatachalam, S.; Fergani, O.; Li, X.; Guo Yang, J.; Chiang, K.N.; Liang, S.Y. Microstructure effects on cutting forces and flow stress in ultra-precision machining of polycrystalline brittle materials. *J. Manuf. Sci. Eng.* **2015**, *137*, 021020. [CrossRef]
18. Jiang, L.; Roos, Å.; Liu, P. The influence of austenite grain size and its distribution on chip deformation and tool life during machining of AISI 304L. *Metall. Mater. Trans. A* **1997**, *28*, 2415–2422. [CrossRef]
19. Wu, X.; Li, L.; He, N.; Zhao, M.; Zhan, Z. Investigation on the influence of material microstructure on cutting force and bur formation in the micro-cutting of copper. *Int. J. Adv. Manuf. Technol.* **2015**, *79*, 321–327. [CrossRef]
20. Popov, K.B.; Dimov, S.S.; Pham, D.T.; Minev, R.M.; Rosochowski, A.; Olejnik, L. Micromilling: Material microstructure effects. *Proc. Inst. Mech. Eng. Part B J. Eng. Manuf.* **2006**, *220*, 1807–1813. [CrossRef]
21. Lai, X.; Li, H.; Li, C.; Lin, Z.; Ni, J. Modelling and analysis of micro-scale milling considering size effect, micro cutter edge radius and minimum chip thickness. *Int. J. Mach. Tools Manuf.* **2008**, *48*, 1–14. [CrossRef]
22. Sun, X.; Yang, Z.; Wei, X.; Tao, Y.; Boczkaj, G.; Yoon, J.Y.; Xuan, X.; Chen, S. Multi-objective optimization of the cavitation generation unit structure of an advanced rotational hydrodynamic cavitation reactor. *Ultrason. Sonochem.* **2021**, *80*, 105771. [CrossRef]
23. Ikawa, N.; Shimada, S.; Tanaka, H. Minimum thickness of cut in micromachining. *Nanotechnology* **1992**, *3*, 6. [CrossRef]
24. Yuan, Z.J.; Zhou, M.; Dong, S. Effect of diamond tool sharpness on minimum cutting thickness and cutting surface integrity in ultraprecision machining. *J. Mater. Process. Technol.* **1996**, *62*, 327–330. [CrossRef]
25. Wang, J.L.; Fu, M.W.; Shi, S.Q. Influences of size effect and stress condition on ductile fracture behavior in micro-scaled plastic deformation. *Mater. Des.* **2017**, *131*, 69–80. [CrossRef]
26. Wang, J.S.; Gong, Y.D.; Abba, G.; Chen, K.; Shi, J.S.; Cai, G.Q. Surface generation analysis in micro end-milling considering the influences of grain. *Microsyst. Technol.* **2008**, *14*, 937–942. [CrossRef]

 *processes* 

Review

A Review of the Preparation, Machining Performance, and Application of Fe-Based Amorphous Alloys

Zexuan Huo [1], Guoqing Zhang [1,2,*], Junhong Han [1], Jianpeng Wang [1], Shuai Ma [1] and Haitao Wang [3]

1 Shenzhen Key Laboratory of High Performance Nontraditional Manufacturing, College of Mechatronics and Control Engineering, Shenzhen University, Nanhai Ave 3688, Shenzhen 518060, China; dg_0769zx@163.com (Z.H.); 2070292131@email.szu.edu.cn (J.H.); 1900291014@email.szu.edu.cn (J.W.); 2070292081@email.szu.edu.cn (S.M.)
2 Guangdong Provincial Key Laboratory of Electromagnetic Control and Intelligent Robots, College of Mechatronics and Control Engineering, Shenzhen University, Nanhai Ave 3688, Shenzhen 518060, China
3 School of Mechanical and Electrical Engineering, Shenzhen Polytechnic, Shenzhen 518055, China; wanghaitao@szpt.edu.cn
* Correspondence: zhanggq@szu.edu.cn; Tel.: +86-755-2653-6306; Fax: +86-755-2655-7471

Abstract: Amorphous alloy is an emerging metal material, and its unique atomic arrangement brings it the excellent properties of high strength and high hardness, and, therefore, have attracted extensive attention in the fields of electronic information and cutting-edge products. Their applications involve machining and forming, make the machining performance of amorphous alloys being a research hotspot. However, the present research on amorphous alloys and their machining performance is widely focused, especially for Fe-based amorphous alloys, and there lacks a systematic review. Therefore, in the present research, based on the properties of amorphous alloys and Fe-based amorphous alloys, the fundamental reason and improvement method of the difficult-to-machine properties of Fe-based amorphous alloys are reviewed and analyzed. Firstly, the properties of amorphous alloys are summarized, and it is found that crystallization and high temperature in machining are the main reasons for difficult-to-machine properties. Then, the unique properties, preparation and application of Fe-based amorphous alloys are reviewed. The review found that the machining of Fe-based amorphous alloys is also deteriorated by extremely high hardness and chemical tool wear. Tool-assisted machining, low-temperature lubrication assisted machining, and magnetic field-assisted machining can effectively improve the machining performance of Fe-based amorphous alloys. The combination of assisted machining methods is the development trend in machining Fe-based amorphous alloys, and even amorphous alloys in the future. The present research provides a systematic summary for the machining of Fe-based amorphous alloys, which would serve as a reference for relevant research.

Keywords: amorphous alloys; Fe-based amorphous alloys; difficult-to-machine; assisted machining

Citation: Huo, Z.; Zhang, G.; Han, J.; Wang, J.; Ma, S.; Wang, H. A Review of the Preparation, Machining Performance, and Application of Fe-Based Amorphous Alloys. *Processes* **2022**, *10*, 1203. https://doi.org/10.3390/pr10061203

Academic Editor: Mehmet Mercangöz

Received: 10 May 2022
Accepted: 7 June 2022
Published: 16 June 2022

1. Introduction

Amorphous alloy is a functional material emerging in the 21st century, also known as metallic glass (MG), which has a unique internal structure with long-range disorder and short-range order in the arrangement of atoms in the 3D space [1], which does not have the grain boundaries in ordinary metallic materials and has the properties of both metal and glass. The arrangement of amorphous alloys can be divided into short-range with regular clusters, medium-range, and long-range with irregular arrangement of clusters, as shown in Figure 1. Amorphous alloys have no fixed melting and boiling points and are isotropic [2], which is distinctly different from crystalline materials. The special atomic structure results in amorphous alloys having excellent properties, such as lower density, high strength, high hardness, and high resistivity, as well as excellent friction resistance, corrosion resistance, and soft magnetic properties, thus, amorphous alloys have received widespread attention.

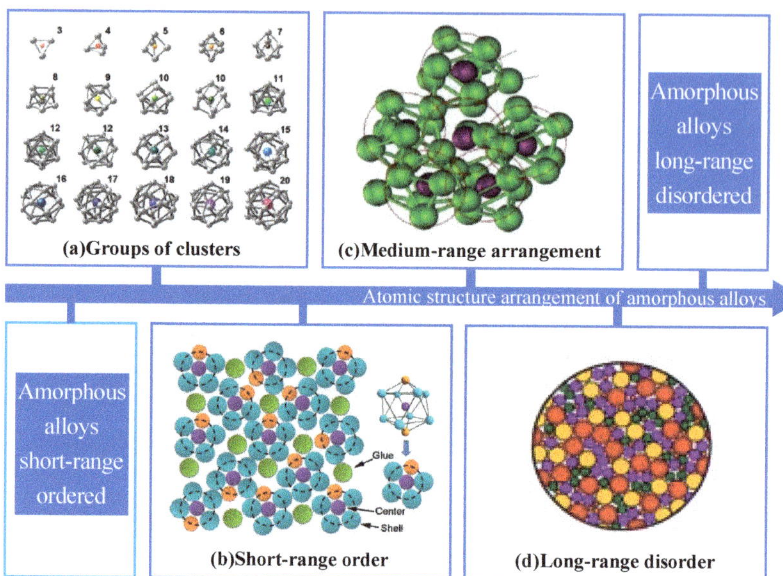

Figure 1. (**a**) The short-range atomic arrangements of amorphous alloys are composed of different kinds of clusters [3]. (**b**) The black line identifies the short-range order characteristic of amorphous alloys. (**c**) The medium-range atomic arrangement of amorphous alloys is randomly distributed by several clusters, adapted with permission from Ref. [4] 2011, Elsevier. (**d**) The long-range atomic arrangement of amorphous alloys becomes disordered.

For a long time, the low glass-forming ability (GFA) limited the preparation of amorphous alloys until Greer [5] was the first to propose the famous confusion principle, which showed that more components could increase the GFA of amorphous alloys. The confusion principle has helped the emergence of numerous amorphous alloys. Amorphous alloys can be divided into different types according to the main forming elements, such as Fe-based, Pd-based, Zr-based, etc. Each type of amorphous alloy has unique properties. Among the many amorphous alloys, Fe-based amorphous alloys have received a lot of attention from researchers because of their excellent soft magnetic properties, high corrosion resistance, high strength and hardness, and high wear resistance [6]. Fe-based amorphous alloys have an elastic modulus similar to conventional stainless steel, yet are 3–4 times stronger and harder than ultra-high-strength steels [7], as shown in Figure 2. The high strength and high hardness of Fe-based amorphous alloys bring many benefits. Meanwhile, it also limits the machining and forming of Fe-based amorphous alloys. General machining methods cannot make Fe-based amorphous alloys obtain ideal machining performance. Therefore, a machining technology with good machining ability is needed to solve the machining and preparation of Fe-based amorphous alloys. In the field of machining, ultra-precision machining technology occupies a very important position in the manufacture of cutting-edge products and modern weapons. As an important ultra-precision machining method, single-point diamond turning (SPDT) can machine some brittle or difficult-to-machine materials into high-hardness workpieces with optical surface quality in the case of assisted machining [8,9]. Machining Fe-based amorphous alloys with SPDT technology is a good choice. However, chemical wear due to the chemical affinity between diamond tools and Fe-based amorphous alloys cannot be ignored.

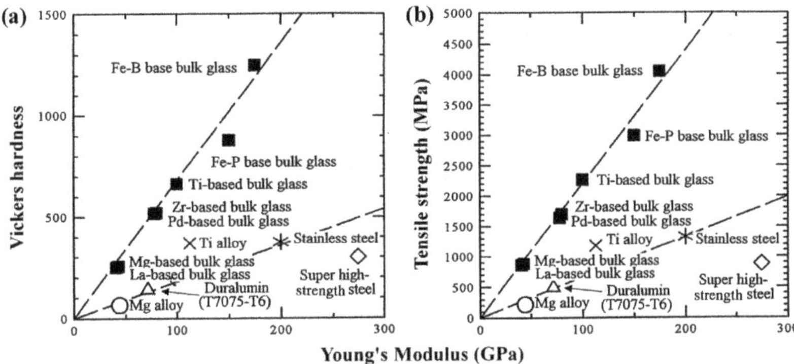

Figure 2. The (**a**) Vickers hardness and (**b**) tensile strength of Fe-based amorphous alloys are much higher than other materials, reprinted with permission from Ref. [7] 2000, Elsevier.

With its excellent soft magnetic properties, Fe-based amorphous alloys have been widely used in various electrical components, and it would be of great importance to use SPDT to achieve good machining performance. Numerous researchers have used many methods to study Fe-based amorphous alloys in an attempt to explore suitable machining methods for Fe-based amorphous alloys. For machining difficult-to-machine materials like Fe-based amorphous alloys, several non-traditional assisted machining methods have demonstrated the ability to improve their machining performance [10]. Although there is much research on Fe-based amorphous alloys, the research on machining Fe-based amorphous alloys is rather scattered, and a systematic review is needed to distill current research progress and future research directions. In the present research, the causes of difficult-to-machine Fe-based amorphous alloys and methods to improve their machining performance are reviewed to guide the subsequent research on the machining of Fe-based amorphous alloys.

2. Preparation and Properties of Amorphous Alloys

2.1. Amorphous Alloys versus Crystalline Materials

Amorphous alloys are glassy substances that are very different from crystals at the atomic level. At the macro level, amorphous alloys are also quite different from crystal metals, which is an important factor affecting the machining performance of amorphous alloys.

Firstly, amorphous alloys are sub-stable, when the temperature or pressure reaches a certain value, the amorphous state crystallizes into a crystalline state. The transition from the amorphous state to the crystalline state has a distinct crystallization peak, which can be measured with a differential thermal analyzer such as differential scanning calorimetric (DSC) [11]. In addition to DSC measurements, X-ray diffraction (XRD) analysis of amorphous alloys can also determine whether the material is in an amorphous state. The diffraction curves of amorphous alloys usually show scattered diffraction peaks (please see Figure 3e), while crystalline alloys show sharp diffraction peaks with different intensities on the diffraction curves (please see Figure 3f) [12]. Amorphous alloys crystallize during machining, which will cause some properties of amorphous alloys to be lost. However, there is a special case, the formation of nanocrystalline or amorphous/nanocrystalline composites by nano-crystallization of amorphous alloys may improve some properties of amorphous alloys [13].

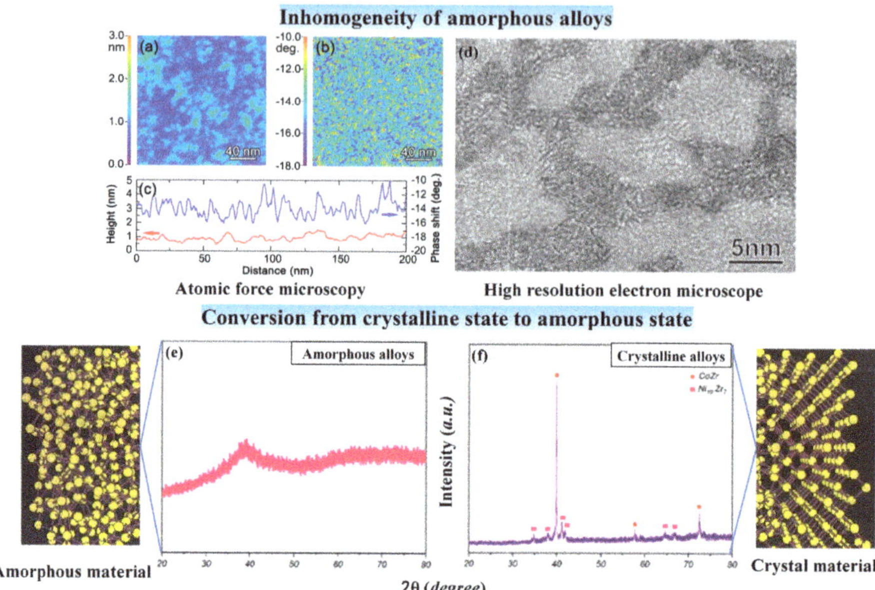

Figure 3. (**a**) The height image with rms roughness of ~0.3 nm and (**b**) phase shift image. (**c**) Height and phase shift profiles taken from the same region (The blue line refers to the phase shift, while the red line refers to the height), adapted with permission from Ref. [14] 2000, APS. (**d**) Amorphous alloys after ultrasonic vibration treatment show the appearance of soft and hard regions, adapted with permission from Ref. [15] 2005, APS. (**e**) XRD pattern of amorphous alloy. (**f**) XRD pattern of the crystalline amorphous alloy, adapted with permission from Ref. [12] 2022, Elsevier.

Secondly, amorphous alloys are isotropic in physical properties, while crystal metals are anisotropic. Although amorphous alloys are isotropic, amorphous alloys are structurally and kinetically inhomogeneous [14–16]. Observation of AFM can find that the amorphous alloy has a phase shift on the nanometer scale (please see Figure 3a–c). After ultrasonic treatment, soft and hard regions appear in the amorphous alloy due to crystallization (please see Figure 3d). These are manifestations of the inhomogeneous structure of amorphous alloys. Meanwhile, amorphous alloys do not have a definite melting point [14], while crystal metals have a definite melting point at room temperature and atmospheric pressure. For example, plastics and glass, as amorphous materials, are heated to gradually soften and eventually become liquid. And amorphous alloys have a glass transition temperature point (Tg), while crystal metals do not [17].

Thirdly, when shear bands appear in the atomic structure of amorphous alloys, it will lead to fatal failure of the material, which is a fatal defect of amorphous alloys compared to crystal metal structures [18]. Therefore, the number of shear bands in amorphous alloys is also considered to be a good indicator of the intrinsic plasticity and fracture toughness of amorphous alloys [19–21]. In machining amorphous alloys, we need to pay attention to shear band changes. Wang et al. [22] found that the bending shear bands of amorphous alloys exhibited asymmetric orientations on the tensile and compressive sides (please see Figure 4a). Gao et al. [23] showed that radial shear bands are the result of the in-plane stress components (please see Figure 4b), while the semicircular shear bands are an artifact of stress relaxation due to the bonded interface (please see Figure 4c). The shear band of amorphous alloys can be simulated using molecular dynamics (MD). Sopu et al. [24] found through MD simulation that the shear band starts at the stress concentration at the root of the notch and extends along the plane of maximum shear stress (please see Figure 4d).

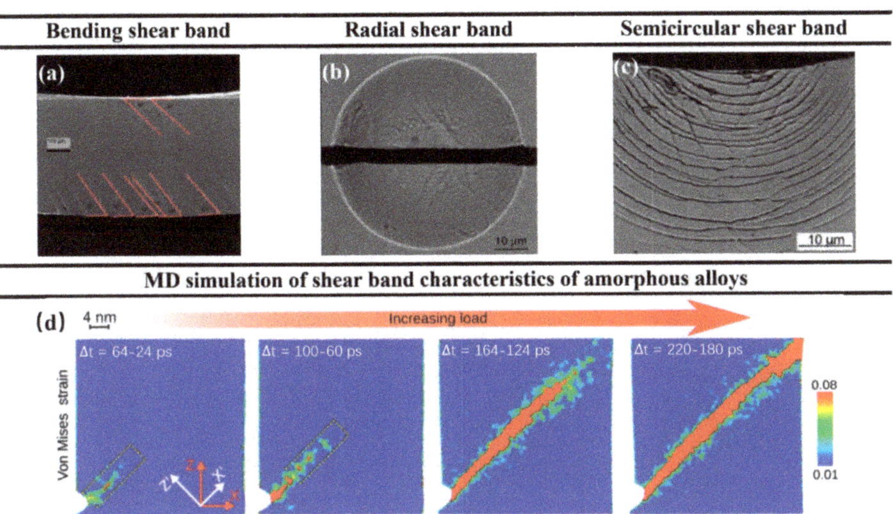

Figure 4. (**a**) Bending shear band, adapted with permission from Ref. [22] 2011, Elsevier. (**b**) Radial shear band. (**c**) Semicircular shear band, adapted with permission from Ref. [23] 2011, Elsevier. (**d**) The shear band of amorphous alloys starts at the stress concentration at the root of the notch and extends along the plane of maximum shear stress, adapted with permission from Ref. [24] 2017, APS.

2.2. Preparation of Amorphous Alloys

In 1959, Klement et al. [25] jointly produced the first amorphous alloy Au75Si25 at Caltech. In 1969, Chen and Turnbull [26] fabricated Pd−based amorphous spheres at a critical cooling rate of 100–1000 °C s^{-1}. Then in the 1980s, Inoue [27] and his colleagues at Tohoku University made a breakthrough in bulk metallic glass (BMG), and discovered a multi-component amorphous alloy with strong GFA. The breakthrough eliminated the need to add noble metal elements to amorphous alloys to enhance its GFA, thus, greatly enhancing the application of amorphous alloys. In 1992, Johnson and Peker [28] discovered the first commercial amorphous alloy Vit1 in some new aerospace materials. A summary of the research progress of amorphous alloys is shown in Table 1.

Amorphous alloys develop an amorphous structure by cooling the metal from the melt at a critical cooling rate below the glass transition zone. In addition to the amorphous structure obtained by rapid cooling of the liquid, it is also possible to form ordered crystalline structures and plastic crystals in the liquid cooling (please see Figure 5a). To ensure that a completely amorphous state is obtained during the preparation, a very high cooling rate (10^6 K/S) is required, as shown in Figure 5b. Thus, limiting the amorphous alloys to often prepare only low dimensional size alloy samples, such as powders and thin strips, and it is difficult to form BMGs with larger critical dimensions, which is the so-called poor GFA of amorphous alloys, Johnson expressed the GFA by the critical cooling rate Rc [46]:

$$R_C = \frac{dT}{dt(K/s)} = \frac{10}{D^2}(cm) \tag{1}$$

where D is the critical size. The GFA is mainly related to two factors: the liquid phase stability of the supercooled liquid and the crystallization resistance. The supercooled liquid is stable or has a high resistance to crystallization, it tends to form glass and has a higher GFA [47]. At the same time, Turnbull [26] proposed that a decrease in the reduction of glass transition temperature Trg (the ratio of Tg to the temperature of the liquid phase line of the alloy Tl) is a good indicator of GFA. Amorphous alloys have superplastic deformability in the supercooled liquid region (between the glass transition temperature Tg and the crystallization onset temperature *Tx*) (please see Figure 5c), which means that viscous flow occurs when the machining temperature of the amorphous alloy exceeds the subcooled liquid region [48].

Table 1. Summary of Amorphous Research Progress.

Time	Events	References
1959	First MG from Cal. Tech. using splat quenching (Au-Si)	[29]
1966	Splat cooling additional findings (MIT)	[30]
1969	MG formation, stability, structure (Pd-Si) (Harvard)	[31]
1977	MG constitutive model	[32,33]
1990	Multi component MG formers using Copper cast (Tohoku University)	[29]
2002	MG tension, compression studies	[34]
2003	MG research with La-Al-Cu-Ni (GFA)	[35]
2004	Application in MEMS, biomedical, sporting goods, and electronics	[36]
2005	MG corrosion wear resistance	[37]
2007	Drilling, machining studies on MG	[38]
2008	MG model into ABAQUS FEA program	[36]
2012	MG cold rolling studies	[39]
2013	MG foam reduce osteopenia in biomedical	[40]
2013	MG by 3D SLM started	[41]
2015	Honeycomb MG	[42]
2016	MG descriptor on GFA best element combination (AFLOW framework)	[43]
2017	MG using 3D SLM with crack-free, complex geometry	[44]
2019	MG measured in a levitation device under microgravity	[45]

Compared with the singleness of the theoretical methods of amorphous formation, the forming methods of amorphous alloys are diverse. The classical methods for their main forming are: end-casting in copper molds, die forging, etc. [28]. Research has shown that heating amorphous alloys in a magnetic field to a softened state is a good method for forming amorphous alloys. The millisecond heating method newly studied by Johnson et al. [50] can easily make the heating rate reach 10^6 K/s. The supercooled liquid can pass the melting point into equilibrium at any temperature above the glass transition (please see Figure 5d), which can intervene well in the crystallization of the supercooled liquid. This forming method may become a promising method for the fabrication of strong metals [50,51]. However, this forming method is not mature at present, and machining is still an unavoidable and important forming method for amorphous alloys. Amorphous alloys are difficult to machine because of their characteristics, and various machining problems need to be solved to obtain better machining performance.

2.3. Properties of Amorphous Alloys

When amorphous alloys were first studied, researchers were drawn to them by their excellent properties, such as strength and hardness. The tensile strength of the Mg80Cu10Y10 amorphous alloy produced by Inoue et al. [52] was as high as 630 MPa at room temperature. Subsequently, Co-based amorphous alloy was found to reach a fracture strength as high as 6.0 GPa [53], which is the highest strength in metallic materials to date. Fe-based amorphous alloys have also been reported to have fracture strengths up to 3.6 GPa, which is several times higher than typical structural steel [54]. The fracture strength of Fe-based amorphous alloys is even higher than that of most materials (please see Figure 6a).

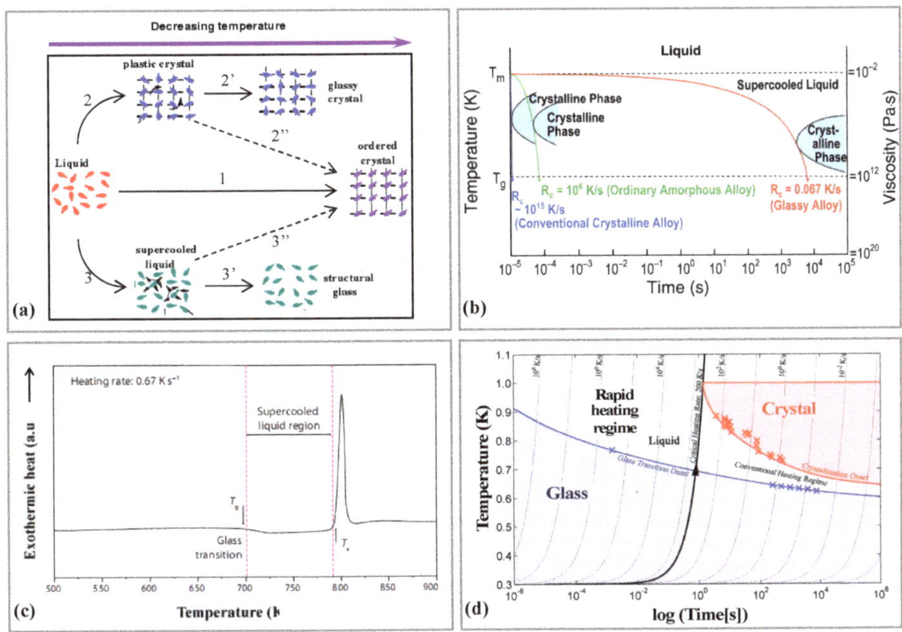

Figure 5. (**a**) Schematic diagram of the structural evolution of the different states of matter obtained by liquid cooling, adapted with permission from Ref. [49] 2019, Wang et al. (**b**) The rapid cooling process of amorphous matter, adapted with permission from Ref. [37] 2017, APS. (**c**) The DSC curve of BMG, the left red line is Tg, the middle of the red lines is the supercooled liquid region, and the right red line is Tx, adapted with permission from Ref. [29] 2011, nature. (**d**) The critical heating rate to completely bypass crystallization on heating from the glass through the liquid is about 200 K/s. Using the millisecond heating method, which enables heating rates on the order of 106 K/s, the undercooled liquid is accessible at any temperature above the glass transition, through the melting point and beyond, where the liquid enters the equilibrium state (upper left quadrant in the diagram), adapted with permission from Ref. [50] 2005, science.

As the research on amorphous alloys progressed, more outstanding properties of amorphous alloys began to be discovered. Based on previous studies, Johnson [56] studied in 2011 to obtain Pd-based amorphous alloys with fracture toughness up to 200 MPa $m^{1/2}$, which is the highest fracture toughness value of the material so far. After research, Fe-based MG has also been clearly shown to successfully remove organic matter from wastewater [57]. Fe-based amorphous alloys are known for its high corrosion resistance, and Fe-based amorphous alloys not only exhibited high corrosion resistance in conventional environments, Pratap et al. [58] further demonstrated that Fe-based amorphous alloys can still have excellent corrosion resistance under simulated body fluids. Although the surface of the Fe-based amorphous alloy is corroded in the simulated body fluids, the surface morphology does not change much (Figure 7). In addition to the excellent properties mentioned above, Fe-based amorphous alloys exhibit excellent soft magnetic properties. Yi [59] even stated that "Fe-(Al, Ga)-(Si-P-B-C) and Fe-Cu-Nb-Si-B amorphous alloy are recognized worldwide as the best comprehensive soft magnetic material".

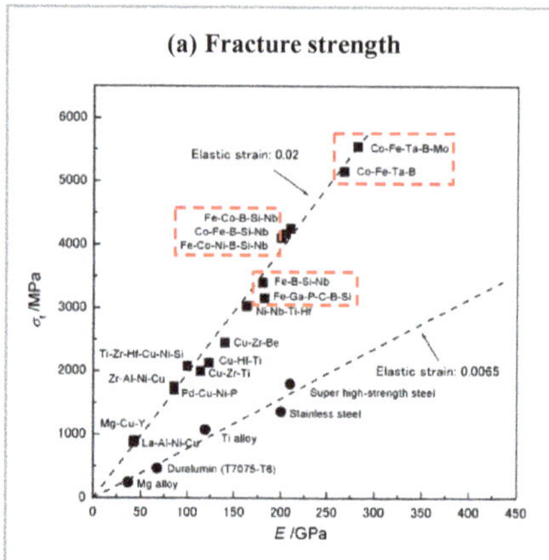

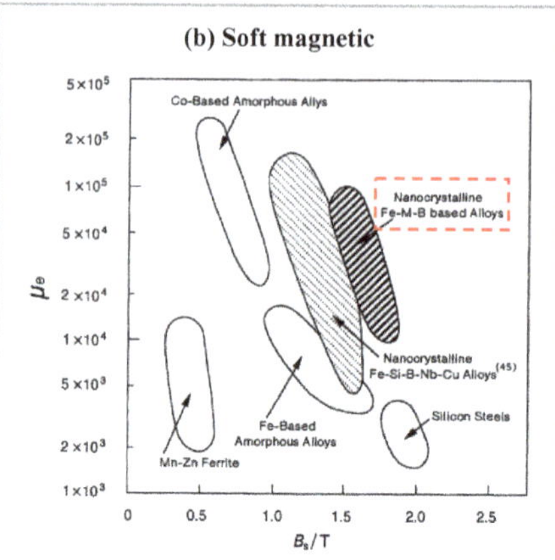

Figure 6. (**a**) Fracture strength of amorphous alloys, adapted with permission from Ref. [7] 2000, Elsevier. (**b**) Comparison of magnetization strength Bs and permeability μc of amorphous alloys with various soft magnetic materials (1 kHz) [55].

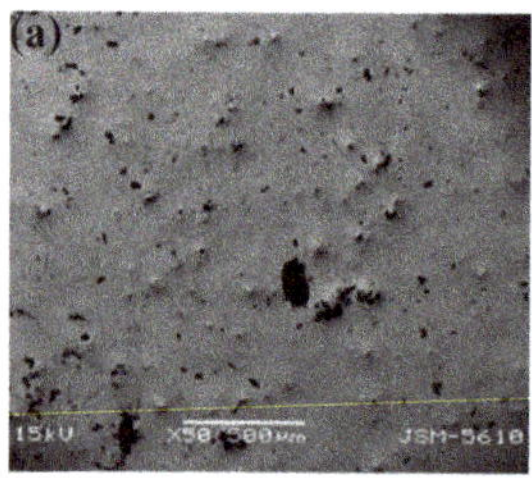
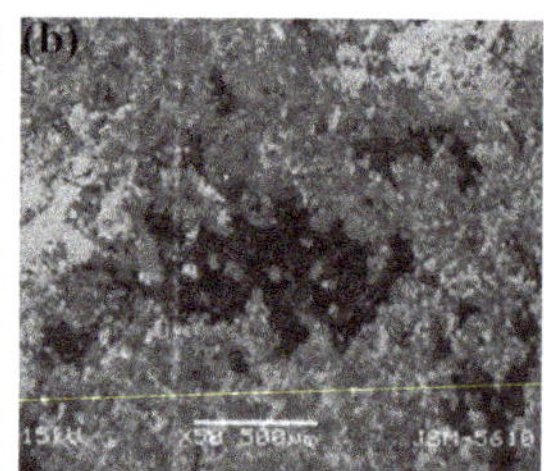

Figure 7. (**a**) Surface corrosion of Fe32Ni36Cr14P12B6 amorphous alloy under artificial saliva. (**b**) Surface corrosion of Fe67Co18B14Si1 amorphous alloy under artificial plasma, reprinted with permission from Ref. [58] 2021, Elsevier. Fe-based amorphous alloys were corroded, but the surface morphology changed little.

Obviously, amorphous alloys have many excellent properties. Amorphous alloys have some record-breaking excellent physical and chemical properties, such as strength, toughness, hardness, and modulus, which break the record for metallic materials [54,56–62], almost every type of amorphous alloy reaches several times the strength of the crystalline material of the same alloy family. Some amorphous alloys also have excellent soft magnetic properties, catalytic properties, and good wear resistance and are widely used in various applications [63,64]. Fe-based nano-amorphous alloys have outstanding magnetization strength and permeability (please see Figure 6b), even better than the comprehensive soft magnetic properties of silicon steel, which is an indispensable soft magnetic material in the power, electronics, and military industries. Different structural configurations exist for amorphous alloys of the same composition [65]. Therefore, the physical properties of amorphous alloys can be regulated by time and process conditions. Meanwhile, amorphous alloys have a wide range of forming compositions [66], and their structure and properties can be modulated by composition. And even amorphous alloys within the same system can possess different properties depending on their composition. For example, there are Fe-based amorphous alloys that can reach a maximum critical size of 16 mm, while there are Fe-based amorphous alloys that have poor GFA but possess excellent soft magnetic properties [67,68].

Conversely, amorphous alloys still have their own unique defects. Although in 1995 Inoue [69] reviewed the gradual appearance of amorphous alloys with great GFA in that era, the critical size of 72 mm for Pd40Cu30Ni10P20 prepared in 1997 [70] is still the largest example of the critical size in amorphous alloys today. The poor GFA of amorphous alloys have largely limited the industrial applications of amorphous alloys. Also limiting the industrial applications of amorphous alloys are their record-breaking strength and hardness, which makes most amorphous alloys have very poor macroscopic room temperature plastic deformation [54,71], making it difficult to obtain the designed shape by conventional methods in applications.

2.4. Machining Performance of Amorphous Alloys

The improved mechanical properties of amorphous alloys compared to crystal metals bring challenges in machining, such as tool wear, high machining temperature, and crystallization on the material surface [72]. Numerous experimental studies and modeling have been performed for the machining of crystalline materials to investigate the chip splitting mechanism during the cutting and machining of crystalline materials [73,74]. The study of these mechanisms plays an important role in the processing of crystal metals. However, the chip splitting mechanism of amorphous alloys is different from that of crystal metals [75], and the systematic knowledge of the machining theory of crystal metals cannot be applied to the machining process of amorphous alloys. Zhu and Fang [76] found that the amorphous alloys removal during the nano-cutting process was achieved by extrusion rather than shear. Some theoretical knowledge of the machining process of amorphous alloys are bound to help improve the machining performance of amorphous alloys. Therefore, there is an urgent need to systematically review the research on a large number of amorphous alloys.

Amorphous alloys are difficult to machine due to their hard-brittleness and sub-stable structure during machining, this does not mean that amorphous alloys cannot be machined to obtain good surface quality. Chong et al. [77] obtained a surface Ra of 6.3 nm when cutting La-based amorphous alloy. What's more, Xiong et al. [78] obtained optical mirror surfaces during SPDT Pd-based BMG. The center area of the machined surface was relatively clean and had no irregular micro/nanostructure exists, as shown in Figure 8a. It can be determined that the Pd-based BMG remains in the amorphous state after turning (please see Figure 8b). This means that amorphous alloys can be machined without causing crystallization of the machined surface by SPDT. Xiong et al. [78] proved that amorphous alloys have the potential to obtain optical mirror surfaces through SPDT, and they also revealed that the high cutting temperature during the machining of amorphous alloys is an important factor for their difficult-to-machine qualities. In contrast, the machined surface quality of Fe-based amorphous alloys is particularly poor [79], as shown in Figure 9.

Xiong et al. [78] further showed that when the temperature of the cutting area exceeds Tg, viscous flow occurs in the cutting area, resulting in additional friction between the tool and the workpiece, and aggravating the temperature rise in the cutting area. If the temperature of the cutting area is higher than Tg for a long time, a surface oxide was generated. Meanwhile, as mentioned above, when the temperature exceeds Tg, the amorphous alloy will crystallize, thus affect the machining. Therefore, the change in temperature of amorphous alloys undercutting and whether the material undergoes crystallization have become the focus of many studies. For example, Maroju and Jin [72] found that the cutting temperature is lower than the crystallization temperature in the cutting experiment of Zr-based BMG. However, such satisfactory results were not obtained under all conditions. Bakkal et al. [80] found that at low turning speeds, the machined surface and chips did not crystallize. At high turning speeds, the BMG of the outer layer is oxidized, and the chips appears to be crystallized to different degrees, as shown in Figure 8c–f. As mentioned in Section 2.1, the formation of nanocrystalline or nanocrystalline/amorphous composites can lead to improvements in material properties. Fu et al. [81] found that high-energy ion milling can make BMG form nanocrystalline phases. When crystallization cannot be avoided in the machining of amorphous alloys, the formation of nanocrystalline phases may be a good choice.

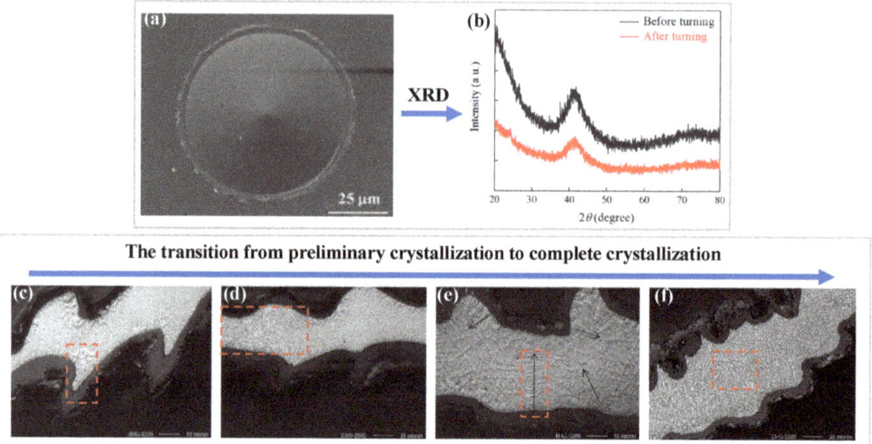

Figure 8. (**a**) The SEM image of the surface of BMG after SPDT. (**b**) XRD pattern of BMG before and after machining, no crystallization was shown [78]. (**c**) Initial crystallization near the oxide layer. (**d**) Region of no crystallization on the right and crystallization on the left. (**e**) Increasing dendritic pattern with increased crystallization. (**f**) Region of complete crystallization of BMG, adapted with permission from Ref. [80] 2004, Elsevier. The red mark is the crystalline region.

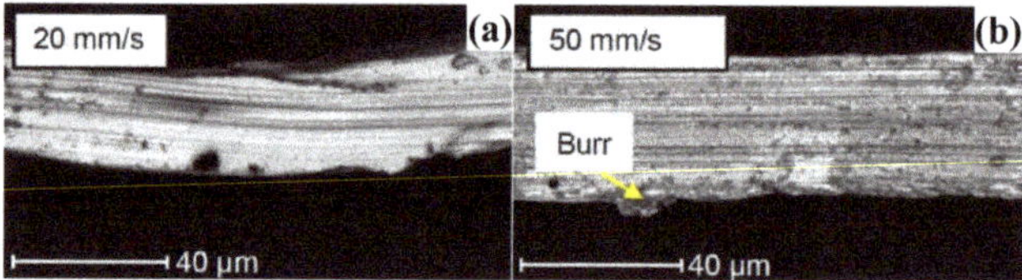

Figure 9. (**a**,**b**) Compared with the excellent machined surface of La-based amorphous alloy, the machined surface of Fe-based amorphous alloy performs extremely poorly, reprinted with permission from Ref. [79] 2021, Elsevier.

The cutting temperature in the machining of amorphous alloys also affects the machining performance of amorphous alloys in terms of the shear band. Zhang et al. [82] expressed that the deformation of BMG after turning was at least partly caused by the adiabatic shear temperature rise. Increased shear speed leads to significant melting of the workpiece surface due to temperature rise in the shear zone. The melting phenomenon caused by the high temperature in the shear zone is also manifested in the chips, and the melting of the chip surface is more serious at higher shear rates, as shown in Figure 10. Although it is confirmed that the temperature rise in the shear zone affects the machining performance of amorphous alloys, in some cases the temperature rise can actually improve the properties of amorphous alloys. Basak and Zhang [83] found that the surface temperature rise of amorphous alloys caused by contact sliding friction heat increases the hardness of MG.

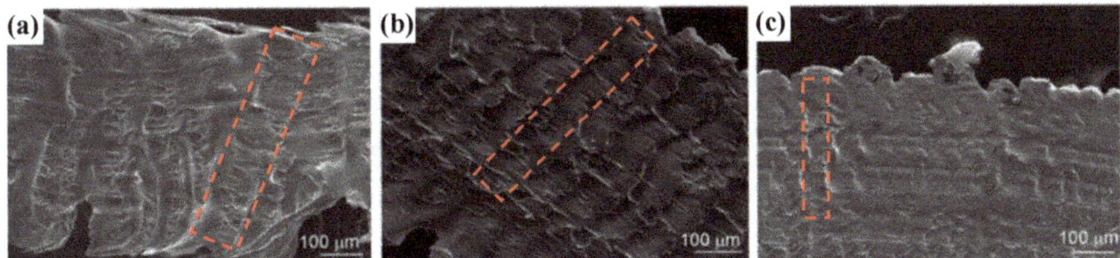

Figure 10. SEM micrographs of the underside of BMGs chips at different shear speeds: (**a**) 1.06 m/s, (**b**) 0.50 m/s and (**c**) 0.25 m/s, reprinted with permission from Ref. [82] 2011, Springer. The red mark is the degree of melting of the chip surface.

3. Preparation, Machining, and Application of Fe-Based Amorphous Alloys

Among the many studies on amorphous alloys mentioned above, it is not difficult to find a special presence among the many amorphous alloy systems, Fe-based amorphous alloy. Fe-based amorphous alloys have far superior chemical properties [84,85], physical and mechanical properties [86,87] than ordinary crystalline alloys, such as high strength, high hardness, low elastic modulus, corrosion resistance, and excellent soft magnetic performance [88], and also has the advantages of low production cost [89], aroused great interest of people [47,90]. However, it is more difficult to obtain ideal surface quality than other amorphous alloys. It is necessary to explore more suitable machining methods for Fe-based amorphous alloys based on the machining of amorphous alloys.

3.1. Preparation and Application of Fe-Based Amorphous Alloys

In 1988, Yoshizawa et al. [91] added a small amount of Cu and Nb to Fe-Si-B amorphous alloy and obtained Fe-Si-B-Nb-C Fe-based nanocrystalline soft magnetic alloy after crystallization and annealing, which has excellent soft magnetic properties and gained wide attention. In late 1989, Prof. A. Inoue [92] proposed the principles contributed greatly to the emergence of many Fe-based amorphous alloys that followed. In 1995, Inoue [93] and coworkers synthesized the first Fe-based BMG with a diameter of 1 mm in the Fe-Al-Ga-P-C-B alloy system by using the copper casting technique. Since the 21st century, research on amorphous alloys has developed rapidly, and Fe-based amorphous alloys have gained a lot of breakthroughs along with the development trend. Although the domestic-related research started late, also made certain achievements, the most representative one is Chen et al. [67] developed Fe41Co7Cr15Mo14C15B6Y2 amorphous alloy with a critical size of 16 mm.

Although under the guidance of Inoue's numerous research achievements, the critical size of Fe-based amorphous alloys have been gradually broken through, and the disadvantage of poor GFA has been alleviated in some applications, there is still a serious problem in the research of Fe-based amorphous alloys, which is room temperature embrittlement [94]. Fe-based amorphous alloys are widely used due to their many excellent properties. For example, Fe-based amorphous strips [95] are used in transformers and cores for motors because of their excellent electromagnetic properties; Fe-based amorphous coatings [96] are widely used in the protection of equipment in the petroleum and electric power industries for their excellent corrosion resistance, wear, and high temperature; Fe-based BMG [67] has been widely used as a special structural material in the military industry and other fields because of its high strength and high hardness mechanical properties; compared to Zr-based and Pd-based BMGs, Fe-based BMGs have superior overall performance, and their high thermal stability and low thermal conductivity can be used in building structures, refrigeration, insulation devices in liquefied petroleum engineering technologies, and highly sensitive precision micro-components [97]. The thermal conductivity of Zr-based amorphous alloys and Pd-based amorphous alloys is already lower than that of most materials (Figure 11). Although the extremely low thermal conductivity of Fe-based amorphous alloys makes it advantageous in some fields, it also makes it difficult to dissipate the machining heat of Fe-based amorphous alloys.

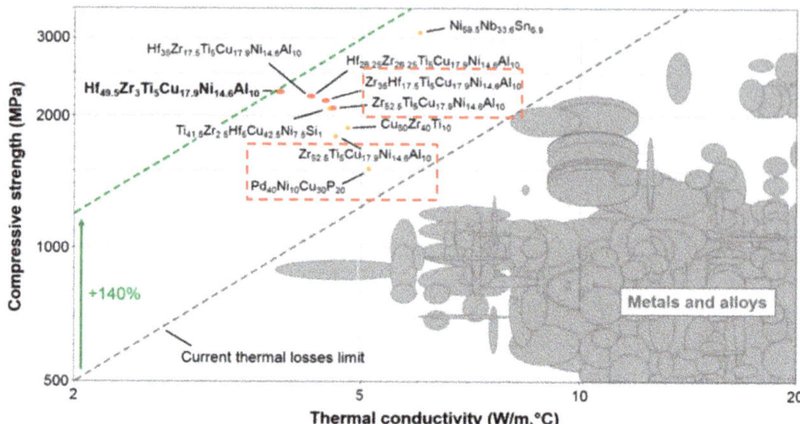

Figure 11. Compressive strength versus room temperature thermal conductivity of metallic glasses, metals, and alloys in Ashby's material space. Conventional crystalline alloys are shaded while other amorphous alloys are in orange and red, reprinted with permission from Ref. [97] 2018, Elsevier.

Figure 12 shows some applications cases of Fe-based amorphous alloy. With excellent soft magnetic properties, powder cores made of Fe-based amorphous alloys with higher efficiency and less heat generation are used for power inductors in laptops (please see Figure 12a). Combined with the excellent soft magnetic properties and extremely high fracture strength of Fe-based amorphous alloys, they are applied as torsion bars to help achieve larger scanning angles (please see Figure 12b). In addition to having extremely high fracture strength, Fe-based amorphous alloys have excellent mechanical properties and can be used in a variety of wires (please see Figure 12d) [93]. Fe-based amorphous alloys also have good catalytic performance and can be used for various wastewater purification applications (please see Figure 12c) [98]. It can be seen that Fe-based amorphous alloys have a very wide range of industrial applications and good prospects for development.

3.2. Machining of Fe-Based Amorphous Alloys

Machining is an important forming method for Fe-based amorphous alloys. Turning and milling are the most common machining methods, and the machining of Fe-based amorphous alloys is often inseparable from these two machining methods. Fang [79] performed two-dimensional cutting of Fe-Si-B amorphous alloy and concluded that the development of shear bands leads to the formation of macroscopic dislocations and cracks in the workpiece, and a work-hardening layer appears in the bottom layer after cutting, making its magnetic properties significantly degraded. Wang [100] conducted cutting experiments on Fe-Al-Cr-B-Si-Nb amorphous alloy and found that the crystallization of chips was up to 52.5% with serious crystallization after increasing the cutting speed and depth of cut. In the above-mentioned cutting experiments on Fe-based amorphous alloys, it was found that the properties of Fe-based amorphous alloys deteriorated after cutting. However, in some experiments of milling, Fe-based amorphous alloys performed well after milling. Fan et al. [101] performed experiments on Fe80B20 by mechanical milling and found that milling induced magnetic anisotropy in the workpiece, which may be attributed to a relief of the induced stresses after bulk crystallization has occurred. Nakao and Fang [102] used their previous experience in milling Fe-Si-B amorphous alloy to take advantage of the brittleness of nanoscale grains in the transition zone for high-speed milling and successfully obtained a better-milled surface. Nakao and Fang found that after increasing the milling speed, the maximum grain and area fraction increased with the rapid growth of nanocrystals as the cutting temperature increased, as shown in Figure 13a. Nakao's case of using the properties of nanocrystals in the transition zone to help to machine and thus obtaining a better-machined surface also confirms the feasibility of obtaining a better-machined surface by removing the surface crystallized material.

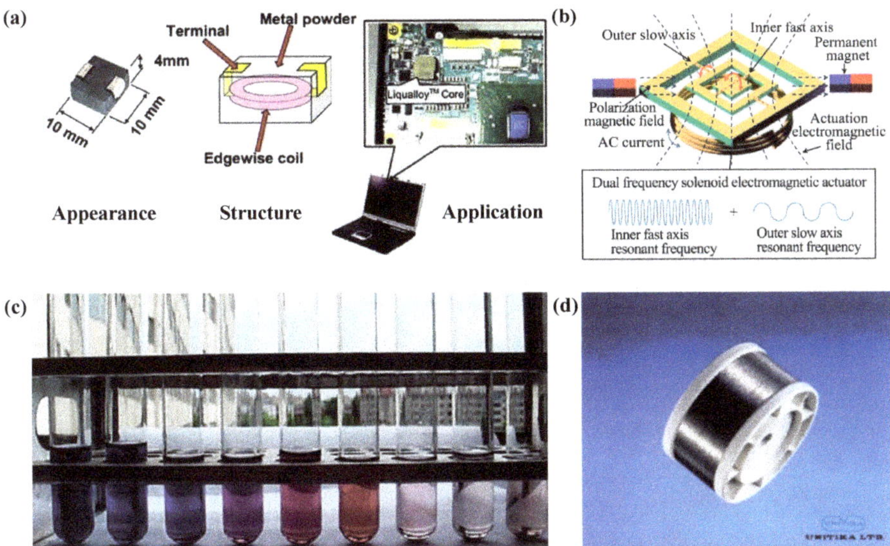

Figure 12. (**a**) Powder cores made of Fe-TM-P-C-B-Si amorphous alloy with high efficiency and low heat generation are used in inductors for laptops, adapted with permission from Ref. [93] 2013, Taylor & Francis. (**b**) A two-dimensional Fe76Si9B10P5 MG microscope was made by Ou, adapted with permission from Ref. [99] 2019, iopscience. (**c**) (Fe0.99Mo0.01)76Si9B13 catalytic purification of wastewater, adapted with permission from Ref. [98] 2010, Elsevier. (**d**) Fe-based amorphous alloy wire.

Although some basic machining methods such as turning and milling are sufficient for most machining needs, in practice, some materials require special machining to meet the requirements. Since the invention of the first laser by C.K.N. Patel in 1964, laser machining has received a lot of attention a long time and is now widely used as a special machining technology in various fields such as micromachining, marking, cutting, and welding. Quintana et al. [104] performed laser machining of Fe81B13.5Si3.5C2 amorphous alloy and the experimental results showed that short-pulse laser machining and long-pulse laser machining are promising techniques for micromachining amorphous alloys. Because they can make amorphous alloys avoid crystallization during machining, thereby ensuring the characteristics of amorphous alloys. Comparing the SEM image of the Fe-based amorphous alloy after laser processing with the SEM image of the polycrystalline Ni-based alloy after laser processing, after testing, it is confirmed that the Fe-based amorphous alloy has no crystallization phenomenon after laser machining. EDM is also a typical special machining technology, which uses thermal energy to remove materials, and its machining ability is not affected by the hardness and strength of the material. The disadvantage of EDM is the poor quality of the machined surface. Tsui et al. [103] used ultrasonic vibration-assisted EDM and added conductive aluminum powder to machine Fe-based amorphous alloys, and the workpiece did not crystallize after machining, as shown in Figure 13c,d. Although there are craters on the machined surface (please see Figure 13c), no crystallization occurred (please see Figure 13d). However, many machining methods cannot make amorphous alloys avoid crystallization during machining like laser machining and EDM. He et al. [105] showed that compression machining of Fe93P10C7 amorphous alloy accelerates its crystallization. Some researchers take advantage of this accelerated crystallization phenomenon.

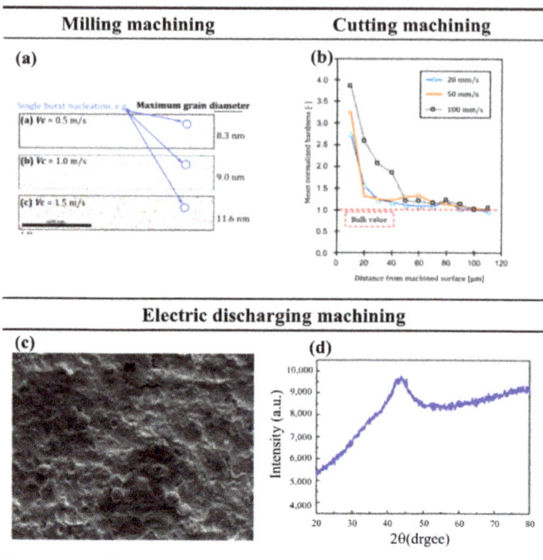

Figure 13. (**a**) The number of nanocrystals increases as the milling speed increases, adapted with permission from Ref. [102]. 2020, Elsevier. (**b**) Changes in microhardness from the machined surface to the inside of the workpiece after two−dimensional cutting of Fe−Si−B by Fang et al., adapted with permission from Ref. [79]. 2021, Elsevier. (**c**) The surface after EDM of the Fe−based BMG. (**d**) The XDR pattern of the Fe−based BMG after EDM [103]. Although there are many pits on the EDM machined surface, the test shows that the Fe-based amorphous alloy does not crystallize.

Nanocrystalline is a special form of amorphous alloy. Mechanical ball milling has been confirmed to make Fe-based amorphous alloys nano-crystallized, thereby improving the properties of Fe-based amorphous alloys. Ramasamy et al. [106] induced the generation of nanocrystalline α-Fe(Nb) after ball milling of Fe80Nb10B10 amorphous alloy. And they showed that the magnetic properties of mechanically induced nanocrystalline α-Fe composites were superior to those of amorphous alloys. The Fe-based amorphous alloy samples hot-pressed at 903 K (please see Figure 14b) have lower porosity and better bonding properties than those hot-pressed at 703 K (please see Figure 14a), which can reflect the good performance of nanocrystals at high temperature. For the two hot-pressed samples, only α-Fe peaks were observed, indicating that α-Fe is stable at temperatures of 703 K and 903 K, as shown in Figure 14c. The saturation magnetization of the samples under hot pressing at 703 and 903 K were 90 and 125 Am^2/kg, respectively, reflecting the excellent magnetic properties of nanocrystals, as shown in Figure 14d. After high-energy ball milling of the Fe-Mo-Si-B amorphous alloy by Guo and Lu [107], the amorphous alloy was also found to generate the nanocrystalline α-Fe. Guo and Lu also showed that α-Fe has remarkable thermal stability at high temperatures. It can be seen that the nanocrystals formed during the machining of Fe-based amorphous alloys can provide some strengthening of their properties.

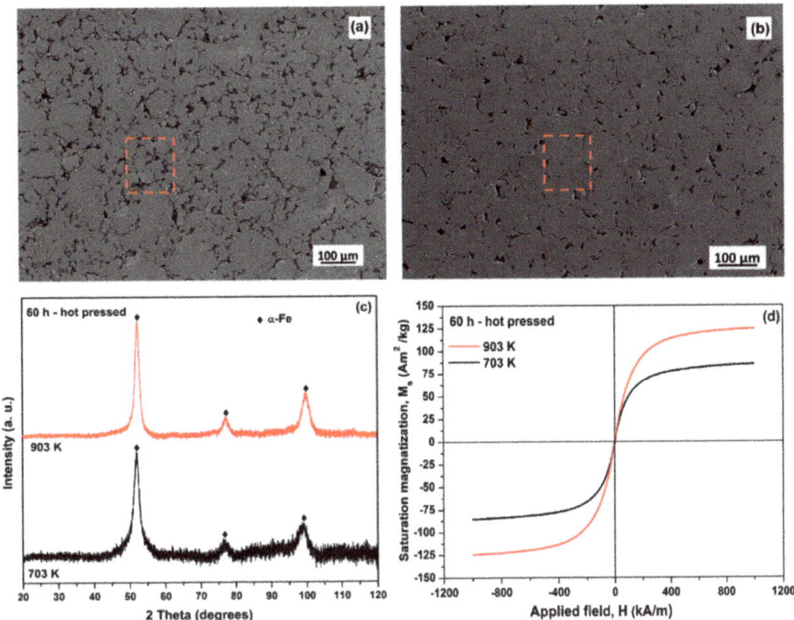

Figure 14. SEM micrographs of the 60−h milled Fe80Nb10B10 metallic glass hot−pressed at (**a**) 703 K with 880 MPa and (**b**) 903 K with 640 MPa. The red mark is the crack on the surface of the material after hot pressing. (**c**) XRD patterns of the 60 h milled samples after hot pressing at 723 K and 903 K. α−Fe is stable at temperatures of 703 K and 903 K. (**d**) Hysteresis loops for the 60 h milled samples hot-pressed at 723 K and 903 K. The saturation magnetization of the samples hot pressed at 703 and 903 K are 90 and 125 Am^2/kg, respectively, reprinted with permission from Ref. [106] 2017, Elsevier.

In most cases, Fe-based amorphous alloy will be crystallized after machining, and some scholars have studied how the crystallization affects them. Lv and Chen [108], in their study of the Fe45Cr15Mo14C15B6Y2Ni3 amorphous alloy, found that the key factor affecting the thermal conductivity of Fe-based BMG was the size and number of nanocrystals, and the thermal conductivity increased significantly after crystallization. In Section 2.4, it is mentioned that cutting temperature is an important factor affecting the machining performance of amorphous alloys. The thermal conductivity can directly affect the cutting temperature. Therefore, it may be possible to reduce the cutting temperature through the crystallization of Fe-based amorphous alloy.

SPDT is expected to achieve good machining of iron-based amorphous alloys. It is not convenient to study the cutting mechanism of amorphous alloys because Fe in Fe-based amorphous alloys can cause graphitization of diamond tools, Han et al. [109] investigated the effects of spindle speed, feed rate, and depth of cut on the machined surface of Zr-based BMGs (Table 2), as shown in Figure 15. The chip morphology at different cutting parameters showed a typical flow pattern, and the cutting morphology changed significantly in sample D with the spindle speed of 500 r/min. Experimental results by Han et al. showed that spindle speed has a greater effect on the BMG machined surface than feed rate and depth of cut, and the larger the spindle speed, the less ductile the surface. In ultra-precision turning, both diamond tools and boron nitride tools are very good and widely used tools. The use of boron nitride tools can effectively avoid chemical wear between diamond tools and Fe-based amorphous alloys. Chen et al. [110] explored the micro-machinability of amorphous alloys by ultra-precision machining of Zr-based amorphous alloys with diamond tools and boron nitride tools. They found that the BMG materials under SPDT could be removed by a ductile mode like metals rather than brittle materials. This mechanism has important implications for the use of SPDT Fe-based amorphous alloy. Fe-based amorphous alloys are machined by SPDT, which is not conducive to direct study of the mechanism, so there is little related research. However, the conclusions drawn by the above-mentioned scholars using other amorphous alloys under SPDT research can provide good help for the ultra-precision machining of Fe-based amorphous alloys.

Table 2. BMG machining conditions [109].

Sample ID	Feed Rate (mm/min)	DOC (μm)	Spindle Rate (r/min)	Groove Depth (nm)	Distance between Fold Lines (μm)	Tool Moving Speed (μm/s)
A	10	1	2000	7	0.35	7
B	20	3	2000	14	1.08	16
C	10	3	2000	20	0.26	8
D	10	3	500	200	1.78	18

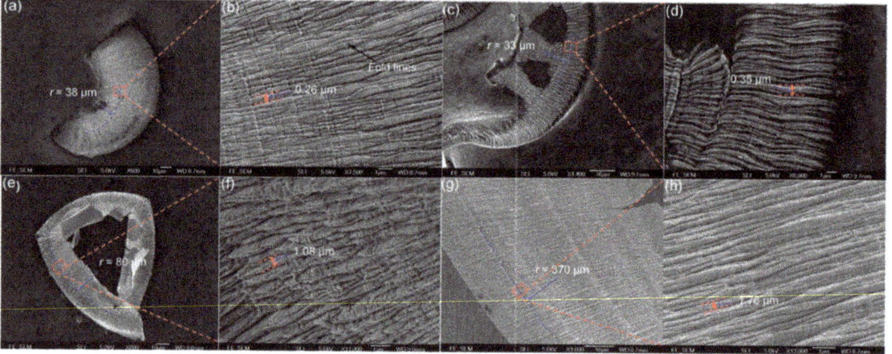

Figure 15. Chip morphologies. (**a**) Overview of chip morphology of sample C. (**b**) Enlarged morphology of the area enclosed in the red rectangle in (**a**). A dark arrow points to the fold line. Distance between neighboring fold lines is also marked. (**c**) Overview of chip morphology of sample A. (**d**) Enlarged morphology of the area enclosed in the red rectangle in (**c**). Distance between neighboring fold lines is also marked. (**e**) Overview of chip morphology of sample B. (**f**) Enlarged morphology of the area enclosed in the red rectangle in (**e**). Distance between neighboring fold lines is also marked. (**g**) Overview of chips' morphology of sample D. (**h**) Enlarged morphology of the area enclosed in the red rectangle in (**g**). Distance between neighboring fold lines is also marked, reprinted with permission from Ref. [109] 2015, Elsevier.

3.3. *Difficult-to-Machine Property of Fe-Based Amorphous Alloys*

The difficult-to-machine property of Fe-based amorphous alloy limits its application, and study is needed of the suppression methods aimed at the causes of its being difficult to machine. Based on the characteristics of Fe-based amorphous alloys, excessive cutting forces, and high cutting temperatures occur during machining. When the cutting temperature exceeds 352 °C, the amorphous alloy will crystallize, which will have a certain negative impact on the machining process [111]. Wang [100] found that the chips eventually crystallized when cutting Fe-based amorphous alloys using cemented carbide tools. Whether it is the Fe-based amorphous alloy itself or the chips at each cutting speed, there is an obvious reduction in the area of the crystallization exothermic peak, which is a signal of crystallization, indicating that the workpiece is crystallized seriously in the machining, as shown in Figure 16. During the cutting process of Fe-based amorphous alloys, the chip temperature rises sharply along with the increase in cutting speed, the chip microstructure undergoes dynamic crystallization. When the machined surface crystallizes, the properties of Fe-based amorphous alloys change, which undoubtedly leads to instability in the machining process, which negatively affects the machining results. However, there is a special case where the nanocrystals produced during crystallization can enhance the properties of Fe-based amorphous alloys [112]. This nano-crystallization phenomenon may be an effective way to solve the problem of crystallization occurring during the machining of Fe-based amorphous alloys.

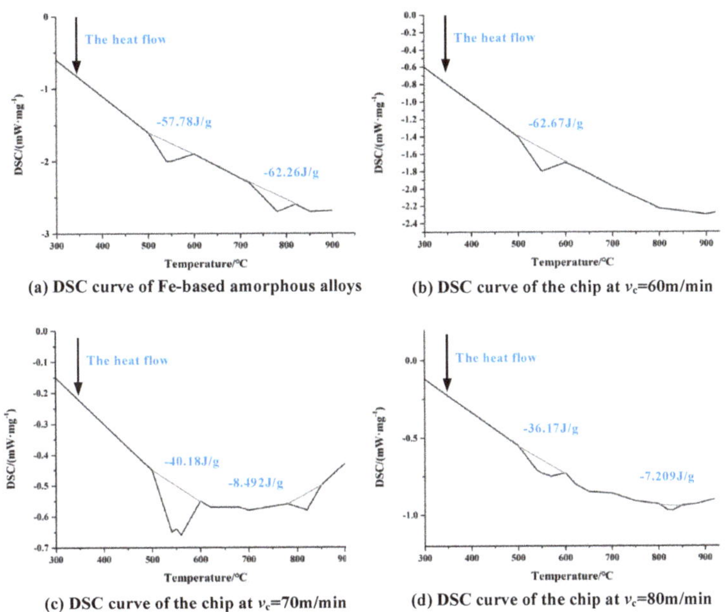

Figure 16. (**a**) DSC curves for Fe-Al-Cr-B-Si-Nb amorphous alloy. (**b**) DSC curves for chips with v_c = 60 m/min, f = 0.09 mm/r, ap = 0.1 mm. (**c**) DSC curves for chips with v_c = 70 m/min, f = 0.09 mm/r, ap = 0.2 mm. (**d**) DSC curves for chips with v_c = 80 m/min, f = 0.09 mm/r, ap = 0.3 mm DSC curve of chips [100]. There is an obvious reduction in the area of the crystallization exothermic peak.

Zhu et al. [48] used MD to demonstrate that the nanoscale cutting temperature of amorphous alloy can reach 600–700 K. Although the actual temperature is difficult to measure, it is also likely that the actual cutting temperature will reach or exceed Tg, which will lead to viscous flow of the workpiece. The temperature in the cutting region is high as shown in Figure 17. Because of the poor heat conductivity of amorphous alloy, the rise of the temperature in the cutting region softens the amorphous alloy. In the experiment of cutting Fe-based amorphous alloy, Wang [100] observed obvious viscous flow caused by high cutting temperature through the SEM images of the cutting surface, as shown in Figure 18a. This indicates that the high cutting temperature will seriously affect its machined surface quality. However, due to the excellent mechanical properties and the low thermal conductivity of Fe-based amorphous alloys, high cutting temperature is inevitably generated during machining, which increases the machining difficulty of Fe-based amorphous alloys. And excessive cutting temperature not only leads to viscous flow in amorphous alloys. Wang et al. [113] found that during cutting experiments on Fe-based amorphous alloys using cemented carbide tools, pyro luminescence phenomenon occurs during machining in some cutting parameters (please see Figure 18b), and the results show burns on the workpiece surface. A similar pyro luminescence phenomenon occurred during the machining of Zr-based amorphous alloy by Ding et al. [114], which led to severe oxidation and crystallization of their machined chips, seriously deteriorate the machined surface quality, as shown in Figure 18c,d. This pyro luminescence phenomenon can cause negative effects such as burns, severe oxidation, and severe crystallization of the workpiece, which is more damaging to the machined surface than viscous flow.

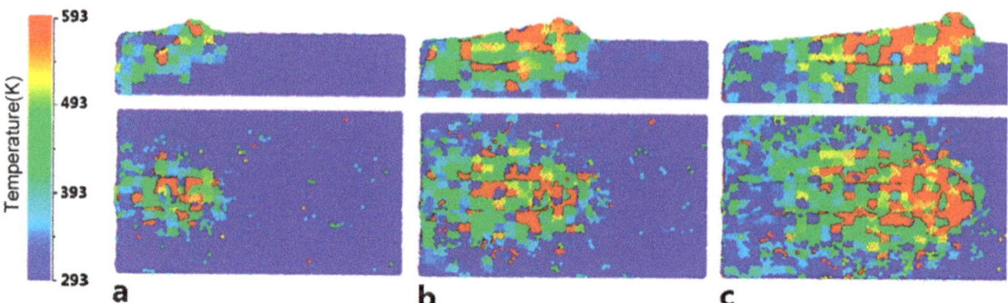

Figure 17. The cutting regional temperature variation diagram of the workpiece during the nanometric cutting process, (**a**–**c**) are cross-section views of x–z plane (the first row) and vertical views (the second row), showing the cutting regional temperature variation at cutting distances of 4, 8, and 14 nm, respectively, reprinted with permission from Ref. [48] 2014, Elsevier.

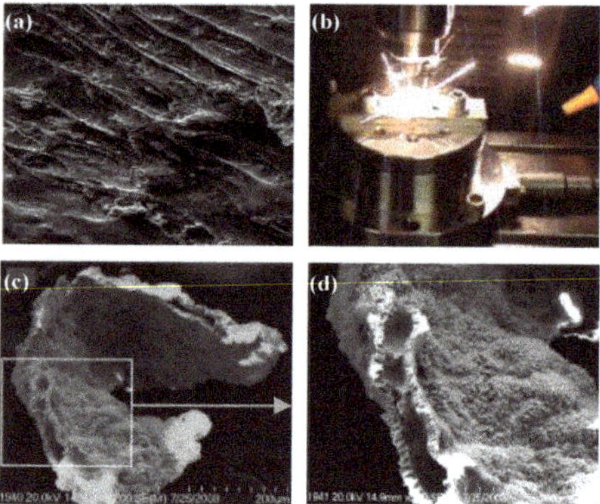

Figure 18. (**a**) Viscous flow on the surface of the chips produced after Fe–Si–B was cut in two dimensions [100]. (**b**) Pyro luminescence phenomenon during cutting of Zr-based BMG. (**c**,**d**) SEM image of the Zr–based BMG after the pyro luminescence, where severe oxidation and crystallization of the BMG can be observed [114].

As a brittle material, the machining of Fe-based amorphous alloys under SPDT is still a difficult research area. Brittle fractures must be avoided in ultra-precision machining [115–117], and machining brittle materials in a ductile mode are essential to achieve a mirror finish in ultra-precision machining [118]. When the feed rate or depth of cut during ultra-precision machining exceeds a critical value, the workpiece will change from definite ductility to indefinite brittleness, and the removal of this brittleness will cause several different cracks in the workpiece [119–121], seriously affecting the machined surface quality. This also causes the problem that Fe-based amorphous alloys are difficult to machine despite under SPDT. The key to solving this problem is that the surface layer of the workpiece cannot be removed by brittle removal, but by plastic removal. Monocrystalline silicon is a typical brittle material, and Ayomoh and Abou-El-Hossein [122] used SPDT monocrystalline silicon to achieve plastic cutting on the surface of monocrystalline silicon under the parameters of a feed rate of 2 μm/rev and depth of cut of 45 μm. The surface roughness *Ra* of the workpiece after machining was only 4.5 nm. This experiment demonstrates that even hard-brittle materials can be machined by SPDT to produce a surface of excellent quality. It is also proved that SPDT may be an important machining method to solve the problem of difficult-to-machine Fe-based amorphous alloys.

In actual production, there are other problems in the machining of hard-brittle materials, such as surface damage, short tool life, and low machining efficiency. Since Fe-based amorphous alloys are ferrous metals, their Fe elements have a chemical affinity with the C elements in diamond tools. When SPDT Fe-based amorphous alloys, diamond tools will also experience graphitic wear similar to that when cutting iron, which seriously damages the life of diamond tools and affects the surface quality. Guo et al. [123] explored the mechanism of diamond graphitization through MD simulations and demonstrated the catalytic effect of Fe-based metals on diamond graphitization, as shown in Figure 19. This also increases the difficulty of machining Fe-based amorphous alloys under SPDT. Although mastering the theoretical knowledge of plastic removal is helpful for machining Fe-based amorphous alloys, as mentioned above, machining Fe-based amorphous alloys still faces various difficulties. The most prominent among them are crystallization during machining, excessive cutting temperature, and severe tool wear. Therefore, it is also necessary to explore efficient assisted machining methods to help solve the difficulties encountered in the machining of Fe-based amorphous alloys and improve the machining performance of Fe-based amorphous alloys.

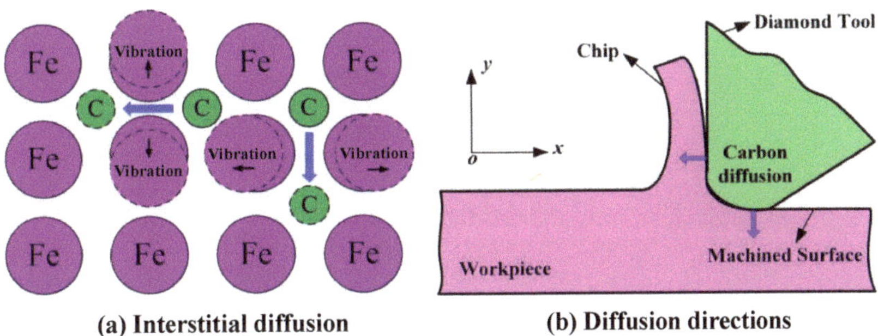

Figure 19. Diffusion process of carbon atoms in cutting ferrous material, reprinted with permission from Ref. [124] 2018, Elsevier. (**a**) Interstitial diffusion of diamond graphitization. (**b**) The direction of diffusion of diamond graphitization to the workpiece.

4. Assisted Machining Methods

The latest research results on the cutting of difficult-to-machine materials abroad show that the improvement of surface quality of parts made of difficult-to-machine materials by optimizing cutting parameters for a given machine tool and tool is very limited [125]. Therefore, the use of assisted machining methods to improve the surface quality of difficult materials is a good option, and SPDT with the use of assisted techniques can also reduce machining time and increase machining efficiency compared to purely mechanical SPDT processes [10]. The introduction of various assisted machining methods will also be expected to improve the machining performance of Fe-based amorphous alloys.

4.1. Tool-Assisted Machining

4.1.1. Ultrasonic Vibration-Assisted Machining

In ultra-precision machining, researchers often use tool-assisted machining to achieve the desired machined surface. Tool-assisted machining mainly apply vibration to the machine tool or workpiece. The vibration systems are divided into two main categories: high-frequency vibration with ultrasonic vibration and low-frequency vibration with tool servo [126].

The high-frequency vibration of ultrasonic vibration is achieved through an ultrasonic vibration system that uses a transducer to convert high-frequency electrical energy into vibration in a mechanical system to drive the diamond tool to produce vibration, usually at frequencies up to 20 kHz. In ultrasonic vibration-assisted SPDT, the ultrasonic vibration system is not affected by the electrical conductivity of the machined workpiece because it affects the formation of the machined surface through high-frequency vibration rather than a thermal process. Therefore, it has a wide range of applications. According to the above, the high cutting temperature in the machining of Fe-based amorphous alloys is a machining challenge that needs to be solved urgently. While Huang et al. [127] used ultrasonic elliptical vibration-assisted SPDT ferrous metals and successfully reduced the cutting temperature of the tool. Because the ultrasonic elliptical vibrator drives the diamond tool and makes the tool achieve periodic intermittent cutting, which can effectively reduce the cutting temperature of the tool and thus reduce the wear of the tool, as shown in Figure 20a–c. Comparing the tool wear diagram of ordinary turning (please see Figure 20a) with the tool wear diagram of ultrasonic elliptical vibration-assisted turning (please see Figure 20b), it can be seen that the tool wear assisted by ultrasonic elliptical vibration is significantly lower than that of ordinary turning. The surface roughness of the workpiece with the aid of ultrasonic elliptical vibration is significantly lower than that of ordinary turning, as shown in Figure 20c. Titanium alloy is a typical difficult-to-machine material, and its thermal conductivity is extremely poor, even resulting in a higher cutting temperature than that of Fe-based amorphous alloys during machining. Zhang et al. [128] introduced ultrasonic vibration assistance when using SPDT titanium alloys, which can significantly improve the surface quality of titanium alloy workpieces by increasing the vibration frequency and amplitude within a certain range. These experimental results show that ultrasonic vibration-assisted machining can improve machining scenarios with excessive cutting temperatures, while reducing tool wear to a certain extent. However, not all application scenarios of ultrasonic vibration-assisted machining can improve this problem. Wang et al. [129] studied ultrasonic vibration-assisted diamond wire saw single crystal silicon, showing that ultrasonic vibration assistance does not significantly change the sawing temperature of diamond wire sawed single crystalline silicon.

In addition to effectively improving tool wear and cutting high temperature, ultrasonic vibration-assisted machining has also been proved to be effective in assisting the machining of hard-brittle materials. Xing et al. [130] used ultrasonic vibration-assisted SPDT single-crystal silicon, and the surface roughness *Ra* of the workpiece edge could reach 1.739 nm and the surface roughness *Ra* of the workpiece center could reach 5.506 nm, which can obtain a very good machined surface quality, as shown in Figure 20d,e. This result indicates that ultrasonic vibration-assisted machining has the potential to improve the machining performance of hard-brittle materials under SPDT. Shen et al. [131] used ultrasonic vibration-assisted diamond grinding wheels to grind ceramics and found that the grinding force did not increase significantly and remained stable as the volume of material removed increased. This indicates that ultrasonic vibration-assisted machining can also reduce cutting force and improve tool wear when assisted machining of hard-brittle materials. In addition, some researches have shown that the use of ultrasonic vibration-assisted SPDT technology can improve the machinability of materials and reduce the wear of diamond tools [132–134]. Most importantly, Tsui et al. [103] successfully used ultrasonic vibration-assisted machining of Fe-based amorphous alloys to obtain the desired machining purpose.

Ultrasonic vibration-assisted machining can improve high cutting temperatures and tool wear. The machining of hard-brittle materials can also be of great help. Therefore, this assisted machining technology is expected to improve the problem of difficult-to-machine Fe-based amorphous alloys. In fact, the application of ultrasonic vibration-assisted technology in SPDT has achieved many remarkable results, such as the improvement of diamond tool life, the generation of optical surfaces, the improvement of tool wear, and cutting force [135–138]. However, ultrasonic vibration-assisted machining has low machining efficiency and is not suitable for large-sized workpieces.

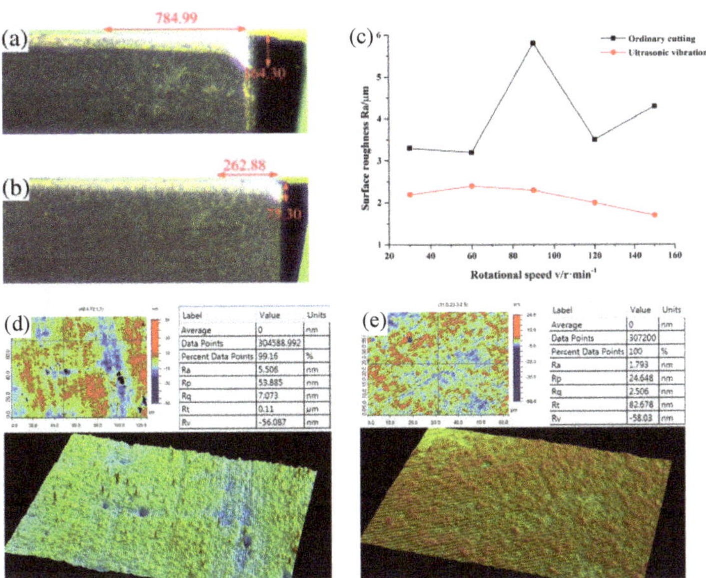

Figure 20. (**a**) Tool wear diagram under normal cutting. (**b**) Tool wear diagram under ultrasonic elliptical vibration assistance. The red mark is the main wear amount of the cutting edge. (**c**) The line graph of the surface roughness under different cutting speeds in ordinary cutting and cutting with the aid of ultrasonic elliptical vibration, adapted with permission from Ref. [127] 2016, Springer. (**d**) The edge roughness of SPDT single crystal silicon with the aid of ultrasonic vibration. (**e**) The center roughness of SPDT single crystal silicon with the aid of ultrasonic vibration [130].

4.1.2. Tool Servo-Assisted Machining

The low-frequency vibration of the tool servo is further divided into fast tool servo (FTS) and slow tool servo (STS). With its high-frequency response and high rigidity [139,140], the fast tool servo is not only considered one of the most promising techniques for machining optical free-form surfaces [141], but also suitable for assisting in cutting some brittle materials [142]. SPDT under the action of a fast tool servo enables high-speed tool motion and high positioning accuracy, while increasing efficiency without compromising machining fidelity and surface integrity [143,144]. FTS-assisted machining has achieved certain success in machining hard-brittle materials. Chen et al. [145] successfully cut micro-grooves in BK7 glass lenses using FTS, showing that the use of FST can effectively assist in machining difficult-to-machine glass materials. Yu et al. [146] used FTS-assisted SPDT MgF2 glass, successfully realized the plastic machining of the entire curved surface. The above experimental results demonstrate that FTS-assisted machining may be an effective way to help achieve plasticity removal in Fe-based amorphous alloys under SPDT.

STS is also often applied to assist SPDT machining and have been reported to be quite effective in machining free-form surfaces and geometries [147]. Nagayama and Yan [148] successfully prepared two-dimensional sine wave grids on typical brittle material monocrystalline silicon in a single cut with a shape accuracy of 8 nm PV and surface roughness less than 1 nm Sa using STS-assisted SPDT, as shown in Figure 21a–c. The SEM image of the cutting chips is shown in Figure 21d. Continuous chips were generated during the machining process, indicating that ductile material removal was realized in this study. Therefore, it is a good choice to use STS-assisted SPDT Fe-based amorphous alloys to achieve ductility removal to improve their machining performance.

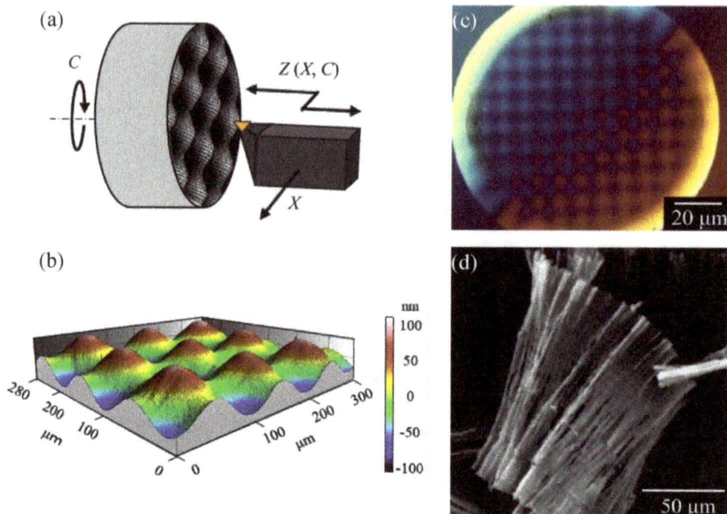

Figure 21. (**a**) STS tuning of the freeform surface. (**b**) 3D image of machined nanometer scaled sine wave grid on single−crystal silicon. (**c**) Microscope image of the machined 2D sine wave grid on single−crystal silicon. (**d**) SEM image of ductile−cut chips, reprinted with permission from Ref. [148] 2021, Elsevier.

Although no researchers have visually demonstrated the usefulness of tool-assisted machining Fe-based amorphous alloys. Based on the results achieved by tool-assisted machining ferrous materials and typical hard-brittle materials, it is reasonable to believe that tool-assisted machining will also have good results in the application of Fe-based amorphous alloys.

4.2. *Low-Temperature Lubrication Assisted Machining*

Difficult-to-machine materials are generally characterized by poor thermal conductivity and high cutting temperatures. The higher cutting temperature generated when cutting difficult-to-machine materials can lead to machining defects such as poor surface quality and severe tool wear [149]. In order to avoid the high cutting temperature and the effect of crystallization on the machining of Fe-based amorphous alloys, it is necessary to use a good coolant to reduce the cutting temperature.

4.2.1. Fe-Based Amorphous Alloys at Low Temperature

Temperature is one of the important factors affecting the way metal materials and engineering structures fracture, some metal materials in low temperature impact absorption work significantly decreased, the material from a ductile state into a brittle state, this phenomenon is called low temperature brittleness. However, amorphous alloys behave exceptionally differently at low temperatures. Maaß et al. [150] showed that at low temperatures, BMG can withstand higher plastic strains than at room temperature. This is especially true for Fe-based amorphous alloys. Daniil et al. [151] found that Fe-Si-Al-Nb-B-Cu nanostructured alloys can still maintain good performance at low temperatures, with a saturation magnetization of 99.3 A m^2/kg, the coercivity is 0.45 A/m, indicating that the soft magnetic properties are not affected. Table 3 shows the parameters of the soft magnetic properties of the Fe-based nanocrystalline alloys at low temperatures. Even some researchers have specifically used low temperature treatment to improve the performance of Fe-based amorphous alloys. Jin et al. [152] concluded that proper deep cooling treatment of Fe-based amorphous alloy coatings can reduce the porosity, and the wear resistance of

Fe-based amorphous alloy coatings gradually increases with longer deep cooling treatment time. Based on the excellent properties exhibited by amorphous alloys at low temperatures, Fan et al. [153] investigated in more depth which amorphous alloys can be benefited by low temperatures and which amorphous alloys are not. They showed that at low temperatures, although the structure of amorphous alloys does not change, amorphous alloys exhibit different mechanical behavior, and in contrast to crystalline materials, the strength and ductility of some amorphous alloys increase significantly with decreasing temperatures in the low-temperature range, such as Ni- and Ni-Fe-based amorphous ribbons. Fe-based amorphous alloys in ribbon forms show ductile to brittle transition, when tested at cryogenic temperatures, as shown in Figure 22a. In this case, the low temperature has a bad effect on the Fe-based amorphous alloy.

Table 3. Magnetic parameters and resistivity of Fe-based nanocrystalline alloys at 77 and 300 K [151].

Composition	M_S (300 K) (A m^2/kg)	M_S (77 K) (A m^2/kg)	H_C (300 K) (A/m)	H_C (300 K) (A/m)
Fe73.5Si13.5Nb3B9Cu1	139.5	151.9	0.52	0.81
Fe68Si15.5Al3.5Nb3B9Cu1	113.3	127.6	0.35	1.26
Fe65.5Si16.5Al5Nb3B9Cu1	98.3	98.3	0.50	0.58
Fe63Si17.5Al6Nb3B9Cu1	80.4	80.4	1.12	0.45
Fe62Si18Al7Nb3B9Cu1	64.1	64.1	1.48	1.45

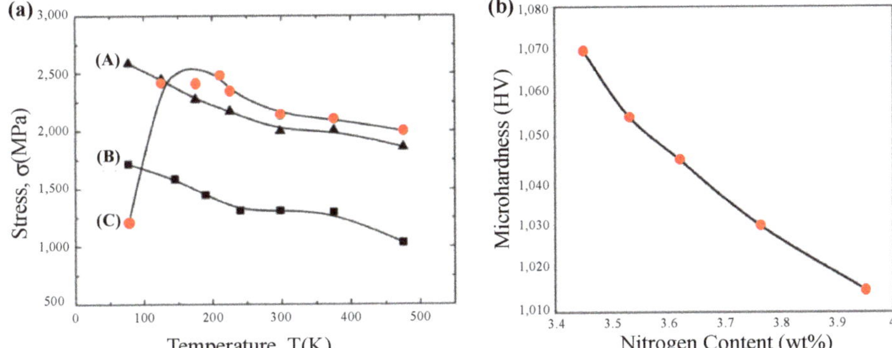

Figure 22. (**a**) Tensile fracture stresses of Ni72P18B7Al3(A), Ni49Fe29P14B6Al2(B), and Fe76P16C4Si2Al2(C) (specimens of 0.025 mm thickness). (**b**) Variation of microhardness values of Fe-18Cr-4Mn-xN amorphous alloy with respect to nitrogen concentration, adapted with permission from Ref. [154] 2009, Elsevier.

In cryogenic-assisted machining, cooling the workpiece with cryogenic gas (CG) is a very common means. In its cooling process, the cooling medium usually uses liquid nitrogen (LN_2) and the workpiece will be under the atmosphere of nitrogen. Salahinejad et al. [154] studied the tissue evolution of Fe-18Cr-8Mn-xN amorphous alloy under a nitrogen atmosphere. They found the number and stability of amorphous phases increased with the increase of nitrogen content. In the study by Fan et al. [153] mentioned above, the microhardness of the Fe-based amorphous alloy decreases with increasing nitrogen content, as shown in Figure 22b. Therefore, reducing the surface hardness of Fe-based amorphous alloys through a nitrogen atmosphere and enhancing the stability of its amorphous phase is helpful for diamond tools to better machine the surface layer of workpieces with reduced hardness. This is beneficial to improving the machining performance of Fe-based amorphous alloys under SPDT. Based on the above effects of low temperature as well as nitrogen atmosphere on Fe-based amorphous alloys, CG-assisted machining might have unexpected effects on the machining of Fe-based amorphous alloys.

4.2.2. Advantages of Low-Temperature Lubrication Assisted Machining

In the cutting process of Fe-based amorphous alloy, the phenomenon appears that the cutting force is large, the temperature of the cutting area is too high, and even produces pyro luminescence phenomenon to burn the workpiece. These phenomena can largely impair the quality of Fe-based amorphous alloy machining, while increasing tool wear. Evans and Bryan [155] concluded that any wear of the diamond tool would increase the force of the tool and thus increase the temperature, which, in turn, would increase the rate of dissolution, diffusion, and catalytic graphitization, then increase the wear of the tool. At low temperatures, both diffusion and catalytic graphitization will slow down. Meanwhile, the high cutting temperature will also lead to the crystallization of Fe-based amorphous alloy, which is not conducive to the study of the machining mechanism of Fe-based amorphous alloy and the tool wear mechanism. In conclusion, effective cooling and lubrication means are urgently needed for SPDT of Fe-based amorphous alloys.

In traditional machining, people mostly use coolant and cutting fluid to achieve the purpose of cooling and lubrication. Although it can play a certain role in most cases, the use of coolant and cutting fluid is not a panacea. In a study by An [156], it was shown that increasing the amount of conventional coolant was not effective in improving the machining performance of difficult-to-machine materials. On the contrary, because there is very little cutting fluid that can effectively penetrate the machining area at room temperature and pressure, it will cause greater waste of resources. Aramcharoen [157] also further proposed that conventional emulsified cutting fluids are actually difficult to penetrate the high temperature cutting area at the tooltip, with limited actual cooling and lubrication, low productivity, and high machining costs. This shows that although the traditional cutting fluid can alleviate a certain cooling and lubricating effect, it is not enough to improve the machining of difficult-to-machine materials such as Fe-based amorphous alloy. In addition, a study by Bordin et al. [158] reported that the use of conventional emulsified cutting fluids can be hazardous to the health of machinists. Therefore, it is urgent to find a cooling and lubricating means to replace cutting fluid and coolant. Bordin et al. also compared the effects of cryogenic turning, dry turning, and wet turning on machining, as shown in Figure 23. The surface roughness under cryogenic turning is generally better than dry turning and wet turning (please see Figure 23a). Frequent chip entanglements occurred during dry turning because the chip breaker exerted no breaking action due to the low depth of cut (please see Figure 23c). In wet turning, the chip radius drastically increased, which damaging both the cutting tool and the machined surface, as shown in Figure 23d. And the chip morphology under cryogenic turning is better, as shown in Figure 23b. In summary, the cryogenic machining process is a good means of cooling and lubrication, and can be a good substitute for cutting fluids and coolants.

LN_2 [159] and carbon dioxide (CO_2) [160] are the most commonly used cooling media in cryogenic machining, and liquid nitrogen is more efficient for cooling at lower temperatures [161]. Both in terms of tool wear and cutting forces, LN_2 assisted machining is better than CO_2 assisted machining and dry machining, as shown in Figure 24. While the tool showed significant wear and built-up-layer (BUL) formation after dry machining (please see Figure 24a) and CO_2 assisted machining (please see Figure 24b), the tool surface after LN_2 assisted machining was smoother and had no BUL formation (please see Figure 24c). At feed rates of 0.1 mm/rev and 0.15 mm/rev, the average flank wear and main cutting force of LN_2 assisted machining were lower than those of dry machining and CO_2 assisted machining, as shown in Figure 24d,e. Experiments by Hong and Ding [162] demonstrated that under the use of LN_2, machining of titanium alloy can effectively reduce the temperature in the cutting area and can effectively replace the cutting fluid to improve the surface quality and improve the tool life. Huang et al. [163] used the combination of CG and minimum quantity lubrication (MQL) to the assisted cutting of titanium alloys when using SPDT titanium alloys, which effectively improved the machining performance of titanium alloys. This shows that this cooling method can effectively replace the traditional cooling cutting fluid at high cutting temperature. The cutting temperature of titanium alloys under

SPDT is even higher than that of Fe-based amorphous alloys. CG+MQL can effectively cool down and lubricate the machining of titanium alloy, it is reasonable to believe that the use of CG+MQL assisted SPDT Fe-based amorphous alloy will also achieve good results.

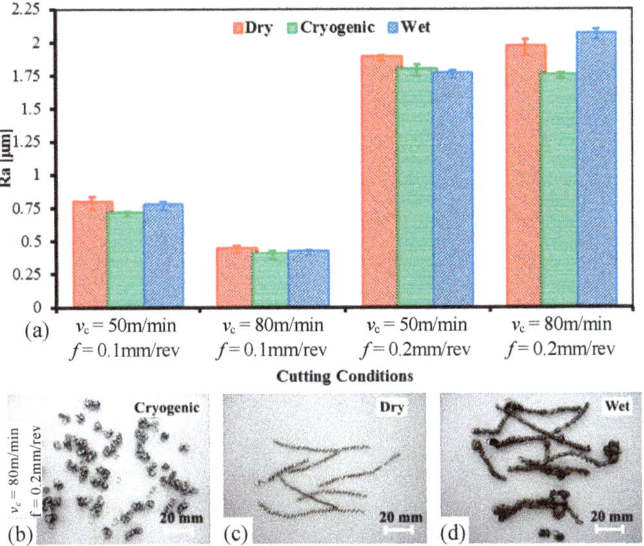

Figure 23. (**a**) Effect of the cooling strategies on the mean surface roughness *Ra* after 8 min of turning. (**b–d**) Chip morphology under cryogenic, dry, and wet turning conditions after 8 min of cutting and for a cutting speed of 80 m/min, adapted with permission from Ref. [158] 2017, Elsevier. Frequent chip entanglements occurred during dry turning. In wet cutting, long tubular chips appear, which damage the tool and workpiece. The chip morphology under cryogenic turning is better.

In addition to having better cooling and lubricating effects than traditional cooling cutting fluids, CG assisted machining also performs better in environmental protection. The use of LN_2 as a cooling medium has the advantages of being non-toxic, harmless, and environmentally friendly. Compared with cutting fluid as a cooling medium, using LN_2 as a low-temperature cooling medium for cutting is a sustainable green machining method [164]. The use of LN_2 is now gaining new focus in the machining industry due to environmental concerns and disposal costs incurred by the use of conventional coolants. The experimental results of Aramcharoen [157] show that the wear resistance of the tool is improved and the cooling efficiency is higher when liquid nitrogen is used as the cryogenic cooling medium compared with the oil-based coolant. Numerous studies have shown that the use of CG as a means of replacing traditional cutting fluids is better than ever, both in terms of effectiveness, safety, and economy.

MQL is a technique that injects a minimal amount of lubricant into the cutting area to provide effective lubricity and improve the machinability of the material [165]. Its disadvantage is that only a small amount of cooling lubricant is used, resulting in unsatisfactory cooling efficiency. The mist lubrication used in MQL has proven to be a good alternative to cutting fluids, the higher pressure of mist lubrication allows the mixture of air and cutting fluid to act effectively on the cutting area [166]. The benefit of MQL over conventional cutting fluids is that it meets the requirements of green machining by optimally spraying a mixture of compressed air and cutting fluid instead of cooling with water injection, from minimizing the use of coolant. MQL has numerous other advantages: its application can reduces cutting forces, cutting zone temperature, tool wear, and friction coefficient compared to dry and wet machining [165]. Kamata and Obikawa [167] found that for different coated tools, the surface finish and tool life obtained with MQL was superior to wet and dry machining. Hadad and Sadeghi [168] observed that MQL improved the turning performance with the flexibility of process parameters, such as nozzle position and orientation. They also found that turning operations using MQL technology required the least cutting forces when compared to dry and wet turning for the entire turning depth. Kishawy et al. [169] explored the cooling effect of MQL and observed that the ability of MQL to improve surface roughness, tool wear, and cutting forces was comparable to that of wet machining. In summary, MQL is effective in reducing cutting forces, lowering cutting temperatures, improving tool wear, and enhancing machined surface quality.

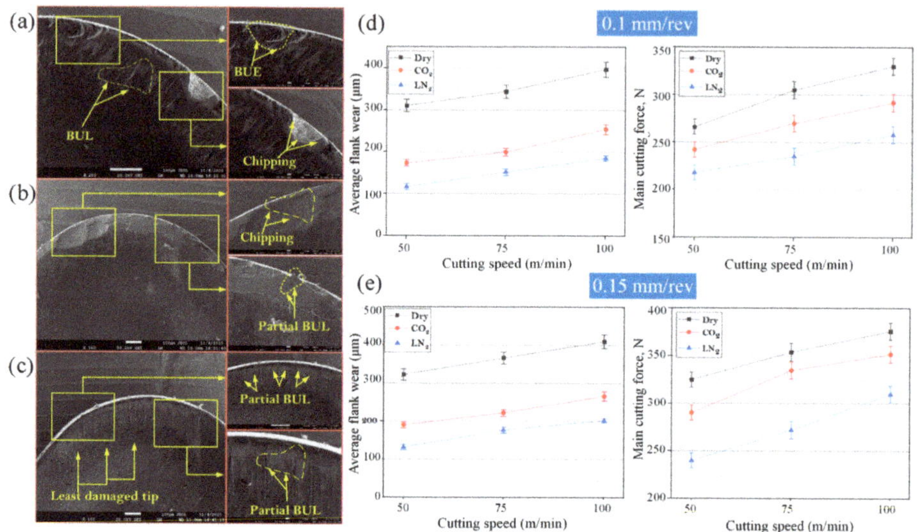

Figure 24. SEM analysis of tool wear under different cooling strategies: (**a**) dry machining. (**b**) CO_2. (**c**) LN_2. The built-up-edge (BUE) and chipping appear on the dry cutting tool surface. CO_2-assisted machining eliminates the formation of BUE, and some sticking in the form of BUL occurs on the rake face. The tool surface under LN_2-assisted machining is relatively smooth, with a small amount of BUL. (**d**) Average flank wear and main cutting forces at feed rate of 0.1 mm/rev. (**e**) Average flank wear and main cutting forces at feed rate of 0.15 mm/rev, adapted with permission from Ref. [161] 2021, Elsevier. The average flank wear and main cutting force were the lowest with LN_2-assisted machining.

CG and MQL are also flawed separately. The cooling efficiency of CG is high, but the lubricating performance is poor. MQL is just the opposite, with good lubrication performance and insufficient cooling efficiency. The combination of CG and MQL will not affect each other's role, even make up for each other's shortcomings. MQL assisted machining achieved less chip adhesion and severe tool wear, CG assisted machining achieved more chip adhesion and less tool wear, while MQL + CG assisted machining achieved moderate chip adhesion and moderate tool wear, as shown in Table 4. The use of CG + MQL hybrid-assisted machining will be a very promising cooling and lubrication process.

Table 4. Tool wear and chip adhesion to the cutting tools under the various cooling condition: MQL; CG; and MQL + CG, adapted with permission from Ref. [163]. 2017, Elsevier.

Cooling Conditions	MQL	CG	MQL + CG
Tool wear	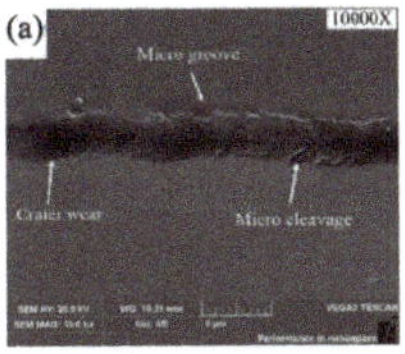 (a) 10000X	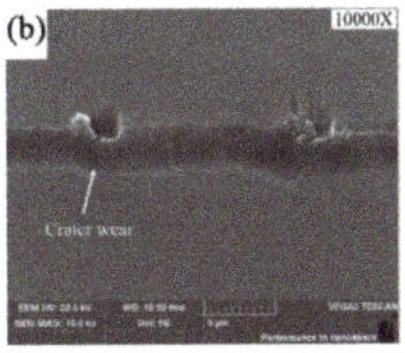 (b) 10000X	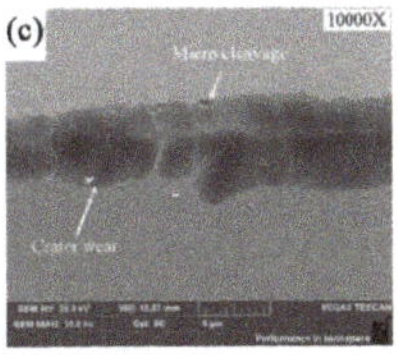 (c) 10000X
Chip adhesion	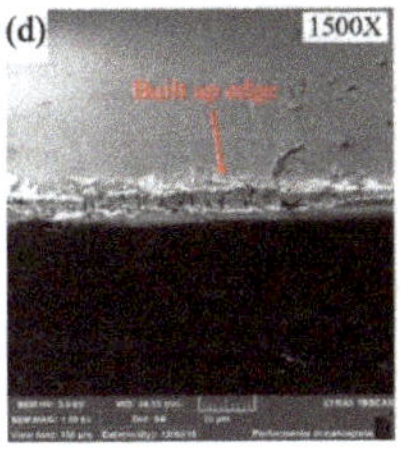 (d) 1500X	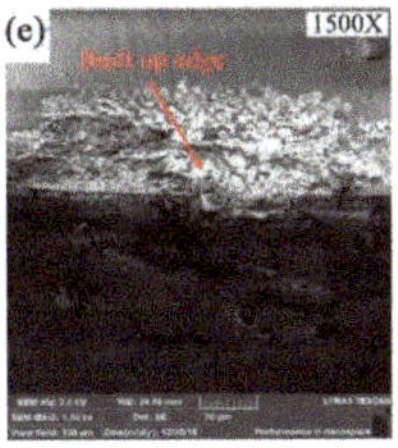 (e) 1500X	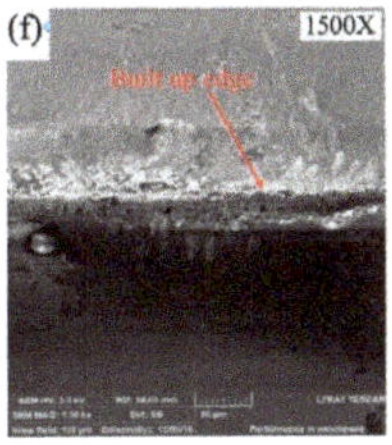 (f) 1500X

4.2.3. Device for Low-Temperature Lubrication Assisted Machining

There are three CG schemes according to how the cold source acts on the workpiece: Scheme 1, using the nitrogen evolved from liquid nitrogen to form an ultra-low temperature environment in which the material is placed so as to achieve a deep cooling treatment [170]; Scheme 2, Immerse the material in LN_2 at a temperature of −196 °C for deep cooling treatment; and Scheme 3, LN_2 is used to enter the machining area under the action of injection pressure, and the cutting temperature is reduced by heat exchange such as heat conduction and heat convection [171]. The volatile nitrogen used in Scheme 1 is not safe enough and it is difficult to control the experimental variables, which affects the meticulousness of the study; Scheme 2 is pretreatment, which does not facilitate the continuous cooling during the machining process; and Scheme 3 is more ideal, while it is more flexible in its use, and can also be divided into liquid nitrogen jets and internal cooling of the turning tool. Gan et al. [164] concluded in their experiments that internal cooling of the turning tool has higher cooling efficiency compared to liquid nitrogen jets, which can effectively reduce surface defects such as material adhesion and pits on the machined surface. However, liquid nitrogen jets also have their outstanding features, and Dhar et al. [172] observed that chip formation was facilitated and cutting forces were significantly reduced when turning was assisted by liquid nitrogen jets. Another advantage of liquid nitrogen jets over internal cooling is flexibility. The cooling area of internal cooling is fixed (please see Figure 25a,b), while liquid nitrogen jets is realized by using external nozzles, and the cooling area can be controlled by artificially changing the position of the nozzles. The experimental results of Khan et al. [173] clearly show that the position of the liquid nitrogen jet nozzle affects tool wear and machined surface quality. According to this situation, the most suitable nozzle position can be experimentally tested according to the specific machining material when using liquid nitrogen jets.

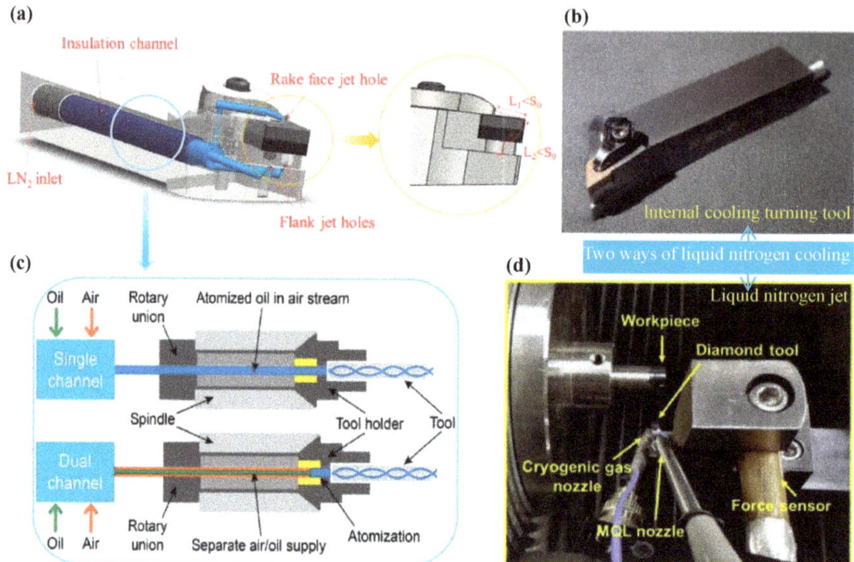

Figure 25. (**a**) 3D modeling picture of the internal cooling device of the turning tool. (**b**) Physical picture of the internal cooling device of the turning tool, adapted with permission from Ref. [164] 2021, Elsevier. (**c**) Internal transfer system of MQL [174]. (**d**) Layout of the CG + MQL assisted SPDT device build, adapted with permission from Ref. [163] 2017, Elsevier.

MQL is also an assisted machining method that uses nozzles to lubricate the machining area. When the cooling scheme of CG is achieved by liquid nitrogen jet, then MQL can be combined with CG to create various cooling and lubrication solutions depending on the nozzle position. MQL technology is delivered internally through specially designed single/dual channels, where air and lubricant then mix with each other in the nozzle and are finally injected into the machining area. The internal channel system of MQL is shown in Figure 25c. As mentioned in Section 4.2.2, the mixed machining of CG + MQL can give full play to the advantages of CG and MQL. Figure 25d shows the liquid nitrogen jets device of Huang et al. [163] using CG and MQL.

4.3. Magnetic Field-Assisted Machining

Magneto-crystallization phenomenon exists in Fe-based amorphous alloys [13], that is, a method of treating Fe-based amorphous alloys with a pulsed magnetic field of a certain frequency to make them undergo nano-crystallization (nano-crystallization of amorphous alloys is a spontaneous process of energy reduction). The magneto-crystallization method is a promising method that can effectively control the amount of amorphous nano-crystallization by controlling the machining conditions. Meanwhile, studies have shown that magnetic field-assisted machining can effectively reduce tool wear of ferromagnetic materials. Therefore, the introduction of magnetic field assistance is not only aimed at improving the machining of Fe-based amorphous alloys by using the magneto-crystallization phenomenon, but also hopes to reduce the tool wear during machining through the magnetic field.

4.3.1. Application of Magnetic Field on SPDT

Gavili et al. [175–179] showed that the thermal conductivity of ferromagnetic metals and ferromagnetic fluids containing ferromagnetic particles can be increased in the presence of a magnetic field. Meanwhile, Gonnet et al. [180–182] found that the thermal conductivity of nanocomposites containing ferrous metals or ferromagnetic fluids can be well improved in the presence of a magnetic field. The basic principle and root cause of using a magnetic field to improve the thermal conductivity of Fe-containing materials can be attributed to the alignment of Fe particles inside nanofluids and nanocomposites under the action of an external magnetic field [183].In the presence of a magnetic field, the magnetic dipole energy is sufficient to exceed the thermal energy, and the Fe particles inside the nanofluid or nanocomposite tend to align along the direction of the applied magnetic field for a positive magnetization of the Fe particles. The well-aligned magnetic particles act as linear chains, which are highly conductive paths for heat transfer, facilitating fast heat transfer in the fluid carrier path [184,185]. Fe-based amorphous alloys suffer from high cutting temperatures under SPDT, resulting in difficult-to-machine. Using the action of the magnetic field to improve the thermal conductivity of the Fe-based amorphous alloy, it is possible to effectively reduce the cutting temperature during machining, thereby improving the machining performance of the Fe-based amorphous alloy. In the application of SPDT, magnetic field as an applied field energy assisted machining has long been of interest to researchers and has achieved some success in many machining scenarios. Yip and To [183] successfully improved the machining performance of titanium alloys under SPDT by increasing the thermal conductivity of titanium alloys through the addition of a magnetic field. The feasibility of improving the high cutting temperature by increasing the thermal conductivity of the workpiece by the magnetic field is effectively confirmed. In another study by Yip and To [186], in SPDT titanium alloy, the addition of a magnetic field caused an eddy current damping effect to occur in the workpiece, which led to a significant reduction in tool wear through the action of the eddy current damping effect. Sodano and Bae [187] reviewed the mechanism of eddy current damping effect generation and concluded that the process of generating vortices and dissipating vibration energy suppresses the vibration in the structure. Sodano et al. [188] built a theoretical model of the eddy current damping system and demonstrated that the eddy current damping system can effectively suppress the vibration of the crossbeam. Bae et al. [189] compared experimental and theoretical models of eddy current dampers using the eddy current damping effect to demonstrate the damping effect of the motion system. These experimental results show that overall turning vibration and tool vibration can be effectively suppressed under the effect of eddy current damping generated by the magnetic field, thus reducing tool wear and improving the machining performance of the workpiece under SPDT. As mentioned in Section 3.3, severe tool wear is an important factor that deteriorates the machining performance of Fe-based amorphous alloys. The eddy current damping effect generated by the magnetic field can reduce the wear of the tool during the SPDT process and effectively improve the machining performance of the Fe-based amorphous alloy. Khalil et al. [190] conducted experiments on magnetic field-assisted SPDT machining of Ti6Al4 and showed that in the process of machining, the magnetic field is conducive to the collection of chips, avoiding edge buildup of chips on the tool and cut instead of the tool cutting edge. Since the chips are easier to discharge, the heat in the chips is prevented from accumulating in the cutting area and affecting the machining process. The chip morphology was also compared in the experiments of Khalil et al. [190] under the applied magnetic field and without the applied magnetic field, as shown in Figure 26. Chips formed when cutting without a magnetic field are serrated, meaning lower cutting performance, as shown in Figure 26a. When 0.01 T and 0.02 T magnetic fields are applied, respectively, the chips formed by cutting are not serrated as before and are long and continuous, implying a relatively high cutting performance, as shown in Figure 26b,c. Fe-based amorphous alloys are used as Fe-containing materials, and the effect of chip discharge will be more obvious.

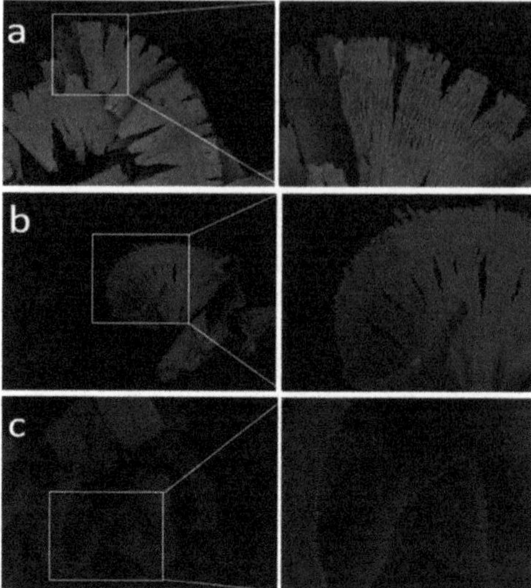

Figure 26. (**a**) Formation of serrated chips when no magnetic field is applied. (**b**) Formation of flat, long and continuous chips when a magnetic field of 0.01 T is applied. (**c**) Formation of flat, long and continuous chips when a magnetic field of 0.02 T is applied, reprinted with permission from Ref. [190] 2022, Elsevier.

Another major advantage of the magnetic field-assisted machining method is that the field energy of the magnetic field can be provided by a simple permanent magnet only, ensuring the economy of the experiment, as shown in Figure 27a–d. In the above experiments where a magnetic field was applied during SPDT, the experimental results obtained many favorable improvements such as increasing the thermal conductivity of the material, effectively reducing tool wear by reducing turning vibration, facilitating cutting collection, and improving cutting performance due to the addition of the magnetic field, respectively. Although there is no systematic theory for the application of magnetic field-assisted machining of difficult-to-machine materials and more areas for optimization, it is well worthwhile to use this method in machining Fe-based amorphous alloy to better improve their machining performance based on the numerous benefits of magnetic field-assisted above.

4.3.2. Application of Magnetic Field on Fe-Based Amorphous Alloys

So far, magnetic field, as a very common applied field energy, has been applied in various fields with certain success, and it is no exception in the case of Fe-based amorphous alloys. Numerous researchers have added magnetic field to their studies on Fe-based amorphous alloys and recorded the changes on Fe-based amorphous alloys after adding magnetic field.

Chao and Zhang [191] optimized the performance of Fe78Si9B13 amorphous strip by treating it with low-frequency pulsed magnetic fields. Their method of treating Fe-based amorphous alloys is the magneto-crystallization method, because the nano-crystallization of amorphous alloys is a spontaneous process with reduced energy, and the magnetic field provides the energy for the transition from the amorphous state to the crystalline state. As mentioned in Section 3.3, nano-crystallization may enable the desired unique properties of Fe-based amorphous alloys and can therefore be used to optimize their performance. Jin et al. [192] also used a pulsed magnetic field of certain intensity to treat Fe52Co34Hf7B6Cu1 samples. Wang [13] also adopted the same method, and investigated how the crystallization and nano-crystallization of Fe-based amorphous alloys would be affected if the intensity of the magnetic field was increased under the applied magnetic field. After treating the Fe-based amorphous alloy with a magnetic field, Wang used transmission electron microscopy to discover nanocrystals, which are about 10 nm in size as shown in Figure 28a,b. And Wang, finally, concluded that the crystallization of the Fe-based amorphous alloy increases monotonically with increasing magnetic field strength. The nature of this magneto-crystallization phenomenon has been studied. Among them, Guo et al. [193] derived from the theory of phase transition kinetics that when a pulsed magnetic field acts on an amorphous alloy, in order to satisfy the minimum free energy, the amorphous alloy will undergo linear magneto-striction. Macroscopically, the workpiece changes in the length direction, and the magneto-striction strengthens the vibration of the internal atomic cycle, provides a driving force for nucleation, and promotes the crystallization of amorphous alloys to produce nanocrystalline. Both Yoshizawa [91] and Suzuki [194] considered this nano-crystallization phenomenon as an effective machining method to improve the soft magnetic properties of Fe-based amorphous alloys.

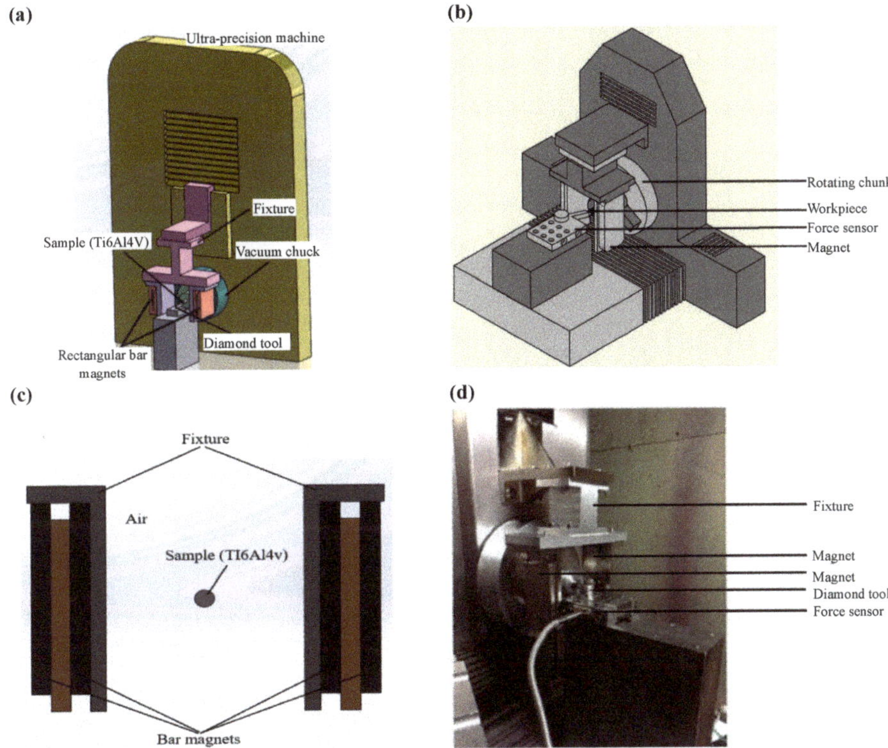

Figure 27. (**a**) Modeling of the machine layout when Khalil et al. [190] use magnetic field assist. (**b**) Modeling of the machine layout when Yip and To use magnetic field assistance. (**c**) Schematic of the distribution of permanent magnet and workpiece positions when using magnetic field assist, adapted with permission from Ref. [190] 2022, Elsevier. (**d**) Actual machine layout when Yip and To use magnetic field assist, adapted with permission from Ref. [186] 2017, Elsevier.

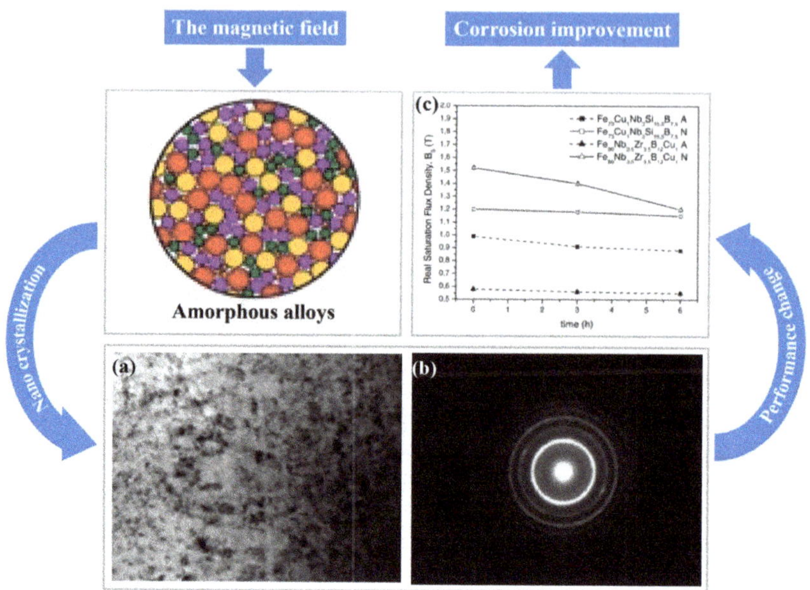

Figure 28. Images of Fe52Co34Hf7B6Cu1 after pulsed magnetic field treatment with nano-crystallization of the workpiece, adapted with permission from Ref. [192] 2006, Elsevier. (**a**) TEM micrograph of Fe52Co34Hf7B6Cu1. (**b**) Distinct diffraction rings are generated. It can be determined that the samples are nano-crystallized after the pulsed magnetic field treatment. (**c**) The values of saturation magnetic flux density, Bs, as function of the immersion time in corrosive environment (0.1 M of H2SO4 solution) for amorphous (A) and nanocrystalline (N) samples, adapted with permission from Ref. [195] 2002, Elsevier. Corrosion enhancement was obtained for both Fe73Nb3Si15.5B7.5Cu1 and Fe73.5Cu1Nb3Si15.5B7.5 after nano-crystallization.

In many cases, nano-crystallization of amorphous alloys results in the strengthening of the workpiece itself. For example, Souza [195] investigated the properties of two Fe-based amorphous alloys, Fe73Nb3Si15.5B7.5Cu1 and Fe73.5Cu1Nb3Si15.5B7.5, after nano-crystallization and found that when they underwent nano-crystallization, their corrosion resistance was somewhat improved, as shown in Figure 28c. Although there is evidence that the formation of nanocrystals enhances the performance of Fe-based amorphous alloys, contrary opinions have been presented. Gostin et al. [196] found that when the matrix precipitation phase of Fe-based amorphous alloys is α-Fe, Fe carbide, and Fe boride or their mixtures, nano-crystallization of Fe-based amorphous alloys will reduce their corrosion resistance. The results show that the addition of a magnetic field can promote the nano-crystallization of Fe-based amorphous alloys, and the intensity of the magnetic field can affect the degree of nano-crystallization. Therefore, magnetic field-assisted machining can make good use of this phenomenon to improve the machining performance of Fe-based amorphous alloys.

4.4. Other Assisted Machining Methods

Amorphous alloys are considered as ideal materials for wear and corrosion resistant coatings due to their excellent properties [197], especially Fe-based amorphous alloys with high strength, high hardness, and excellent wear and corrosion resistance. The protection mechanism of this coating may also be a good way to improve the machining performance of Fe-based amorphous alloys, as coating technology has been an important way to improve tool wear resistance in the machining field [198]. Chemical wear of Fe and C deteriorates the machining performance of Fe-based amorphous alloys under SPDT. Brinksmeier and

Glabe [199] demonstrated the potential of TIC and TIN coatings on diamond tools to eliminate chemical wear. Xiao [200] et al. showed that the nano-SiC/Ni composite coating can further protect the diamond from graphitization and can result in higher bending strength and wear resistance of the diamond turning tool bit. It can be seen that coating-assisted technology has the potential to improve the machining performance of Fe-based amorphous alloys under SPDT. However, the hardness of the coating is often lower than that of diamond, and the machined surface quality is not ideal. Therefore, the application of coating technology to improve the surface quality of Fe-based amorphous alloys and other difficult-to-machine materials still has a lot of room for development.

Diamond tools used in SPDT are the hardest known material and are widely used in ultra-precision machining, but diamond suffers from severe tool wear affecting the surface quality when machining Fe and other transition metal alloys. Implanting the near-surface of diamond with ion implantation of other elements to modify its surface mechanical and chemical behavior is considered as a promising assisted method to address this wear [201]. Already in 1999, Klocke and Krieg [198] suggested applying a protective coating to diamond tools to create a diffusion barrier. As an Fe-based amorphous alloy with Fe as the main element, it is also promising to improve the machining performance of Fe-based amorphous alloy under SPDT with the aid of ion implanted modified diamond technology. Wear occurrences are compared between ion implanted diamond and unmodified diamond by Lee et al. [202], as shown in Figure 29. The cutting tool is considered to be worn when it is incapable of achieving the desired surface finish of the work material. After machining a distance of 350 m with the ion implanted tool, a clear reflection of the emblem can still be observed, signifying that the ion implanted tool can operate at a further distance in comparison to the unmodified tool (please see Figure 29c). Correspondingly, the surface roughness measurements showed a similar magnitude of 2.6 times increase in *Ra* and *Rq* for the surface produced by the unmodified cutting tool (please see Figure 29b). However, there are still obstacles that must be overcome in the technique of diamond ion implantation such as the removal of radiation damage after ion implantation without causing graphitization of the diamond [203]. These obstacles also greatly affect the widespread application of ion implantation modified diamond tools as an assisted technology.

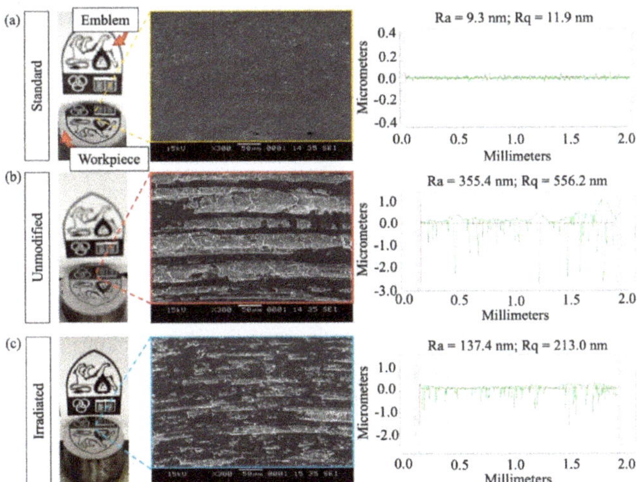

Figure 29. Images of the machined iron surface quality with the magnified observation using a SEM and the respective surface roughness measurements: (**a**) standard requirements achieved after machining 50 m, (**b**) after machining with an unmodified diamond tool over a distance of 350 m, and (**c**) after machining with a gallium irradiated diamond tool over a distance of 350 m, reprinted with permission from Ref. [202] 2019, Elsevier.

In addition to the above—mentioned diamond tool coating techniques and ion implantation diamond modification techniques. Inert gas-assisted machining methods and electric field-assisted machining methods [204] are also favored by researchers often applied to improve the machining performance of difficult-to-machine materials.

5. Summary and Outlook

This paper systematically reviews the properties and machining performance of amorphous alloys. As a special case of amorphous alloy, the preparation, application and machining of Fe-based amorphous alloy are systematically summarized in this review. It is found that single-point diamond turning (SPDT) is a promising machining method to overcome the extremely high hardness of Fe-based amorphous alloys. However, under SPDT, the problems of high machining temperature, machining crystallization and chemical wear still greatly deteriorate the machining performance of Fe-based amorphous alloys. Assisted machining methods such as tool-assisted machining, low-temperature lubrication assisted machining and magnetic field assisted machining et al. are found effective improving the machining performance of Fe-based amorphous alloys.

Ultrasonic vibration assisted machining is expected to reduce the high cutting temperature of Fe-based amorphous alloys through periodic intermittent machining, thereby reducing the impact of high cutting temperatures on the crystallization and oxidation. Meanwhile, ultrasonic vibration assisted machining can effectively reduce tool wear and cutting force, whereby effectively improve the machining performance of Fe-based amorphous alloys. However, the machining efficiency of ultrasonic vibration assisted machining is low. In addition, fast tool servo (FTS) and slow tool servo (STS) are expected to help Fe-based amorphous alloys to achieve ductile removal, and they can effectively assist machining of the microstructure of Fe-based amorphous alloys.

Low-temperature lubrication assisted machining can greatly reduce the cutting temperature of Fe-based amorphous alloys. It is expected to control the cutting temperature of Fe-based amorphous alloys not to exceed glass transition temperature point (Tg) through low-temperature lubrication assisted machining, thereby eliminating the effects of crystallization and high-temperature oxidation on machining. Meanwhile, the nitrogen atmosphere can effectively reduce the surface hardness of the Fe-based amorphous alloys, which is helpful for better machining. The form of cryogenic gas (CG)+minimum quantity lubrication (MQL) has good flexibility and can effectively adapt to various cooling and lubrication requirements in the machining of Fe-based amorphous alloys. However, surface rebound of the workpiece caused by temperature changes has enormous deteriorated the finish machining accuracy.

The presence of magneto-crystallization in Fe-based amorphous alloys has shown promise for improving machining performance by promoting nano-crystallization on workpiece surfaces through magnetic fields. The magnetic field promotes the thermal conductivity of Fe-containing materials and improves the excessive machining temperature of Fe-based amorphous alloys caused by low thermal conductivity, and also facilitates the collection of chips from magnetic materials. The eddy current damping effect caused by the magnetic field at the workpiece can effectively suppress overall machining vibration and tool vibration. Magnetic field-assisted machining has achieved good results in the machining of many ferromagnetic materials because the implementation of simple equipment also has good economic benefits. Unfortunately, magnetic field-assisted machining has not yet formed a systematic theory, and the experimental process is not easy to control.

In addition to the above three highlighted assisted machining methods, traditional coating protection methods and novel ion implantation modified diamond tools are also effective ways of improving the machining performance of Fe-based amorphous alloys. The use of coating protection can alleviate tool wear to a certain extent. However, since the hardness of the coating is not high enough, the effect is often not ideal when machining high hardness materials such as Fe-based amorphous alloys. Direct machining with diamond tools can well overcome the problem that the tool hardness is not high enough. However, the chemical affinity of diamond and Fe-based amorphous alloys can cause chemical wear and aggravate tool wear. Ion-injected diamond modification can form a wear-resistant and inert barrier layer for the cutting edge of the tool, which helps to improve the wear of this tool and improve the surface quality. However, the ion implantation technology is not mature enough and can radiate damage to the diamond.

The combination of SPDT and assisted machining methods is a promising method for machining Fe-based amorphous alloys. However, assisted machining methods also cannot fully provide favorable factors, combined with the machining process problems, a reasonable combination of different assisted machining technology may be able to achieve better results. For example, instead of degrading the effect, the combined use of CG and MQL can compensate each other. Therefore, when machining Fe-based amorphous alloys or even amorphous alloys, it is advisable to improve their machining performance more often in the form of combinations based on assisted machining methods. The critical dimension of Fe48Cr15Mo14C15B6Er2 is 12 mm, and the critical dimension of Fe41Co7Cr15Mo14C15B6Y2 is the largest 16 mm. In the study of the machining of Fe−based amorphous alloys, it is suggested that these two samples can be used to facilitate the analysis of the machining mechanism. Meanwhile, Fe77.5Si17.5B15 has been proved that its glass removal not only reduces the magnitude of internal stress, but also significantly reduces the magnetostriction, which is suitable as an object for machining.

Author Contributions: Z.H.: Writing-Original Draft, Writing-Review and Editing, Conceptualization. G.Z.: Supervision, Writing—Review and Editing, Funding acquisition. J.H.: Methodology, Software. J.W.: Investigation, Supervision. S.M.: Data Curation, Resources. H.W.: Funding acquisition. All authors have read and agreed to the published version of the manuscript.

Funding: The work described in this paper was supported by the National Natural Science Foundation of China (Grant No. U2013603, 51827901), the Shenzhen Natural Science Foundation University Stability Support Project (Grant No. 20200826160002001, 20200821110721002), and the Postgraduate Innovation Development Fund Project of Shenzhen University (Grant No. 315-0000470813).

Institutional Review Board Statement: Not applicable.

Informed Consent Statement: Not applicable.

Data Availability Statement: Not applicable.

Conflicts of Interest: No conflict of interest exists in this submitted manuscript, and the manuscript is approved by all authors for publication. I would like to declare on behalf of my co-authors that the work described was original research that has not been published previously, and not under consideration for publication elsewhere, in whole, or in part.

References

1. Shen, G.D.; Li, J.P.; Zhou, C.W.; Yang, F. Fe-based Amorphous Soft Magnetic Alloy and Its Crystallization. *J. Nanjing Univ. Sci. Technol.* **1998**, *22*, 544–547. (In Chinese) [CrossRef]
2. Pei, Y.; Zhou, G.; Luan, N.; Zong, B.; Qiao, M.; Tao, F.F. Synthesis and catalysis of chemically reduced metal–metalloid amorphous alloys. *Chem. Soc. Rev.* **2012**, *41*, 8140–8162. [CrossRef] [PubMed]
3. Miracle, D.B.; Lord, E.A.; Ranganathan, S. Candidate atomic cluster configurations in metallic glass structures. *Mater. Trans.* **2006**, *47*, 1737–1742. [CrossRef]
4. Cheng, Y.; Ma, E. Atomic-level structure and structure–property relationship in metallic glasses. *Prog. Mater. Sci.* **2011**, *56*, 379–473. [CrossRef]
5. Greer, A.L. Confusion by design. *Nature* **1993**, *366*, 303–304. [CrossRef]
6. Souza, C.; Ribeiro, D.; Kiminami, C. Corrosion resistance of Fe-Cr-based amorphous alloys: An overview. *J. Non-Cryst. Solids* **2016**, *442*, 56–66. [CrossRef]
7. Inoue, A. Stabilization of metallic supercooled liquid and bulk amorphous alloys. *Acta Mater.* **2000**, *48*, 279–306. [CrossRef]
8. Jain, V.K. *Nanofinishing Science and Technology: Basic and Advanced Finishing and Polishing Processes*; CRC Press: Boca Raton, FL, USA, 2016.
9. Li, R.T. *Fundamental Study on the Ultra-Precision Machining of Brittle Material and Construction of Micro-Cutting System*; Tianjin University: Tianjin, China, 2018. (In Chinese) [CrossRef]

10. Hatefi, S.; Abou-El-Hossein, K. Review of non-conventional technologies for assisting ultra-precision single-point diamond turning. *Int. J. Adv. Manuf. Technol.* **2020**, *111*, 2667–2685. [CrossRef]
11. Wang, W.H.; Dong, C.; Shek, C. Bulk metallic glasses. *Mater. Sci. Eng. R Rep.* **2004**, *44*, 45–89. [CrossRef]
12. Jalali, A.; Malekan, M.; Park, E.S.; Rashidi, R.; Bahmani, A.; Yoo, G.H. Thermal behavior of newly developed Zr33Hf8Ti6Cu32Ni10 Co5Al6 high-entropy bulk metallic glass. *J. Alloys Compd.* **2022**, *892*, 162220. [CrossRef]
13. Wang, L. *The Research on Preparation and Properties of Fe-Based Amorphous Alloys*; Northeastern University: Shenyang, China, 2014. (In Chinese)
14. Liu, Y.; Wang, D.; Nakajima, K.; Zhang, W.; Hirata, A.; Nishi, T.; Inoue, A.; Chen, M. Characterization of nanoscale mechanical heterogeneity in a metallic glass by dynamic force microscopy. *Phys. Rev. Lett.* **2011**, *106*, 125504. [CrossRef] [PubMed]
15. Ichitsubo, T.; Matsubara, E.; Yamamoto, T.; Chen, H.; Nishiyama, N.; Saida, J.; Anazawa, K. Microstructure of fragile metallic glasses inferred from ultrasound-accelerated crystallization in Pd-based metallic glasses. *Phys. Rev. Lett.* **2005**, *95*, 245501. [CrossRef] [PubMed]
16. Wagner, H.; Bedorf, D.; Kuechemann, S.; Schwabe, M.; Zhang, B.; Arnold, W.; Samwer, K. Local elastic properties of a metallic glass. *Nat. Mater.* **2011**, *10*, 439–442. [CrossRef] [PubMed]
17. Wang, W.H. The nature and properties of amorphous matter. *Prog. Phys.* **2013**, *33*, 177–351. (In Chinese)
18. Park, E.S. Understanding of the shear bands in amorphous metals. *Appl. Microsc.* **2015**, *45*, 63–73. [CrossRef]
19. Lewandowski, J.; Shazly, M.; Nouri, A.S. Intrinsic and extrinsic toughening of metallic glasses. *Scr. Mater.* **2006**, *54*, 337–341. [CrossRef]
20. Liu, Y.H.; Wang, G.; Wang, R.J.; Zhao, D.Q.; Pan, M.X.; Wang, W.H. Super plastic bulk metallic glasses at room temperature. *Science* **2007**, *315*, 1385–1388. [CrossRef]
21. Conner, R.; Li, Y.; Nix, W.; Johnson, W. Shear band spacing under bending of Zr-based metallic glass plates. *Acta Mater.* **2004**, *52*, 2429–2434. [CrossRef]
22. Wang, L.; Bei, H.; Gao, Y.; Lu, Z.P.; Nieh, T. Effect of residual stresses on the hardness of bulk metallic glasses. *Acta Mater.* **2011**, *59*, 2858–2864. [CrossRef]
23. Gao, Y.; Wang, L.; Bei, H.; Nieh, T.-G. On the shear-band direction in metallic glasses. *Acta Mater.* **2011**, *59*, 4159–4167. [CrossRef]
24. Şopu, D.; Stukowski, A.; Stoica, M.; Scudino, S. Atomic-level processes of shear band nucleation in metallic glasses. *Phys. Rev. Lett.* **2017**, *119*, 195503. [CrossRef] [PubMed]
25. Klement, W.; Willens, R.; Duwez, P. Non-crystalline structure in solidified gold–silicon alloys. *Nature* **1960**, *187*, 869–870. [CrossRef]
26. Turnbull, D. Under what conditions can a glass be formed? *Contemp. Phys.* **1969**, *10*, 473–488. [CrossRef]
27. Inoue, A.; Zhang, T.; Masumoto, T. Zr–Al–Ni amorphous alloys with high glass transition temperature and significant supercooled liquid region. *Mater. Trans. JIM* **1990**, *31*, 177–183. [CrossRef]
28. Axinte, E.; Bofu, A.; Wang, Y.; Abdul-Rani, A.M.; Aliyu, A.A.A. An overview on the conventional and nonconventional methods for manufacturing the metallic glasses. *MATEC Web Conf.* **2017**, *112*, 03003. [CrossRef]
29. Chen, M. A brief overview of bulk metallic glasses. *NPG Asia Mater.* **2011**, *3*, 82–90. [CrossRef]
30. Ruhl, R.C. Cooling rates in splat cooling. *Mater. Sci. Eng.* **1967**, *1*, 313–320. [CrossRef]
31. Chen, H.S.; Turnbull, D. Formation, stability and structure of palladium-silicon based alloy glasses. *Acta Metall.* **1969**, *17*, 1021–1031. [CrossRef]
32. Spaepen, F. A microscopic mechanism for steady state inhomogeneous flow in metallic glasses. *Acta Metall.* **1977**, *25*, 407–415. [CrossRef]
33. Argon, A.S. Plastic deformation in metallic glasses. *Acta Metall.* **1979**, *27*, 47–58. [CrossRef]
34. Huang, R.; Suo, Z.; Prevost, J.H.; Nix, W.D. Inhomogeneous deformation in metallic glasses. *J. Mech. Phys. Solids* **2002**, *50*, 1011–1027. [CrossRef]
35. Tan, H.; Zhang, Y.; Ma, D.; Feng, Y.; Li, Y. Optimum glass formation at off-eutectic composition and its relation to skewed eutectic coupled zone in the La based La–Al–(Cu, Ni) pseudo ternary system. *Acta Mater.* **2003**, *51*, 4551–4561. [CrossRef]
36. Ekambaram, R.; Thamburaja, P.; Nikabdullah, N. On the evolution of free volume during the deformation of metallic glasses at high homologous temperatures. *Mech. Mater.* **2008**, *40*, 487–506. [CrossRef]
37. Inoue, A.; Takeuchi, A. Recent development and application products of bulk glassy alloys. *Acta Mater.* **2011**, *59*, 2243–2267. [CrossRef]
38. Xingchao, Z.; Yong, Z.; Hao, T.; Yong, L.; Xiaohua, C.; Guoliang, C. Micro-electro-discharge machining of bulk metallic glasses. In Proceedings of the 2007 International Symposium on High Density packaging and Microsystem Integration, Shanghai, China, 26–28 June 2007; pp. 1–4. [CrossRef]
39. Rizzi, P.; Habib, A.; Castellero, A.; Battezzati, L. Ductility and toughness of cold-rolled metallic glasses. *Intermetallics* **2013**, *33*, 38–43. [CrossRef]
40. Li, J.B.; Lin, H.C.; Jang, J.S.C.; Kuo, C.N.; Huang, J.C. Novel open-cell bulk metallic glass foams with promising characteristics. *Mater. Lett.* **2013**, *105*, 140–143. [CrossRef]
41. Li, X.; Kang, C.; Huang, H.; Sercombe, T. The role of a low-energy–density re-scan in fabricating crack-free $Al_{85}Ni_5Y_6Co_2Fe_2$ bulk metallic glass composites via selective laser melting. *Mater. Des.* **2014**, *63*, 407–411. [CrossRef]

42. Liu, Z.; Chen, W.; Carstensen, J.; Ketkaew, J.; Ojeda Mota, R.M.; Guest, J.K.; Schroers, J. 3D metallic glass cellular structures. *Acta Mater.* **2016**, *105*, 35–43. [CrossRef]
43. Perim, E.; Lee, D.; Liu, Y.; Toher, C.; Gong, P.; Li, Y.; Simmons, W.N.; Levy, O.; Vlassak, J.J.; Schroers, J. Spectral descriptors for bulk metallic glasses based on the thermodynamics of competing crystalline phases. *Nat. Commun.* **2016**, *7*, 12315. [CrossRef] [PubMed]
44. Yang, C.; Zhang, C.; Xing, W.; Liu, L. 3D printing of Zr-based bulk metallic glasses with complex geometries and enhanced catalytic properties. *Intermetallics* **2018**, *94*, 22–28. [CrossRef]
45. Mohr, M.; Wunderlich, R.K.; Zweiacker, K.; Prades-Rödel, S.; Sauget, R.; Blatter, A.; Logé, R.; Dommann, A.; Neels, A.; Johnson, W.L. Surface tension and viscosity of liquid Pd43Cu27Ni10P20 measured in a levitation device under microgravity. *Npj Microgravity* **2019**, *5*, 4. [CrossRef] [PubMed]
46. Lin, X.; Johnson, W. Formation of Ti–Zr–Cu–Ni bulk metallic glasses. *J. Appl. Phys.* **1995**, *78*, 6514–6519. [CrossRef]
47. Li, H.; Lu, Z.; Wang, S.; Wu, Y.; Lu, Z. Fe-based bulk metallic glasses: Glass formation, fabrication, properties and applications. *Prog. Mater. Sci.* **2019**, *103*, 235–318. [CrossRef]
48. Zhu, P.-Z.; Qiu, C.; Fang, F.-Z.; Yuan, D.-D.; Shen, X.-C. Molecular dynamics simulations of nanometric cutting mechanisms of amorphous alloy. *Appl. Surf. Sci.* **2014**, *317*, 432–442. [CrossRef]
49. Cao, T.F. *Studies of Relationship between Melting Entropy and Melting Point Viscosity of Zr2ni-Ti2ni Alloys*; Yanshan University: Qinhuangdao, China, 2019. (In Chinese) [CrossRef]
50. Johnson, W.L.; Kaltenboeck, G.; Demetriou, M.D.; Schramm, J.P.; Liu, X.; Samwer, K.; Kim, C.P.; Hofmann, D.C. Beating crystallization in glass-forming metals by millisecond heating and processing. *Science* **2011**, *332*, 828–833. [CrossRef] [PubMed]
51. Kaltenboeck, G.; Harris, T.; Sun, K.; Tran, T.; Chang, G.; Schramm, J.P.; Demetriou, M.D.; Johnson, W.L. Accessing thermoplastic processing windows in metallic glasses using rapid capacitive discharge. *Sci. Rep.* **2014**, *4*, 6441. [CrossRef]
52. Inoue, A.; Nakamura, T.; Nishiyama, N.; Masumoto, T. Mg–Cu–Y bulk amorphous alloys with high tensile strength produced by a high-pressure die casting method. *Mater. Trans. JIM* **1992**, *33*, 937–945. [CrossRef]
53. Wang, J.; Li, R.; Hua, N.; Zhang, T. Co-based ternary bulk metallic glasses with ultrahigh strength and plasticity. *J. Mater. Res.* **2011**, *26*, 2072–2079. [CrossRef]
54. Wang, W.H. The elastic properties, elastic models and elastic perspectives of metallic glasses. *Prog. Mater. Sci.* **2012**, *57*, 487–656. [CrossRef]
55. Makino, A.; Inoue, A.; Masumoto, T. Nanocrystalline soft magnetic Fe–M–B (M = Zr, Hf, Nb) alloys produced by crystallization of amorphous phase (overview). *Mater. Trans. JIM* **1995**, *36*, 924–938. [CrossRef]
56. Demetriou, M.D.; Launey, M.E.; Garrett, G.; Schramm, J.P.; Hofmann, D.C.; Johnson, W.L.; Ritchie, R.O. A damage-tolerant glass. *Nat. Mater.* **2011**, *10*, 123–128. [CrossRef] [PubMed]
57. Lin, B.; Bian, X.; Wang, P.; Luo, G. Application of Fe-based metallic glasses in wastewater treatment. *Mater. Sci. Eng. B* **2012**, *177*, 92–95. [CrossRef]
58. Pratap, A.; Kasyap, S.; Prajapati, S.; Upadhyay, D. Bio-corrosion studies of Fe-based metallic glasses. *Mater. Today Proc.* **2021**, *42*, 1669–1672. [CrossRef]
59. Yi, H.Q. *Research of the Fe76al4p12b4si4 Fe-Based Amorphous Alloy*; Shanghai Jiao Tong University: Shanghai, China, 2009. (In Chinese)
60. Gong, P.; Deng, L.; Jin, J.; Wang, S.; Wang, X.; Yao, K. Review on the research and development of Ti-based bulk metallic glasses. *Metals* **2016**, *6*, 264. [CrossRef]
61. Li, F.; Zhang, T.; Guan, S.; Shen, N. A novel dual-amorphous-phased bulk metallic glass with soft magnetic properties. *Mater. Lett.* **2005**, *59*, 1453–1457. [CrossRef]
62. Schroers, J. Bulk metallic glasses. *Phys. Today* **2013**, *66*, 32. [CrossRef]
63. Parisi, G.; Sciortino, F. Flying to the bottom. *Nat. Mater.* **2013**, *12*, 94–95. [CrossRef]
64. Suryanarayana, C. Mechanical behavior of emerging materials. *Mater. Today* **2012**, *15*, 486–498. [CrossRef]
65. Wang, W. Bulk metallic glasses with functional physical properties. *Adv. Mater.* **2009**, *21*, 4524–4544. [CrossRef]
66. Highmore, R.; Greer, A. Eutectics and the formation of amorphous alloys. *Nature* **1989**, *339*, 363–365. [CrossRef]
67. Shen, J.; Chen, Q.; Sun, J.; Fan, H.; Wang, G. Exceptionally high glass-forming ability of an FeCoCrMoCBY alloy. *Appl. Phys. Lett.* **2005**, *86*, 151907. [CrossRef]
68. Inoue, A.; Shinohara, Y.; Gook, J.S. Thermal and magnetic properties of bulk Fe-based glassy alloys prepared by copper mold casting. *Mater. Trans. JIM* **1995**, *36*, 1427–1433. [CrossRef]
69. Inoue, A. High strength bulk amorphous alloys with low critical cooling rates (overview). *Mater. Trans. JIM* **1995**, *36*, 866–875. [CrossRef]
70. Inoue, A.; Nishiyama, N.; Kimura, H. Preparation and thermal stability of bulk amorphous Pd40Cu30Ni10P20 alloy cylinder of 72 mm in diameter. *Mater. Trans. JIM* **1997**, *38*, 179–183. [CrossRef]
71. Chen, M. Mechanical behavior of metallic glasses: Microscopic understanding of strength and ductility. *Annu. Rev. Mater. Res.* **2008**, *38*, 445–469. [CrossRef]
72. Maroju, N.K.; Jin, X. Mechanism of chip segmentation in orthogonal cutting of Zr-based bulk metallic glass. *J. Manuf. Sci. Eng.* **2019**, *141*, 081003. [CrossRef]

73. Komanduri, R.; Schroeder, T.; Hazra, J.; Von Turkovich, B.; Flom, D. On the catastrophic shear instability in high-speed machining of an AISI 4340 steel. *J. Eng. Ind. May* **1982**, *102*, 121–131. [CrossRef]
74. Molinari, A.; Musquar, C.; Sutter, G. Adiabatic shear banding in high speed machining of Ti–6Al–4V: Experiments and modeling. *Int. J. Plast.* **2002**, *18*, 443–459. [CrossRef]
75. Zhang, L.; Huang, H. Micro machining of bulk metallic glasses: A review. *Int. J. Adv. Manuf. Tech.* **2019**, *100*, 637–661. [CrossRef]
76. Zhu, P.; Fang, F. On the mechanism of material removal in nanometric cutting of metallic glass. *Appl. Phys. A* **2014**, *116*, 605–610. [CrossRef]
77. Chong, F.; To, S.; Chan, K.C. Cutting characteristics of lanthanum base metallic glass in single point diamond turning. In *Key Engineering Materials*; Trans Tech Publ.: Zurich, Switzerland, 2012. [CrossRef]
78. Xiong, J.; Wang, H.; Zhang, G.; Chen, Y.; Ma, J.; Mo, R. Machinability and surface generation of Pd40Ni10Cu30P20 bulk metallic glass in single-point diamond turning. *Micromachines* **2020**, *11*, 4. [CrossRef] [PubMed]
79. Fang, Z.; Nagato, K.; Liu, S.; Sugita, N.; Nakao, M. Investigation into surface integrity and magnetic property of FeSiB metallic glass in two-dimensional cutting. *J. Manuf. Process* **2021**, *64*, 1098–1104. [CrossRef]
80. Bakkal, M.; Liu, C.T.; Watkins, T.R.; Scattergood, R.O.; Shih, A.J. Oxidation and crystallization of Zr-based bulk metallic glass due to machining. *Intermetallics* **2004**, *12*, 195–204. [CrossRef]
81. Fu, E.; Carter, J.; Martin, M.; Xie, G.; Zhang, X.; Wang, Y.; Littleton, R.; McDeavitt, S.; Shao, L. Ar-ion-milling-induced structural changes of Cu50Zr45Ti5 metallic glass. *Nucl. Instrum. Meth. B* **2010**, *268*, 545–549. [CrossRef]
82. Zhang, W.; Ma, M.; Song, A.; Liang, S.; Hao, Q.; Tan, C.; Jing, Q.; Liu, R. Temperature rise and flow of Zr-based bulk metallic glasses under high shearing stress. *Sci. China Phys. Mech.* **2011**, *54*, 1972–1976. [CrossRef]
83. Basak, A.; Zhang, L. Deformation of Ti-Based bulk metallic glass under a cutting tip. *Tribol. Lett.* **2018**, *66*, 1–8. [CrossRef]
84. Wang, J.Q.; Liu, Y.H.; Chen, M.W.; Xie, G.Q.; Louzguine-Luzgin, D.V.; Inoue, A.; Perepezko, J.H. Rapid degradation of azo dye by Fe-based metallic glass powder. *Adv. Funct. Mater.* **2012**, *22*, 2567–2570. [CrossRef]
85. Sun, H.; Zheng, H.; Yang, X. Efficient degradation of orange II dye using Fe-based metallic glass powders prepared by commercial raw materials. *Intermetallics* **2021**, *129*, 107030. [CrossRef]
86. Zhou, J.; Di, S.-Y.; Sun, B.-A.; Zhao, R.; Zeng, Q.-S.; Wang, J.-G.; Sun, Z.-Z.; Wang, W.-H.; Shen, B.-L. Pronounced β-relaxation in plastic FeNi-based bulk metallic glasses and its structural origin. *Intermetallics* **2021**, *136*, 107234. [CrossRef]
87. Cheng, Y.; Hao, Q.; Pelletier, J.; Pineda, E.; Qiao, J. Modelling and physical analysis of the high-temperature rheological behavior of a metallic glass. *Int. J. Plast.* **2021**, *146*, 103107. [CrossRef]
88. Yang, W.; Liu, H.; Xue, L.; Li, J.; Dun, C.; Zhang, J.; Zhao, Y.; Shen, B. Magnetic properties of (Fe1− xNix) 72B20Si4Nb4 (x= 0.0–0.5) bulk metallic glasses. *J. Magn. Magn. Mater.* **2013**, *335*, 172–176. [CrossRef]
89. Gao, J.; Chen, Z.; Du, Q.; Li, H.; Wu, Y.; Wang, H.; Liu, X.; Lu, Z. Fe-based bulk metallic glass composites without any metalloid elements. *Acta Mater.* **2013**, *61*, 3214–3223. [CrossRef]
90. Lu, Z.; Liu, C.; Thompson, J.; Porter, W. Structural amorphous steels. *Phys. Rev. Lett.* **2004**, *92*, 245503. [CrossRef] [PubMed]
91. Yoshizawa, Y.a.; Oguma, S.; Yamauchi, K. New Fe-based soft magnetic alloys composed of ultrafine grain structure. *J. Appl. Phys.* **1988**, *64*, 6044–6046. [CrossRef]
92. Inoue, A.; Nishiyama, N.; Amiya, K.; Zhang, T.; Masumoto, T. Ti-based amorphous alloys with a wide supercooled liquid region. *Mater. Lett.* **1994**, *19*, 131–135. [CrossRef]
93. Suryanarayana, C.; Inoue, A. Iron-based bulk metallic glasses. *Int. Mater. Rev.* **2013**, *58*, 131–166. [CrossRef]
94. Zhao, Y.Y. *Molecular Dynamic Simulation of Fe-Based Amorphous Alloys*; Shijiazhuang Tiedao University: Shijiazhuang, China, 2014. (In Chinese)
95. Wang, W.M.; Gebert, A.; Roth, S.; Kuehn, U.; Schultz, L. Effect of Si on the glass-forming ability, thermal stability and magnetic properties of Fe–Co–Zr–Mo–W–B alloys. *J. Alloys Compd.* **2008**, *459*, 203–208. [CrossRef]
96. Huang, Y.; Guo, Y.; Fan, H.; Shen, J. Synthesis of Fe–Cr–Mo–C–B amorphous coating with high corrosion resistance. *Mater. Lett.* **2012**, *89*, 229–232. [CrossRef]
97. Lenain, A.; Blandin, J.; Kapelski, G.; Volpi, F.; Gravier, S. Hf-rich bulk metallic glasses as potential insulating structural material. *Mater. Des.* **2018**, *139*, 467–472. [CrossRef]
98. Zhang, C.; Zhang, H.; Lv, M.; Hu, Z. Decolorization of azo dye solution by Fe–Mo–Si–B amorphous alloy. *J. Non-Cryst Solids* **2010**, *356*, 1703–1706. [CrossRef]
99. Ou, C.-H.; Lin, Y.-C.; Keikoin, Y.; Ono, T.; Esashi, M.; Tsai, Y.-C. Two-dimensional MEMS Fe-based metallic glass micromirror driven by an electromagnetic actuator. *Jpn. J. Appl. Phys.* **2019**, *58*, SDDL01. [CrossRef]
100. Wang, M. Experimental Study on Machineability of Fe-based Amorphous Alloy. *Mach. Tool Hydraul.* **2019**, *47*, 112–115. [CrossRef]
101. Fan, G.; Quan, M.; Hu, Z. Induced magnetic anisotropy in Fe80B20 metallic glass by mechanical milling. *Appl. Phys. Lett.* **1996**, *68*, 1159–1161. [CrossRef]
102. Fang, Z.; Nakao, M. Local magnetic deterioration on work-hardening layer of FeSiB metallic glass by milling. *CIRP Ann.* **2020**, *69*, 501–504. [CrossRef]
103. Tsui, H.-P.; Lee, P.-H.; Yeh, C.-C.; Hung, J.-C. Ultrasonic vibration-assisted electrical discharge machining on Fe-based metallic glass by adding conductive powder. *Procedia CIRP* **2020**, *95*, 425–430. [CrossRef]

104. Quintana, I.; Dobrev, T.; Aranzabe, A.; Lalev, G. Laser micromachining of metallic glasses: Investigation of the material response to machining with micro-second and pico-second lasers. In *Laser Applications in Microelectronic and Optoelectronic Manufacturing XV*; International Society for Optics and Photonics: Washington, DC, USA, 2010. [CrossRef]
105. He, Y.; Shiflet, G.; Poon, S. Ball milling-induced nanocrystal formation in aluminum-based metallic glasses. *Acta Metal. Mater.* **1995**, *43*, 83–91. [CrossRef]
106. Ramasamy, P.; Shahid, R.N.; Scudino, S.; Eckert, J.; Stoica, M. Influencing the crystallization of Fe80Nb10B10 metallic glass by ball milling. *J. Alloys Compd.* **2017**, *725*, 227–236. [CrossRef]
107. Guo, F.; Lu, K. Ball-milling-induced crystallization and ball-milling effect on thermal crystallization kinetics in an amorphous FeMoSiB alloy. *Metall. Mater. Trans. A* **1997**, *28*, 1123–1131. [CrossRef]
108. Lv, Y.; Chen, Q. Internal friction behavior and thermal conductivity of Fe-based bulk metallic glasses with different crystallization. *Thermochim. Acta* **2018**, *666*, 36–40. [CrossRef]
109. Han, D.; Wang, G.; Li, J.; Chan, K.C.; To, S.; Wu, F.; Gao, Y.; Zhai, Q. Cutting characteristics of Zr-based bulk metallic glass. *J. Mater. Sci. Technol.* **2015**, *31*, 153–158. [CrossRef]
110. Chen, X.; Xiao, J.; Zhu, Y.; Tian, R.; Shu, X.; Xu, J. Micro-machinability of bulk metallic glass in ultra-precision cutting. *Mater. Des.* **2017**, *136*, 1–12. [CrossRef]
111. Wang, D.D. *Experomental and Simulation Study on Ultrasonic Torsional Vibration Assisted Milling of Amorphous Alloys*; Yanshan University: Qinhuangdao, China, 2019. (In Chinese) [CrossRef]
112. Zhou, S.X.; Lu, Z.C.; Chen, J.C. Amorphous state physics and the manufacture of soft magnetic materials. *Physical* **2002**, *31*, 430–436. (In Chinese)
113. Wang, M.; Xu, B.; Dong, S.; Zhang, J.; Wei, S. Experimental investigations of cutting parameters influence on cutting forces in turning of Fe-based amorphous overlay for remanufacture. *Int. J. Adv. Manuf. Technol.* **2013**, *65*, 735–743. [CrossRef]
114. Ding, F.; Wang, C.; Zhang, T.; Zheng, L.; Zhu, X. High performance cutting of Zr-based bulk metallic glass: A review of chip formation. *Procedia Cirp.* **2018**, *77*, 421–424. [CrossRef]
115. Komanduri, R.; Lucca, D.; Tani, Y. Technological advances in fine abrasive processes. *CIRP Ann.* **1997**, *46*, 545–596. [CrossRef]
116. Venkatesh, V.; Inasaki, I.; Toenshof, H.; Nakagawa, T.; Marinescu, I. Observations on polishing and ultraprecision machining of semiconductor substrate materials. *CIRP Ann.* **1995**, *44*, 611–618. [CrossRef]
117. Gee, A.E.; Spragg, R.; Puttick, K.E.; Rudman, M. Single-point diamond form-finishing of glasses and other macroscopically brittle materials. *Commer. Appl. Precis. Manuf. Sub-Micron Level* **1992**, *1573*, 39–48. [CrossRef]
118. Fang, F.; Chen, L. Ultra-precision cutting for ZKN7 glass. *CIRP Ann.* **2000**, *49*, 17–20. [CrossRef]
119. Fang, F.; Venkatesh, V. Diamond cutting of silicon with nanometric finish. *CIRP Ann.* **1998**, *47*, 45–49. [CrossRef]
120. Shimada, S.; Ikawa, N.; Inamura, T.; Takezawa, N.; Ohmori, H.; Sata, T. Brittle-ductile transition phenomena in microindentation and micromachining. *CIRP Ann.* **1995**, *44*, 523–526. [CrossRef]
121. Lucca, D.; Brinksmeier, E.; Goch, G. Progress in assessing surface and subsurface integrity. *CIRP Ann.* **1998**, *47*, 669–693. [CrossRef]
122. Ayomoh, M.; Abou-El-Hossein, K. Surface finish in ultra-precision diamond turning of single-crystal silicon. In *Optifab 2015*; International Society for Optics and Photonics: Washington, DC, USA, 2015. [CrossRef]
123. Guo, X.; Zhai, C.; Jin, Z.; Guo, D. The study of diamond graphitization under the action of iron-based catalyst. *Chin. J. Mech. Eng.* **2015**, *51*, 162–168. [CrossRef]
124. Zou, L.; Yin, J.; Huang, Y.; Zhou, M. Essential causes for tool wear of single crystal diamond in ultra-precision cutting of ferrous metals. *Diamond Relat. Mater.* **2018**, *86*, 29–40. [CrossRef]
125. Biček, M.; Dumont, F.; Courbon, C.; Pušavec, F.; Rech, J.; Kopač, J. Cryogenic machining as an alternative turning process of normalized and hardened AISI 52100 bearing steel. *J. Mater. Process. Technol.* **2012**, *212*, 2609–2618. [CrossRef]
126. Hatefi, S.; Abou-El-Hossein, K. Review of single-point diamond turning process in terms of ultra-precision optical surface roughness. *Int. J. Adv. Manuf. Tech.* **2020**, *106*, 2167–2187. [CrossRef]
127. Huang, S.; Liu, X.; Chen, F.; Zheng, H.; Yang, X.; Wu, L.; Song, J.; Xu, W. Diamond-cutting ferrous metals assisted by cold plasma and ultrasonic elliptical vibration. *Int. J. Adv. Manuf. Tech.* **2016**, *85*, 673–681. [CrossRef]
128. Zhang, Y.; Zhou, Z.; Wang, J.; Li, X. Diamond tool wear in precision turning of titanium alloy. *Mater. Manuf. Process.* **2013**, *28*, 1061–1064. [CrossRef]
129. Wang, Y.; Song, L.-X.; Liu, J.-G.; Wang, R.; Zhao, B.-C. Investigation on the sawing temperature in ultrasonic vibration assisted diamond wire sawing monocrystalline silicon. *Mater. Sci. Semicond Process.* **2021**, *135*, 106070. [CrossRef]
130. Xing, Y.; Liu, Y.; Li, C.; Yang, C.; Xue, C. Ductile-brittle coupled cutting of a single-crystal silicon by ultrasonic assisted diamond turning. *Opt. Express* **2021**, *29*, 23847–23863. [CrossRef]
131. Shen, J.; Wang, J.; Jiang, B.; Xu, X. Study on wear of diamond wheel in ultrasonic vibration-assisted grinding ceramic. *Wear* **2015**, *332*, 788–793. [CrossRef]
132. Zhong, Z.; Lin, G. Diamond turning of a metal matrix composite with ultrasonic vibrations. *Mater. Manuf. Process.* **2005**, *20*, 727–735. [CrossRef]
133. Zhong, Z.; Lin, G. Ultrasonic assisted turning of an aluminium-based metal matrix composite reinforced with SiC particles. *Int. J. Adv. Manuf. Technol.* **2006**, *27*, 1077–1081. [CrossRef]

134. Zhong, Z.; Hung, N.P. Diamond turning and grinding of aluminum-based metal matrix composites. *Mater. Manuf. Process.* **2000**, *15*, 853–865. [CrossRef]
135. Kumar, J. Ultrasonic machining—A comprehensive review. *Mach. Sci. Technol.* **2013**, *17*, 325–379. [CrossRef]
136. Singh, R.; Khamba, J. Ultrasonic machining of titanium and its alloys: A review. *J. Mater. Process Technol.* **2006**, *173*, 125–135. [CrossRef]
137. Koshimizu, S. Ultrasonic vibration-assisted cutting of titanium alloy. In *Key Engineering Materials*; Trans Tech Publ.: Zurich, Switzerland, 2009. [CrossRef]
138. Zhu, Z.; To, S.; Xiao, G.; Ehmann, K.F.; Zhang, G. Rotary spatial vibration-assisted diamond cutting of brittle materials. *Precis. Eng.* **2016**, *44*, 211–219. [CrossRef]
139. Tian, F.; Yin, Z.; Li, S. A novel long range fast tool servo for diamond turning. *Int. J. Adv. Manuf. Technol.* **2016**, *86*, 1227–1234. [CrossRef]
140. Feng, H.; Xia, R.; Li, Y.; Chen, J.; Yuan, Y.; Zhu, D.; Chen, S.; Chen, H. Fabrication of freeform progressive addition lenses using a self-developed long stroke fast tool servo. *Int. J. Adv. Manuf. Technol.* **2017**, *91*, 3799–3806. [CrossRef]
141. Fang, F.; Zhang, X.; Weckenmann, A.; Zhang, G.; Evans, C. Manufacturing and measurement of freeform optics. *CIRP Ann.* **2013**, *62*, 823–846. [CrossRef]
142. Zhu, L.; Li, Z.; Fang, F.; Huang, S.; Zhang, X. Review on fast tool servo machining of optical freeform surfaces. *Int. J. Adv. Manuf. Technol.* **2018**, *95*, 2071–2092. [CrossRef]
143. Liu, Q.; Zhou, X.; Xu, P. A new tool path for optical freeform surface fast tool servo diamond turning. *Proc. Inst. Mech. Eng. Pt. B J. Eng. Manuf.* **2014**, *228*, 1721–1726. [CrossRef]
144. Yuan, J.; Lyu, B.; Hang, W.; Deng, Q. Review on the progress of ultra-precision machining technologies. *Front. Mech. Eng.* **2017**, *12*, 158–180. [CrossRef]
145. Chen, Y.-L.; Cai, Y.; Tohyama, K.; Shimizu, Y.; Ito, S.; Gao, W. Auto-tracking single point diamond cutting on non-planar brittle material substrates by a high-rigidity force controlled fast tool servo. *Precis. Eng.* **2017**, *49*, 253–261. [CrossRef]
146. Yu, D.P.; Wong, Y.S.; Hong, G.S. Ductile-regime machining for fast tool servo diamond turning of micro-structured surfaces on brittle materials. In *Advanced Materials Research*; Trans Tech Publ.: Zurich, Switzerland, 2012. [CrossRef]
147. Yin, Z.; Dai, Y.; Li, S.; Guan, C.; Tie, G. Fabrication of off-axis aspheric surfaces using a slow tool servo. *Int. J. Mach. Tool Manuf.* **2011**, *51*, 404–410. [CrossRef]
148. Nagayama, K.; Yan, J. Deterministic error compensation for slow tool servo-driven diamond turning of freeform surface with nanometric form accuracy. *J. Manuf. Process.* **2021**, *64*, 45–57. [CrossRef]
149. Singh, R.; Dureja, J.; Dogra, M.; Gupta, M.K.; Jamil, M.; Mia, M. Evaluating the sustainability pillars of energy and environment considering carbon emissions under machining ofTi-3Al-2.5 V. *Sustain. Energy Technol.* **2020**, *42*, 100806. [CrossRef]
150. Maaß, R.; Klaumünzer, D.; Preiß, E.; Derlet, P.; Löffler, J.F. Single shear-band plasticity in a bulk metallic glass at cryogenic temperatures. *Scr. Mater.* **2012**, *66*, 231–234. [CrossRef]
151. Daniil, M.; Osofsky, M.S.; Gubser, D.U.; Willard, M.A. (Fe, Si, Al)-based nanocrystalline soft magnetic alloys for cryogenic applications. *Appl. Phys. Lett.* **2010**, *96*, 162504. [CrossRef]
152. Jin, X.J.; Zhu, X.B.; Xue, W.C.; Wang, G. Effect of Cryogenic Treatment on Microstructure and Properties of Fe -Based Amorphous Alloy Coating. *Mater. Prot.* **2019**, *52*, 90–93. [CrossRef]
153. Fan, C.; Liu, C.; Yan, H. Mechanical properties of bulk metallic glasses at cryogenic temperatures. *Mod. Phys. Lett. B* **2009**, *23*, 2703–2722. [CrossRef]
154. Salahinejad, E.; Amini, R.; Marasi, M.; Sritharan, T.; Hadianfard, M. The effect of nitrogen on the glass-forming ability and micro-hardness of Fe–Cr–Mn–N amorphous alloys prepared by mechanical alloying. *Mater. Chem. Phys.* **2009**, *118*, 71–75. [CrossRef]
155. Evans, C.; Bryan, J. Cryogenic diamond turning of stainless steel. *CIRP Ann.* **1991**, *40*, 571–575. [CrossRef]
156. An, Q.L. *Cryogenic Mist Jet Impinging Cooling and Its Application in Machining of Titanium Alloy*; Nanjing University of Aeronautics Astronautics: Nanjing, China, 2006. (In Chinese)
157. Aramcharoen, A. Influence of cryogenic cooling on tool wear and chip formation in turning of titanium alloy. *Procedia CIRP.* **2016**, *46*, 83–86. [CrossRef]
158. Bordin, A.; Sartori, S.; Bruschi, S.; Ghiotti, A. Experimental investigation on the feasibility of dry and cryogenic machining as sustainable strategies when turning Ti6Al4V produced by Additive Manufacturing. *J. Clean. Prod.* **2017**, *142*, 4142–4151. [CrossRef]
159. Khanna, N.; Agrawal, C.; Gupta, M.K.; Song, Q. Tool wear and hole quality evaluation in cryogenic Drilling of Inconel 718 superalloy. *Tribol. Int.* **2020**, *143*, 106084. [CrossRef]
160. Pereira, O.; Celaya, A.; Urbikaín, G.; Rodríguez, A.; Fernández-Valdivielso, A.; de Lacalle, L.N.L. CO_2 cryogenic milling of Inconel 718: Cutting forces and tool wear. *J. Mater. Res. Technol.* **2020**, *9*, 8459–8468. [CrossRef]
161. Gupta, M.K.; Song, Q.; Liu, Z.; Sarikaya, M.; Mia, M.; Jamil, M.; Singla, A.K.; Bansal, A.; Pimenov, D.Y.; Kuntoğlu, M. Tribological performance based machinability investigations in cryogenic cooling assisted turning of α-β titanium alloy. *Tribol. Int.* **2021**, *160*, 107032. [CrossRef]
162. Hong, S.Y.; Ding, Y. Cooling approaches and cutting temperatures in cryogenic machining of Ti-6Al-4V. *Int. J. Mach. Tool Manuf.* **2001**, *41*, 1417–1437. [CrossRef]

163. Huang, P.; Li, H.; Zhu, W.-L.; Wang, H.; Zhang, G.; Wu, X.; To, S.; Zhu, Z. Effects of eco-friendly cooling strategy on machining performance in micro-scale diamond turning of Ti–6Al–4V. *J. Clean. Prod.* **2020**, *243*, 118526. [CrossRef]
164. Gan, Y.; Wang, Y.; Liu, K.; Wang, S.; Yu, Q.; Che, C.; Liu, H. The development and experimental research of a cryogenic internal cooling turning tool. *J. Clean. Prod.* **2021**, *319*, 128787. [CrossRef]
165. Sharma, A.K.; Tiwari, A.K.; Dixit, A.R. Effects of Minimum Quantity Lubrication (MQL) in machining processes using conventional and nanofluid based cutting fluids: A comprehensive review. *J. Clean. Prod.* **2016**, *127*, 1–18. [CrossRef]
166. Sharma, V.S.; Singh, G.; Sørby, K. A Review on Minimum Quantity Lubrication for Machining Processes. *Mater. Manuf. Process.* **2015**, *30*, 935–953. [CrossRef]
167. Kamata, Y.; Obikawa, T. High speed MQL finish-turning of Inconel 718 with different coated tools. *J. Mater. Process. Technol.* **2007**, *192*, 281–286. [CrossRef]
168. Hadad, M.; Sadeghi, B. Minimum quantity lubrication-MQL turning of AISI 4140 steel alloy. *J. Clean. Prod.* **2013**, *54*, 332–343. [CrossRef]
169. Kishawy, H.; Dumitrescu, M.; Ng, E.-G.; Elbestawi, M. Effect of coolant strategy on tool performance, chip morphology and surface quality during high-speed machining of A356 aluminum alloy. *Int. J. Mach. Tool Manu.* **2005**, *45*, 219–227. [CrossRef]
170. Zhou, J.W. *Effect of Cryogenic and Magnetic Pulse Treatment on Fe-Based Amorphous Microstructure and Magnetic Properties*; Jiangsu University: Zhenjiang, China, 2019. [CrossRef]
171. Wang, Y.; Guo, D.M.; Guo, L.J.; Liu, K.; Ren, F.; Liu, H.B.; Jiang, S.W.; Wang, S.Q. Research Status and Development Trend of Cryogenic Machining Technology. *Aerosp. Shanghai (Chin. Engl.)* **2020**, *37*, 11–21. (In Chinese) [CrossRef]
172. Dhar, N.; Kishore, N.S.; Paul, S.; Chattopadhyay, A. The effects of cryogenic cooling on chips and cutting forces in turning AISI 1040 and AISI 4320 steels. *Proc. Inst. Mech. Eng. Pt. B J Eng. Manuf.* **2002**, *216*, 713–724. [CrossRef]
173. Khan, A.; Ali, M.Y.; Haque, M. A New Approach of Applying Cryogenic Coolant in Turning AISI 304 Stainless Steel. *Int. J. Mech. Mater. Eng.* **2010**, *5*, 171–174.
174. Sen, B.; Mia, M.; Krolczyk, G.; Mandal, U.K.; Mondal, S.P. Eco-friendly cutting fluids in minimum quantity lubrication assisted machining: A review on the perception of sustainable manufacturing. *Int. J. Precis. Eng. Manuf.-Green Technol.* **2021**, *8*, 249–280. [CrossRef]
175. Gavili, A.; Zabihi, F.; Isfahani, T.D.; Sabbaghzadeh, J. The thermal conductivity of water base ferrofluids under magnetic field. *Exp. Therm. Fluid Sci.* **2012**, *41*, 94–98. [CrossRef]
176. Altan, C.L.; Elkatmis, A.; Yüksel, M.; Aslan, N.; Bucak, S. Enhancement of thermal conductivity upon application of magnetic field to Fe3O4 nanofluids. *J. Appl. Phys.* **2011**, *110*, 093917. [CrossRef]
177. Philip, J.; Shima, P.; Raj, B. Enhancement of thermal conductivity in magnetite based nanofluid due to chainlike structures. *Appl. Phys. Lett.* **2007**, *91*, 203108. [CrossRef]
178. Younes, H.; Christensen, G.; Luan, X.; Hong, H.; Smith, P. Effects of alignment, p H, surfactant, and solvent on heat transfer nanofluids containing Fe2O3 and CuO nanoparticles. *J. Appl. Phys.* **2012**, *111*, 064308. [CrossRef]
179. Sundar, L.S.; Singh, M.K.; Sousa, A.C. Investigation of thermal conductivity and viscosity of Fe3O4 nanofluid for heat transfer applications. *Int. Commun. Heat Mass* **2013**, *44*, 7–14. [CrossRef]
180. Gonnet, P.; Liang, Z.; Choi, E.S.; Kadambala, R.S.; Zhang, C.; Brooks, J.S.; Wang, B.; Kramer, L. Thermal conductivity of magnetically aligned carbon nanotube buckypapers and nanocomposites. *Curr. Appl. Phys.* **2006**, *6*, 119–122. [CrossRef]
181. Han, Z.; Fina, A. Thermal conductivity of carbon nanotubes and their polymer nanocomposites: A review. *Prog. Polym. Sci.* **2011**, *36*, 914–944. [CrossRef]
182. Horton, M.; Hong, H.; Li, C.; Shi, B.; Peterson, G.; Jin, S. Magnetic alignment of Ni-coated single wall carbon nanotubes in heat transfer nanofluids. *J. Appl. Phys.* **2010**, *107*, 104320. [CrossRef]
183. Yip, W.; To, S. Reduction of material swelling and recovery of titanium alloys in diamond cutting by magnetic field assistance. *J. Alloys Compd.* **2017**, *722*, 525–531. [CrossRef]
184. Philip, J.; Shima, P.; Raj, B. Evidence for enhanced thermal conduction through percolating structures in nanofluids. *Nanotechnology* **2008**, *19*, 305706. [CrossRef]
185. Nkurikiyimfura, I.; Wang, Y.; Pan, Z. Effect of chain-like magnetite nanoparticle aggregates on thermal conductivity of magnetic nanofluid in magnetic field. *Exp. Therm. Fluid. Sci.* **2013**, *44*, 607–612. [CrossRef]
186. Yip, W.; To, S. Tool life enhancement in dry diamond turning of titanium alloys using an eddy current damping and a magnetic field for sustainable manufacturing. *J. Clean. Prod.* **2017**, *168*, 929–939. [CrossRef]
187. Sodano, H.A.; Bae, J.-S. Eddy current damping in structures. *Shock. Vib. Dig.* **2004**, *36*, 469. [CrossRef]
188. Sodano, H.A.; Bae, J.-S.; Inman, D.J.; Belvin, W.K. Improved concept and model of eddy current damper. *J. Vib. Acoust. Jun* **2006**, *128*, 294–302. [CrossRef]
189. Bae, J.S.; Kwak, M.K.; Inman, D.J. Vibration suppression of a cantilever beam using eddy current damper. *J. Sound Vib.* **2005**, *284*, 805–824. [CrossRef]
190. Khalil, A.K.; Yip, W.; To, S. Theoretical and experimental investigations of magnetic field assisted ultra-precision machining of titanium alloys. *J. Mater. Process Technol.* **2022**, *300*, 117429. [CrossRef]
191. Zhang, Y.H.; Chao, Y.S. Nano-Crystallization of Amorphous Alloy Fe78Si9B13 by Treatment of Low-Frequency Pulsating Magnetic Field. *J. Northeast. Univ. (Nat. Sci.)* **2003**, *24*, 1018–1020. (In Chinese)

192. Jin, Y.; Chao, Y.; Liu, F.; Wang, J.; Sun, M. Nanocrystallization and magnetostriction coefficient of Fe52Co34Hf7B6Cu1 amorphous alloy treated by medium-frequency magnetic pulse. *J. Magn. Magn. Mater.* **2018**, *468*, 181–184. [CrossRef]
193. Guo, H.; Chao, Y.S.; Zhang, l. Nano-Crystallization of Amorphous Alloy by Magnetic Pulsing and Optimization of Soft Magnetic Properties. *Rare Met. Mater. Eng.* **2013**, *42*, 1236–1240. (In Chinese)
194. Suzuki, K.; Makino, A.; Kataoka, N.; Inoue, A.; Masumoto, T. High saturation magnetization and soft magnetic properties of bcc Fe–Zr–B and Fe–Zr–B–M (M= transition metal) alloys with nanoscale grain size. *Mater. Trans. JIM* **1991**, *32*, 93–102. [CrossRef]
195. Souza, C.; May, J.; Carlos, I.; de Oliveira, M.; Kuri, S.; Kiminami, C. Influence of the corrosion on the saturation magnetic density of amorphous and nanocrystalline Fe73Nb3Si15. 5B7. 5Cu1 and Fe80Zr3. 5Nb3. 5B12Cu1 alloys. *J. Non-Cryst. Solids* **2002**, *304*, 210–216. [CrossRef]
196. Gostin, P.; Gebert, A.; Schultz, L. Comparison of the corrosion of bulk amorphous steel with conventional steel. *Corros Sci.* **2010**, *52*, 273–281. [CrossRef]
197. Guo, R.; Zhang, C.; Chen, Q.e.; Yang, Y.; Li, N.; Liu, L. Study of structure and corrosion resistance of Fe-based amorphous coatings prepared by HVAF and HVOF. *Corros Sci* **2011**, *53*, 2351–2356. [CrossRef]
198. Klocke, F.; Krieg, T. Coated tools for metal cutting–features and applications. *CIRP Ann.* **1999**, *48*, 515–525. [CrossRef]
199. Brinksmeier, E.; Gläbe, R. Advances in precision machining of steel. *CIRP Ann.* **2001**, *50*, 385–388. [CrossRef]
200. Xiao, C. Properties of nano-SiC/Ni composite coating on diamond surfaces. *Surf. Eng.* **2018**, *34*, 832–837. [CrossRef]
201. Stock, H.-R.; Schlett, V.; Kohlscheen, J.; Mayr, P. Characterization and mechanical properties of ion-implanted diamond surfaces. *Surf. Coat. Technol.* **2001**, *146*, 425–429. [CrossRef]
202. Lee, Y.J.; Hao, L.; Lüder, J.; Chaudhari, A.; Wang, S.; Manzhos, S.; Wang, H. Micromachining of ferrous metal with an ion implanted diamond cutting tool. *Carbon* **2019**, *152*, 598–608. [CrossRef]
203. Prawer, S. Ion implantation of diamond and diamond films. *Diamond Relat. Mater.* **1995**, *4*, 862–872. [CrossRef]
204. Zhang, G. Method for Extending Diamond Tool Life in Diamond Machining of Materials that Chemically React with Diamond. U.S. Patent 7198043B1, 3 April 2007.

MDPI
St. Alban-Anlage 66
4052 Basel
Switzerland
Tel. +41 61 683 77 34
Fax +41 61 302 89 18
www.mdpi.com

Processes Editorial Office
E-mail: processes@mdpi.com
www.mdpi.com/journal/processes

www.ingramcontent.com/pod-product-compliance
Lightning Source LLC
LaVergne TN
LVHW071027170726
843515LV00006B/1218

* 9 7 8 3 0 3 6 5 6 0 3 4 2 *